Quantitative Seismic Interpretation
Applying Rock Physics Tools to Reduce Interpretation Risk

Seismic data analysis is one of the key technologies for characterizing reservoirs and monitoring subsurface pore fluids. While there have been great advances in 3D seismic data processing, the *quantitative* interpretation of the seismic data for rock properties still poses many challenges.

Quantitative Seismic Interpretation demonstrates how rock physics can be applied to predict reservoir parameters, such as lithologies and pore fluids, from seismically derived attributes. It shows how the multidisciplinary combination of rock physics models with seismic data, sedimentological information and stochastic techniques can lead to more powerful results than can be obtained from a single technique. The authors provide an integrated methodology and practical tools for quantitative interpretation, uncertainty assessment, and characterization of subsurface reservoirs using well-log and seismic data. They illustrate the advantages of these new methodologies, while providing advice about limitations of the methods and traditional pitfalls.

This book is aimed at graduate students, academics and industry professionals working in the areas of petroleum geoscience and exploration seismology. It will also interest environmental geophysicists seeking a quantitative subsurface characterization from shallow seismic data. The book includes problem sets and a case-study, for which seismic and well-log data, and Matlab codes are provided on a website (http://www.cambridge.org/9780521816014). These resources will allow readers to gain a hands-on understanding of the methodologies.

Per Avseth is a geophysical consultant at Odin Petroleum in Bergen, Norway, and adjunct professor in reservoir geophysics at Norwegian University of Science and Technology (NTNU) in Trondheim, Norway. He received his M.Sc. in Applied Petroleum Geosciences from NTNU, and his Ph.D. in Geophysics from Stanford University, California. Dr Avseth worked at Norsk Hydro Research Center in Bergen from 2001–2006, and at Rock Physics Technology from 2006–2007. His research interests include applied rock physics, and AVO analysis, for quantitative seismic exploration and reservoir characterization. He has taught applied rock physics courses for several oil companies and has served as course instructor at EAGE Educational Days in London and Moscow, and was the SEG Honorary Lecturer in Europe in 2009.

Tapan Mukerji received his Ph.D. in Geophysics from Stanford University in 1995 and is now an Associate Professor (Research) in Energy Resources Engineering and a member of the Stanford Rock Physics Project at Stanford University. Professor Mukerji co-directs the Stanford Center for Reservoir Forecasting (SCRF) focussing on problems related to uncertainty and data integration for reservoir modeling. His research interests include wave propagation and statistical rock physics, and he specializes in applied rock physics and geostatistical methods for seismic reservoir characterization, fracture detection, 4-D monitoring, and shallow subsurface environmental applications. Professor Mukerji is also a co-author of *The Rock Physics Handbook* (Cambridge University Press, second edition 2009), and has taught numerous industry courses. He received the Karcher award from the Society of Exploration Geophysicists in 2000.

Gary Mavko received his Ph.D. in geophysics from Stanford University in 1977 where he is now Professor (Research) of Geophysics. Professor Mavko co-directs the Stanford Rock Physics and Borehole Geophysics Project (SRB), a group of approximately 25 researchers working on problems related to wave propagation in earth materials. Professor Mavko is also a co-author of *The Rock Physics Handbook*, and has been an invited instructor for numerous industry courses on rock physics for seismic reservoir characterization. He received the Honorary Membership award from the Society of Exploration Geophysicists in 2001, and was the SEG Distinguished Lecturer in 2006.

Quantitative Seismic Interpretation

Applying Rock Physics Tools to Reduce Interpretation Risk

Per Avseth
Norsk Hydro Research Centre, Bergen

Tapan Mukerji
Stanford University

Gary Mavko
Stanford University

CAMBRIDGE UNIVERSITY PRESS
Cambridge, New York, Melbourne, Madrid, Cape Town, Singapore,
São Paulo, Delhi, Dubai, Tokyo

Cambridge University Press
The Edinburgh Building, Cambridge CB2 8RU, UK

Published in the United States of America by Cambridge University Press, New York

www.cambridge.org
Information on this title: www.cambridge.org/9780521151351

© P. Avseth, T. Mukerji and G. Mavko 2005, 2010

This book is in copyright. Subject to statutory exception
and to the provisions of relevant collective licensing agreements,
no reproduction of any part may take place without
the written permission of Cambridge University Press.

First published 2005
Sixth printing 2008
Paperback edition 2010

Printed in the United Kingdom at the University Press, Cambridge

A catalog record for this book is available from the British Library

ISBN 978-0-521-81601-4 Hardback
ISBN 978-0-521-151351 Paperback

Cambridge University Press has no responsibility for the persistence or accuracy of URLs for external or third-party internet websites referred to in this book, and does not guarantee that any content on such websites is, or will remain, accurate or appropriate.

Do not believe in anything simply because it is found written in your books. But after observation and analysis, when you find that anything agrees with reason and is conducive to the good and benefit of one and all, then accept it and live up to it. **The Buddha**

Contents

	Preface	*page* xi

1 Introduction to rock physics — 1

1.1	Introduction	1
1.2	Velocity–porosity relations for mapping porosity and facies	2
1.3	Fluid substitution analysis	15
1.4	Pressure effects on velocity	24
1.5	The special role of shear-wave information	30
1.6	Rock physics "*What ifs?*": fluid and lithology substitution	42
1.7	All models are wrong … some are useful	43

2 Rock physics interpretation of texture, lithology and compaction — 48

2.1	Introduction	48
2.2	The link between rock physics properties and sedimentary microstructure: theory and models	51
2.3	Example: rock physics interpretation of microstructure in North Sea turbidite systems	70
2.4	Relating rock physics to lithofacies and depositional environments	81
2.5	Example: seismic lithofacies in a North Sea turbidite system	83
2.6	Rock physics depth trends	90
2.7	Example: rock physics depth trends and anomalies in a North Sea field	96
2.8	Rock physics templates: a tool for lithology and fluid prediction	101
2.9	Discussion	107
2.10	Conclusions	109

3 Statistical rock physics: Combining rock physics, information theory, and statistics to reduce uncertainty — 111

3.1	Introduction	111
3.2	Why quantify uncertainty?	112
3.3	Statistical rock physics workflow	123
3.4	Information entropy: some simple examples	132
3.5	Monte Carlo simulation	136
3.6	Statistical classification and pattern recognition	138
3.7	Discussion and summary	165

4 Common techniques for quantitative seismic interpretation — 168

4.1	Introduction	168
4.2	Qualitative seismic amplitude interpretation	168
4.3	AVO analysis	180
4.4	Impedance inversion	230
4.5	Forward seismic modeling	252
4.6	Future directions in quantitative seismic interpretation	256

5 Case studies: Lithology and pore-fluid prediction from seismic data — 258

5.1	Case 1: Seismic reservoir mapping from 3D AVO in a North Sea turbidite system	258
5.2	Case 2: Mapping lithofacies and pore-fluid probabilities in a North Sea reservoir using seismic impedance inversions and statistical rock physics	278
5.3	Case 3: Seismic lithology prediction and reservoir delineation using statistical AVO in the Grane field, North Sea	295
5.4	Case 4: AVO depth trends for lithology and pore fluid classification in unconsolidated deep-water systems, offshore West Africa	306
5.5	Case 5: Seismic reservoir mapping using rock physics templates. Example from a North Sea turbidite system	312

6 Workflows and guidelines 317

6.1	AVO reconnaissance	318
6.2	Rock physics "*What ifs*" and AVO feasibility studies	320
6.3	RPT analysis	322
6.4	AVO classification constrained by rock physics depth trends	323
6.5	Seismic reservoir characterization constrained by lithofacies analysis and statistical rock physics	325
6.6	Why and when should we do quantitative seismic interpretation?	328

7 Hands-on 332

7.1	Introduction	332
7.2	Problems	332
7.3	Project	336

| *References* | 340 |
| *Index* | 356 |

The color plates are situated between pages 176 and 177.

Preface

Every year finding new oil is harder, riskier, and more expensive – a natural consequence of its finiteness. As dictated by M. King Hubbert's "peak," declines in discoveries and production are inevitable. Yet demand continues, forcing us to deeper water, more complex reservoirs, and smaller, more subtle oil fields.

A key to managing this complexity and risk has always been effective integration of the diverse petroleum technologies. Workstations, visualization software, and geostatistics have contributed to integrating the vast amounts of *data* that we sometimes drown in. Perhaps more important are the asset teams that exploit diverse data by *integrating expertise*. Our goal, in preparing Quantitative Seismic Interpretation, is to help illustrate the powerful role that rock physics can play in integrating both the data and expertise of geophysics and geology for reservoir characterization.

Our objective for this book is to help make the links between seismic and reservoir properties more *quantitative*. Most of our examples use amplitude signatures and impedances, but we consider quantitative seismic interpretation to include the use of any seismic attributes for which there are specific models relating them to the rock properties. Our approach is to introduce fundamental rock physics relations, which help to quantify the geophysical signatures of rock and fluid properties. Since rock properties are a consequence of geologic processes, we begin to quantify the seismic signatures of various geologic trends. We also fully embrace probabilistic and geostatistical tools, as quantitative means for managing the inevitable uncertainty that accompanies all quantitative methods. Quantifying, managing, and understanding the uncertainties are critical for survival in a risky environment.

For many years, rock physics focused on *physics*. We carefully measured wave propagation under a variety of laboratory conditions, and we developed marvelously clever acoustic analogs of rocks, finding ways to model grains and pores, and the fluids that sit inside them. We know how to parameterize seismic velocities in terms of mineralogy, porosity, aspect ratios, and grain contacts. We understand how pore pressure and stress affect velocity, attenuation, and their anisotropies. We have a sense for why (high-frequency) laboratory velocities differ from (low-frequency) field velocities. And we can make excellent predictions of how velocities change when pore fluids change.

Surprisingly, some of the most important breakthroughs in rock physics during the past decade have come not from additional mathematics, but from rediscovering the physics of rock geology. Our rock textural parameters that control elastic response can now be related to depositional maturity, and the overprint of compaction and diagenesis. Pore aspect ratios have given way to parameters such as grain sorting; linear impedance–porosity trends have given way to sand-shale "boomerang" plots in the velocity–porosity plane, reflecting depositional cycles. Quantitative geologic constraints can define the relevant trajectories through geophysical planes (velocity versus porosity; V_P versus V_S), which physics-based models can only parameterize.

One of the most powerful uses of rock physics is for *extrapolation*. At a well – assuming that data quality is good – we pretty much know "the answer." Cuttings, cores, and logs tell us about the lithology, porosity, permeability, and fluids. The problem is, often, knowing what happens as we move away from the well. This is the role of the rock physics "*What if?*" Using rock physics, we can extrapolate to geologically plausible conditions that might exist away from the well, exploring how the seismic signatures might change. This is particularly useful when we wish to understand the seismic signatures of fluids and facies that are not represented in the well. For statistical methods, such as clustering analysis or neural networks, such extrapolations are critical for extending the training data. What if the pore fluids change? What if the lithology changes? What if the depositional environment changes?

Another exciting development is the appearance of *statistical rock physics*. Simulation-based quantitative interpretation is one of the main messages of statistical rock physics. Geophysicists and geologists have tended to shy away from (even scorn) statistics. We have somehow felt that statistical methods were giving up the physics, even getting sloppy. But stochastic methods do not throw away the physics. They just put in some of the realities and heterogeneities that are not modeled by the idealized physics. When was the last time you saw a seismic section with error bars? Not long ago one of our colleagues, after hearing a presentation on stochastic simulation, remarked "You mean that you just make up random numbers?" Thankfully, these misconceptions are (slowly) melting away. Just because we do not yet have the perfect imaging and velocity estimation algorithm does not mean we should stop making interpretations and wait for perfect data. Decisions need to be made in the face of uncertainty, with imperfect and incomplete data. As better-quality data become available, one can update prior interpretations and reduce the associated uncertainty. One of the complaints about statistical methods is that they require lots of data. It is true that more data help the statistics. But scenarios with scarce data are the ones where the uncertainty is the greatest. It is these situations with few data that benefit the most from stochastic methods for quantifying and reducing the uncertainty.

"Quantitative" does not mean without uncertainty. We also stress that uncertainty estimates and probabilities are always subjective. Subjective information plays an important role in quantitative interpretation. "Subjective" and "quantitative" are not mutually exclusive.

Uncertainty and risk pervade our decisions on reservoirs. One source of uncertainty is model approximations of a hopelessly complex Earth. Rocks are neither linear nor elastic nor isotropic. Yet much of seismic analysis assumes so, leaving imperfections in our seismic images. Another source of uncertainty is the fundamental nonuniqueness of interpretation. The most perfect seismic inversion assuming isotropic linear elasticity yields at best three parameters: V_P, V_S, and density. We're still struggling to get even these three. In addition to V_P, V_S and density, perhaps we might be able to estimate something about Q and anisotropy with appropriate models. The wave equation that we base most of our work on depends only on these few parameters. Yet there are many more rock unknowns: mineralogy, porosity, pore shapes, grain size distributions, angularity, packing, pore fluids, saturations, temperature, pore pressure, stress, etc. So even with perfect data we have a tremendous uncertainty that needs to be described and reduced by optimum use of geology.

Chapter 1 gives a brief introduction to rock physics, the science aimed at discovering and understanding the relations between seismic observables (velocity, impedance, amplitude) and rock properties (lithology, porosity, permeability, pore fluids, temperature, and stress). We introduce the concepts of bounds on elastic properties, and show how they also can serve as powerful interpolators when describing depositional and diagenetic trends in the velocity–porosity plane. We give an extensive discussion of fluid substitution, and explore the special role that shear wave information plays when separating lithologic, pressure, and saturation effects. We also discuss some of the effects that pore pressure has on seismic velocities.

Chapter 2 focuses on the rock physics link to depositional and diagenetic trends of sands, shales, and shaly sands. We introduce a number of specialized models that describe the velocity–porosity behavior of clastics, and illustrate these with a number of field examples. We establish important links between depositional facies and rock physics properties, investigate depth trends in the rock physics of sands and shales as a function of diagenesis, and finally put it all together in basin-specific rock physics templates (RPTs) which can be used both for well-log and seismic data analysis.

Chapter 3 focuses on statistical rock physics. It gives brief introductions to various statistical classification techniques, and shows how combining rock physics models with modern computational statistics helps us to go beyond what is possible using either statistics or physics alone. We show how Monte Carlo simulations help us to quantify uncertainties in rock physics interpretation of seismic attributes. We also discuss the concept of derived distributions to extend and extrapolate training data. MATLAB™ functions for Monte Carlo simulation and statistical classification techniques described in Chapter 3 may be downloaded from the website for our book. Two excellent texts that we recommend for elaborate discussions on statistical classification techniques are *The Elements of Statistical Learning: Data Mining, Inference, and Prediction* (Hastie, Tibshirani and Freidman, 2001) and *Pattern Classification* (Duda, Hart and Stork, 2001). *Decision Making with Insight* (Savage, 2003) highlights in an entertaining manner the pitfalls of ignoring uncertainty in quantitative modeling.

Chapter 4 provides a compilation of the most common techniques used for quantitative seismic interpretation, including the new contributions made by the authors of this book. We start with explaining some common pitfalls in qualitative seismic interpretation, and how quantitative techniques can solve important ambiguities, and improve the detectability of hydrocarbons. Amplitude variation with offset (AVO) analysis is the most common quantitative technique used in the industry today, and we give an overview of the many aspects of AVO, ranging from wave-propagation theory, processing and acquisition effects, and different ways to interpret the AVO information. This chapter also includes an overview on various methodologies to extract rock properties from near and far impedance inversions. We stress the many pitfalls associated with the various techniques, but also the great potential to obtain rock and fluid properties. We extend the discussion from deterministic techniques to probabilistic AVO analysis as a technique for seismic prediction of reservoir properties. The new techniques of AVO constrained by rock physics depth trends and seismic applications of RPT analysis are also presented. Finally, we give a brief overview on forward seismic modeling as a technique to quantify subsurface reservoir properties.

Chapter 5 describes different case studies where the concepts described in the previous chapters are used systematically for quantitative prediction of lithology and pore fluids from seismic data. Although our examples are drawn from siliciclastic depositional systems, the methods and workflows can be applied to other problems, such as carbonates, gas hydrates, fractured reservoirs, and shallow hydrologic site characterization. Moreover, we discuss only static reservoir characterization, but the methods can be extended to include time-lapse seismic.

Chapter 6 recommends specific workflows for applying the methodologies of quantitative seismic interpretation at various stages of reservoir exploration, appraisal, development and management. By including these workflows, we hope to make the methodology appealing to everyone who routinely interprets geophysical data.

Chapter 7 provides problem sets and an extended reservoir characterization project based on an example seismic data set and well logs provided at the website. We emphasize the value of working through the problems. The best way to learn is by doing. We hope the exercises, example data set, and MATLAB functions will help the reader to understand the techniques better by providing practical hands-on experience. We believe the resources at the website (*http://srb.stanford.edu/books*) will make this book suitable for teaching.

Quantitative Seismic Interpretation is complementary to other works. For in-depth discussions of specific rock physics topics, we recommend *The Rock Physics Handbook* (Mavko *et al.*, 1998); *Acoustics of Porous Media* (Bourbié *et al.*, 1987); and *Introduction to the Physics of Rocks* (Guéguen and Palciauskas, 1994). We also draw your attention to *3-D Seismic Interpretation* (Bacon *et al.*, 2003), and *Interpretation of Three-Dimensional Seismic Data* (Brown, 1992). More geologic discussions can be found in *Principles of Sedimentology and Stratigraphy* (Boggs, 1987). Excellent discussions of AVO technology can be found in *Offset-Dependent Reflectivity: Theory and Practice of AVO Analysis* (Castagna and Backus (eds), 1993) and discussions of inversion methods in *Global Optimization Methods in Geophysical Inversion* (Sen and Stoffa, 1995). We found especially useful the works of Yoram Rubin, including *Applied Stochastic Hydrogeology* (Rubin, 2003).

We wish to thank Norsk Hydro, Statoil and Total for permission to publish many of the field data illustrated in this book, and we acknowledge Norsk Hydro for their generous support of Per Avseth while working on this book. Special thanks to Aart-Jan van Wijngaarden, Harald Flesche, Susanne Lund Jensen, Erik Ødegaard, Johannes Rykkje, and Jorunn Aune Tyssekvam at Norsk Hydro, and Tor Veggeland at DONG, for contributions to this book. We also thank Jon Gjelberg, Tom Dreyer, Ivar Sandø, Erik Holtar, Toril Dyreng, Ragnhild Ona, Hans Helle, and Torbjørn Fristad at Norsk Hydro, for valuable feedback and discussions on the techniques and examples shown in this book. We are happy to thank the faculty, students, industrial affiliates, and friends of the Stanford Rock Physics and Borehole Geophysics (SRB) project for many valuable comments and insights. We found particularly useful discussions with Jack Dvorkin, Jef Caers, Biondo Biondi, Henning Omre, Mario Gutierrez, Ran Bachrach, Jo Eidsvik, Nizar Chemingui, Ezequiel Gonzalez, and Youngseuk Keehm. Arild Jørstad worked with us on one of the early statistical rock physics projects when we developed methods of applying Monte Carlo simulation techniques in rock physics modeling and statistical classification. And as always, we are indebted to Amos Nur whose work, past and present, has helped to make the field of rock physics what it is today.

We hope you find this book useful.

1 Introduction to rock physics

Make your theory as simple as possible, but no simpler. *Albert Einstein*

1.1 Introduction

The sensitivity of seismic velocities to critical reservoir parameters, such as porosity, lithofacies, pore fluid type, saturation, and pore pressure, has been recognized for many years. However, the practical need to quantify seismic-to-rock-property transforms and their uncertainties has become most critical over the past decade, with the enormous improvement in seismic acquisition and processing and the need to interpret amplitudes for hydrocarbon detection, reservoir characterization, and reservoir monitoring. Discovering and understanding the seismic-to-reservoir relations has been the focus of rock physics research.

One of our favorite examples of the need for rock physics is shown in Plate 1.1. It is a seismic P–P reflectivity map over a submarine fan, or turbidite system. We can begin to interpret the image without using much rock physics, because of the striking and recognizable shape of the feature. A sedimentologist would tell us that the main feeder channel (indicated by the high amplitude) on the left third of the image is likely to be massive, clean, well-sorted sand – good reservoir rock. It is likely to be cutting through shale, shown by the low amplitudes. So we might propose that high amplitudes correspond to good sands, while the low amplitudes are shales.

Downflow in the lobe environment, however, the story changes. Well control tells us that on the right side of the image, the low amplitudes correspond to *both* shale and clean sand – the sands are transparent. In this part of the image the bright spots are the poor, shale-rich sands. So, what is going on?

We now understand many of these results in terms of the interplay of sedimentologic and diagenetic influences. The clean sands on the left (Plate 1.1) are very slightly cemented, causing them to have higher acoustic impedance than the shales. The clean sands on the right are uncemented, and therefore have virtually the same impedance as the shales. However, on the right, there are more facies associated with lower energy deposition, and these tend to be more poorly sorted and clay-rich. We know from laboratory work and theory that poor sorting can also influence impedance. In the turbidite system in Plate 1.1 both the clean, slightly cemented sand and the clean uncemented sand are oil-saturated. These sands have essentially the same porosity and composition, yet they have very different seismic signatures.

> This example illustrates the need to incorporate rock physics principles into seismic interpretation, and reservoir geophysics in general. Despite the excellent seismic quality and well control, the correct interpretation required quantifying the connection between geology and seismic data. A purely correlational approach, for instance using neural networks or geostatistics, would not have been so successful.

Our goal in this first chapter is to review some of the basic rock physics concepts that are critical for reservoir geophysics. Although the discussion is not exhaustive, we assess the strengths, weaknesses, and common pitfalls of some currently used methods, and we make specific recommendations for seismic-to-rock-property transforms for mapping of lithology, porosity, and fluids. Several of these rock physics methods are further discussed and applied in Chapters 2, 3, and 5.

1.2 Velocity–porosity relations for mapping porosity and facies

Rock physics models that relate velocity and impedance to porosity and mineralogy (e.g. shale content) form a critical part of seismic analysis for porosity and lithofacies. In this section we illustrate how to recognize the appropriate velocity–porosity relation when approaching a new reservoir geophysics problem.

> **Pitfall**
>
> One of the most serious and common mistakes that we have observed in industry practice is the use of inappropriate velocity–porosity relations for seismic mapping of porosity and lithofacies. The most common error is to use overly stiff velocity–porosity relations, such as the classical empirical trends of Wyllie *et al.* (1956), Raymer, Hunt, and Gardner (Raymer *et al.*, 1980), Han (1986), or Raiga-Clemenceau *et al.* (1988), the critical porosity model (Nur, 1992), or penny-shaped crack models. "Sonic porosity," derived from sonic logs using the Wyllie time average, is perhaps the worst example. Implicit in these relations is that porosity is controlled by variations in diagenesis, which is not always the case. Hence, critical sedimentologic variations are ignored.

1.2 Velocity–porosity relations for mapping porosity and facies

> **Solution**
>
> Rock physics diagnostic analysis of well logs and cores, coupled to the geologic model, usually leads to more rational velocity–porosity relations. Certain aspects are highlighted in this section.

The importance of velocity–porosity relations applies to other rock physics problems, as well. Even seismic pore fluid analysis, which we discuss in the next section, depends on the velocity–porosity relation. We can start to see this by looking at the Gassmann (1951) relation, which can be represented in the form (Zimmerman, 1991; Mavko and Mukerji, 1995; Mavko et al., 1998):

$$\frac{1}{K_{\text{rock}}} = \frac{1}{K_{\text{mineral}}} + \frac{\phi}{\tilde{K}_\phi}$$

where K_{rock}, K_{mineral}, and $\tilde{K}_\phi$ are the bulk moduli of the saturated rock, the mineral, and the saturated pore space, respectively, and ϕ is the porosity. The pore space modulus is approximately the sum of the dry pore modulus and the fluid modulus: $\tilde{K}_\phi \approx K_\phi + K_{\text{fluid}}$. (We will define these more carefully later.) Hence, we can see that the sensitivity of rock modulus (and velocity) to pore fluid changes depends directly on the ratio of pore space stiffness to porosity, K_ϕ/ϕ. Rocks that are relatively stiff have a small seismic sensitivity to pore fluids, and rocks that are soft have a large sensitivity to pore fluids.

> We encounter the link between fluid substitution and velocity–porosity relations in several common ways:
> - When first analyzing well logs to derive a velocity–porosity relation, it is essential first to map the data to a common fluid. Otherwise, the effects of the rock frame and pore fluid become mixed.
> - When interpreting 3D seismic data for hydrocarbon detection, the Gassmann analysis requires a good estimate of porosity, which also must be mapped from the seismic data.
> - When populating reservoir models with acoustic properties (V_P and V_S) for 4D feasibility studies, we often need to map from porosity to velocity. Beginning the exercise with the incorrect mapping quickly makes the fluid substitution analysis wrong.

1.2.1 Background on elastic bounds

We begin with a discussion of upper and lower bounds on the elastic moduli of rocks. The bounds provide a useful and elegant framework for velocity–porosity relations.

Many "effective-medium" models have been published, attempting to describe theoretically the effective elastic moduli of rocks and sediments. (For a review, see

Introduction to rock physics

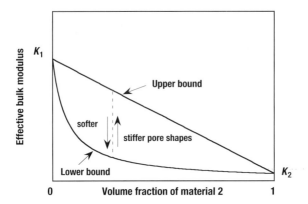

Figure 1.2 Conceptual illustration of bounds for the effective elastic bulk modulus of a mixture of two materials.

Mavko *et al.*, 1998.) Some models approximate the rock as an elastic block of mineral perturbed by holes. These are often referred to as "inclusion models." Others try to describe the behavior of the separate elastic grains in contact. These are sometimes called "granular-medium models" or "contact models." Regardless of the approach, the models generally need to specify three types of information: (1) the volume fractions of the various constituents, (2) the elastic moduli of the various phases, and (3) the geometric details of how the phases are arranged relative to each other.

In practice, the geometric details of the rock and sediment have never been adequately incorporated into a theoretical model. Attempts always lead to approximations and simplifications, some better than others.

When we specify only the volume fractions of the constituents and their elastic moduli, without geometric details of their arrangement, then we can predict only the upper and lower *bounds* on the moduli and velocities of the composite rock. However, the elastic bounds are extremely reliable and robust, and they suffer little from the approximations that haunt most of the geometry-specific effective-medium models. Furthermore, since well logs yield information on constituents and their volume fractions, but relatively little about grain and pore microstructure, the bounds turn out to be extremely valuable rock physics tools.

Figure 1.2 illustrates the concept for a simple mixture of two constituents. These might be two different minerals or a mineral plus fluid (water, oil, or gas). At any given volume fraction of constituents the effective modulus of the mixture will fall between the bounds (somewhere along the vertical dashed line in the figure), but its precise value depends on the geometric details. We use, for example, terms like "stiff pore shapes" and "soft pore shapes" to describe the geometric variations. Stiffer grain or pore shapes cause the value to be higher within the allowable range; softer grain or pore shapes cause the value to be lower.

1.2 Velocity–porosity relations for mapping porosity and facies

The Voigt and Reuss bounds

The simplest, but not necessarily the best, bounds are the Voigt (1910) and Reuss (1929) bounds. The *Voigt upper bound* on the effective elastic modulus, M_V, of a mixture of N material phases is

$$M_V = \sum_{i=1}^{N} f_i M_i \qquad (1.1)$$

with

f_i the volume fraction of the ith constituent
M_i the elastic modulus of the ith constituent

There is no way that nature can put together a mixture of constituents (i.e., a rock) that is elastically *stiffer* than the simple arithmetic average of the constituent moduli given by the Voigt bound. The Voigt bound is sometimes called the *isostrain* average, because it gives the ratio of average stress to average strain when all constituents are assumed to have the same strain.

The *Reuss lower bound* of the effective elastic modulus, M_R, is

$$\frac{1}{M_R} = \sum_{i=1}^{N} \frac{f_i}{M_i} \qquad (1.2)$$

There is no way that nature can put together a mixture of constituents that is elastically *softer* than this harmonic average of moduli given by the Reuss bound. The Reuss bound is sometimes called the *isostress* average, because it gives the ratio of average stress to average strain when all constituents are assumed to have the same stress.

Mathematically the M in the Voigt and Reuss formulas can represent any modulus: the bulk modulus K, the shear modulus μ, Young's modulus E, etc. However, it makes most sense to compute the Voigt and Reuss averages of only the shear modulus, $M = \mu$, and the bulk modulus, $M = K$, and then compute the other moduli from these, using the rules of isotropic linear elasticity.

Figure 1.3 shows schematically the bounds for elastic bulk and shear moduli, when one of the constituents is a liquid or gas. In this case, the lower bound corresponds to a suspension of the particles in the fluid, which is an excellent model for very soft sediments at low effective stress. Note that the lower bound on shear modulus is zero, as long as the volume fraction of fluid is nonzero.

> The Reuss average describes exactly the effective moduli of a suspension of solid grains in a fluid. This will turn out to be the basis for describing certain types of clastic sediments. It also describes the moduli of "shattered" materials where solid fragments are completely surrounded by the pore fluid.

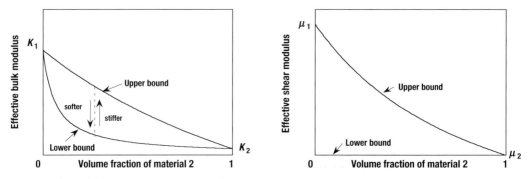

Figure 1.3 Conceptual illustration of upper and lower bounds to bulk and shear moduli for a mixture of two materials, one of which is a fluid.

> When all constituents are gases or liquids with zero shear modulus, then the Reuss average gives the effective moduli of the mixture, exactly.
>
> In contrast to the Reuss average which describes a number of real physical systems, real isotropic mixtures can never be as stiff as the Voigt bound (except for the single-phase end members).

Hashin–Shtrikman bounds

The best bounds for an *isotropic* elastic mixture, defined as giving the narrowest possible range of elastic moduli without specifying anything about the geometries of the constituents, are the *Hashin–Shtrikman bounds* (Hashin and Shtrikman, 1963). For a mixture of two constituents, the Hashin–Shtrikman bounds are given by

$$K^{HS\pm} = K_1 + \frac{f_2}{(K_2 - K_1)^{-1} + f_1(K_1 + 4\mu_1/3)^{-1}}$$

$$\mu^{HS\pm} = \mu_1 + \frac{f_2}{(\mu_2 - \mu_1)^{-1} + 2f_1(K_1 + 2\mu_1)/[5\mu_1(K_1 + 4\mu_1/3)]}$$

(1.3)

with

K_1, K_2 bulk moduli of individual phases
μ_1, μ_2 shear moduli of individual phases
f_1, f_2 volume fractions of individual phases

Upper and lower bounds are computed by interchanging which material is subscripted 1 and which is subscripted 2. Generally, the expressions give the upper bound when the *stiffest* material is subscripted 1 in the expressions above, and the lower bound when the *softest* material is subscripted 1.

The physical interpretation of a material whose bulk modulus would fall on one of the Hashin–Shtrikman bounds is shown schematically in Figure 1.4. The space is filled by an assembly of spheres of material 2, each surrounded by a spherical shell of material 1. Each sphere and its shell have precisely the volume fractions f_1 and f_2.

1.2 Velocity–porosity relations for mapping porosity and facies

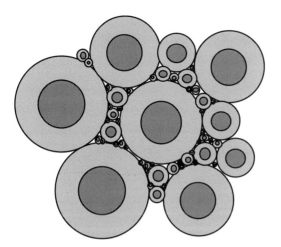

Figure 1.4 Physical interpretation of the Hashin–Shtrikman bounds for bulk modulus of a two-phase material.

The upper bound is realized when the stiffer material forms the shell; the lower bound, when it is in the core.

A more general form of the Hashin–Shtrikman bounds, which can be applied to more than two phases (Berryman, 1995), can be written as

$$K^{HS+} = \Lambda(\mu_{max}), \quad K^{HS-} = \Lambda(\mu_{min})$$
$$\mu^{HS+} = \Gamma(\zeta(K_{max}, \mu_{max})), \quad \mu^{HS-} = \Gamma(\zeta(K_{min}, \mu_{min})) \quad (1.4)$$

where

$$\Lambda(z) = \left\langle \frac{1}{K(r) + 4z/3} \right\rangle^{-1} - \frac{4}{3}z$$

$$\Gamma(z) = \left\langle \frac{1}{\mu(r) + z} \right\rangle^{-1} - z$$

$$\zeta(K, \mu) = \frac{\mu}{6}\left(\frac{9K + 8\mu}{K + 2\mu}\right)$$

The brackets $\langle \cdot \rangle$ indicate an average over the medium, which is the same as an average over the constituents, weighted by their volume fractions.

The separation between the upper and lower bounds (Voigt–Reuss or Hashin–Shtrikman) depends on how elastically different the constituents are. As shown in Figure 1.5, the bounds are often fairly similar when mixing solids, since the elastic moduli of common minerals are usually within a factor of two of each other. Since many effective-medium models (e.g., Biot, 1956; Gassmann, 1951; Kuster and Toksöz, 1974) assume a homogeneous mineral modulus, it is often useful (and adequate) to represent a mixed mineralogy with an "average mineral" modulus, equal either to one of the bounds computed for the mix of minerals or to their average $(M^{HS+} + M^{HS-})/2$.

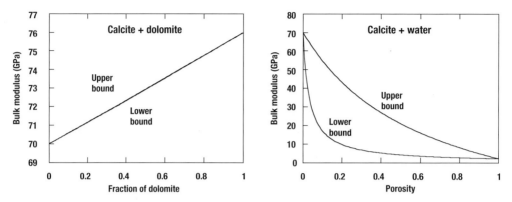

Figure 1.5 On the left, a mixture of two minerals. The upper and lower bounds are close when the constituents are elastically similar. On the right, a mixture of mineral and water. The upper and lower bounds are far apart when the constituents are elastically different.

On the other hand, when the constituents are quite different – such as minerals and pore fluids – then the bounds become quite separated, and we lose some of the predictive value.

> Note that when $\mu_{min} = 0$, then K^{HS-} is the same as the Reuss bound. In this case, the Reuss or Hashin–Shtrikman lower bounds describe exactly the moduli of a suspension of grains in a pore fluid. These also describe the moduli of a mixture of fluids and/or gases.

1.2.2 Generalized velocity–porosity models for clastics

Brief "life story" of a clastic sediment

The bounds provide a framework for understanding the acoustic properties of sediments. Figure 1.6 shows P-wave velocity versus porosity for a variety of water-saturated sediments, ranging from ocean-bottom suspensions to consolidated sandstones. The Voigt and Reuss bounds, computed for mixtures of quartz and water, are shown for comparison. (Strictly speaking, the bounds describe the allowable range for elastic moduli. When the corresponding P- and S-wave velocities are derived from these moduli, it is common to refer to them as the "upper and lower bounds on velocity.")

Before deposition, sediments exist as particles suspended in water (or air). As such, their acoustic properties must fall on the Reuss average of mineral and fluid. When the sediments are first deposited on the water bottom, we expect their properties still to lie on (or near) the Reuss average, as long as they are weak and unconsolidated. Their porosity position along the Reuss average is determined by the geometry of the particle packing. Clean, well-sorted sands will be deposited with porosities near 40%. Poorly sorted sands will be deposited along the Reuss average at lower porosities. Chalks will be deposited at

1.2 Velocity–porosity relations for mapping porosity and facies

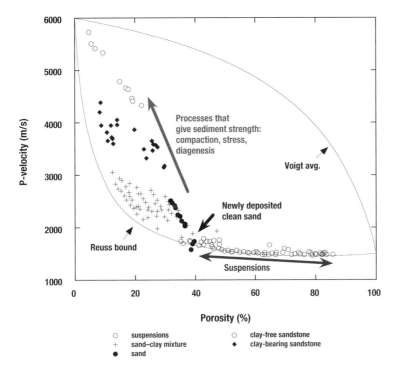

Figure 1.6 P-wave velocity versus porosity for a variety of water-saturated sediments, compared with the Voigt–Reuss bounds. Data are from Yin (1992), Han (1986) and Hamilton (1956).

high initial porosities, 55–65%. We sometimes call this porosity of the newly deposited sediment the *critical porosity* (Nur, 1992). Upon burial, the various processes that give the sediment strength – effective stress, compaction, and cementing – must move the sediments off the Reuss bound. We observe that with increasing diagenesis, the rock properties fall along steep trajectories that extend upward from the Reuss bound at critical porosity, toward the mineral end point at zero porosity. We will see below that these diagenetic trends can be described once again using the bounds.

Han's empirical relations

Figure 1.7 shows typical plots of seismic V_P and V_S vs. porosity for a large set of laboratory ultrasonic data for water-saturated sandstones (Han, 1986). All of the data points shown are at 40 MPa effective pressure. In both plots, we see the usual general trend of decreasing velocity with increasing porosity. There is a great deal of scatter around the trend, which we know from Han's work is well correlated with the clay content. Han described this velocity–porosity–clay behavior with the empirical relations:

$$V_P = 5.59 - 6.93\phi - 2.13C$$
$$V_S = 3.52 - 4.91\phi - 1.89C$$
(1.5)

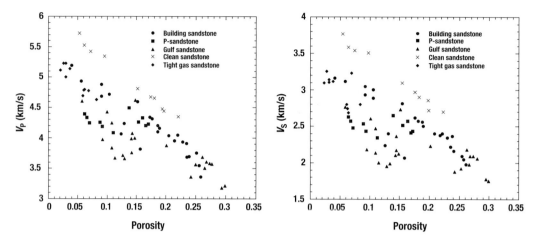

Figure 1.7 Velocity versus porosity for water-saturated sandstones at 40 MPa. Data are ultrasonic measurements from Han (1986).

where the velocities are in km/s, ϕ is the porosity, and C is the clay volume fraction. These relations can be rewritten slightly in the form

$$V_P = (5.59 - 2.13C) - 6.93\phi$$
$$V_S = (3.52 - 1.89C) - 4.91\phi \tag{1.6}$$

which can be thought of as a series of parallel velocity–porosity trends, whose zero-porosity intercepts depend on the clay content. These contours of constant clay content are illustrated in Figure 1.8, and are essentially the steep diagenetic trends mentioned in Figure 1.6. Han's clean (clay-free) line mimics the diagenetic trend for clean sands, while Han's more clay-rich contours mimic the diagenetic trends for dirtier sands. Vernik and Nur (1992) and Vernik (1997) found similar velocity–porosity relations, and were able to interpret the Han-type contours in terms of petrophysical classifications of siliciclastics. Klimentos (1991) also obtained similar empirical relations between velocity, porosity, clay content and permeability for sandstones.

As with any empirical relations, equations (1.5) and (1.6) are most meaningful for the data from which they were derived. It is dangerous to extrapolate them to other situations, although the concepts that porosity and clay have large impacts on P- and S-wave velocities are quite general for clastic rocks.

When using relations like these, it is very important to consider the coupled effects of porosity and clay. If two rocks have the same porosity, but different amounts of clay, then chances are good that the high clay rock has lower velocity. But if porosity decreases as clay volume increases, then the high clay rock might have a higher velocity. (See also Section 2.2.3.)

1.2 Velocity–porosity relations for mapping porosity and facies

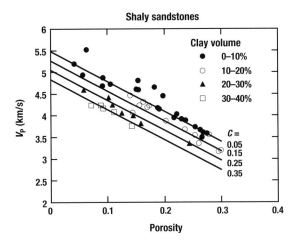

Figure 1.8 A subset of Han's data from Figure 1.7, sorted by clay content. The empirical relations, equations (1.5) and (1.6), can be thought of as a series of parallel lines as shown, all with the same slope, but with different clay contents.

Distinction between cementing and sorting trends

A number of workers (e.g., Dvorkin and Nur, 1996) have recognized that the slope of the velocity–porosity trend (or impedance–porosity trend) in sandstones is highly variable, and depends largely on the geologic process that is controlling porosity. The steep velocity–porosity trends shown in Figures 1.7 and 1.8 for sandstones are representative of porosity variations controlled by *diagenesis*, i.e. porosity reduction due to pressure solution, compaction, and cementation. Hence, we often see steep velocity–porosity trends when examining data spanning a great range of depths or ages. The classical empirical trends of Wyllie *et al*. (1956), Raymer–Hunt–Gardner (Raymer *et al*., 1980), Han (1986) and Raiga-Clemenceau *et al*. (1988), all show versions of the steep, diagenetically controlled velocity–porosity trend.

On the other hand, porosity variations resulting from variations in sorting and clay content tend to yield much flatter velocity–porosity trends. That is, porosity controlled by *sedimentation* is generally expected to yield flatter trends, which we sometimes refer to as *depositional trends*. Data sets from narrow depth ranges or individual reservoirs often (though not always) show this behavior.

This distinction of diagenetic vs. depositional trends as a generalized velocity–porosity model for clastics is illustrated schematically in Figure 1.9. We have found that the diagenetic trends, which connect the newly deposited sediment on the Reuss bound with the mineral point, can often be described well using an *upper bound*. In fact, we sometimes refer to it as a *modified upper bound*, because we use it to describe a mixture of the newly deposited sediment at critical porosity with additional mineral, instead of describing a mixture of mineral and pore fluid. The *modified upper Hashin–Shtrikman bound* approximating the diagenetic trend for clean sands is shown by the heavy black curve in Figure 1.9. The thinner black curves below (and parallel to) the clean sand

Introduction to rock physics

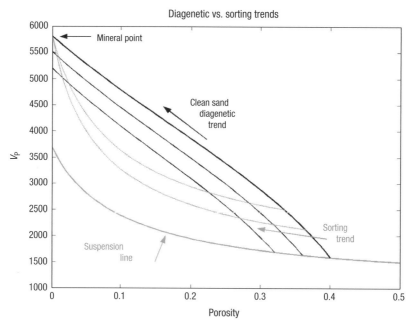

Figure 1.9 Generalized clastic model. Sediments are deposited along the suspension line. Clean, well-sorted sands will have initial (critical) porosity of ~0.4. Poorly sorted sediments will have a smaller critical porosity. Burial, compaction and diagenesis move data off the suspension line. Sediments of constant shaliness or sorting and variable age (or degree of diagenesis) fall along the (black) cementing trends. Sediments of constant age but variable shaliness or sorting will fall along the (gray) sorting trends.

line represent the diagenetic trends for more clay-rich sands. They are computed again using the Hashin–Shtrikman upper bound, connecting the lower critical porosities for more clay-rich sands with the elastically softer mineral moduli for quartz–clay mixtures. These parallel trends are essentially the same as Han's empirical lines, shown in Figure 1.8.

> We observe empirically that the modified upper Hashin–Shtrikman bound describes fairly well the variation of velocity with porosity during compaction and diagenesis of sandstones. While it is difficult to derive from first principles, a heuristic argument for the result is that diagenesis is the stiffest way to mix a young sediment with additional mineral (i.e., the stiffest way to reduce porosity); an upper bound describes the stiffest way to mix two constituents.

A slight improvement over the modified upper Hashin–Shtrikman bound as a diagenetic trend for sands can be obtained by steepening the high-porosity end. An effective way to do this is to append Dvorkin's model (Dvorkin and Nur, 1996) for cementing of grain contacts, as illustrated in Figure 1.10 (discussed more in Chapter 2). The

1.2 Velocity–porosity relations for mapping porosity and facies

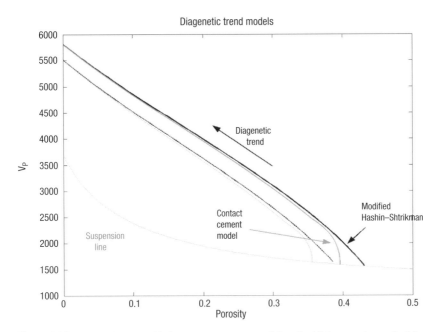

Figure 1.10 Appending Dvorkin's contact cement model at the high-porosity end of the modified Hashin–Shtrikman bound improves the agreement with sands.

contact-cement model captures the rapid increase in elastic stiffness of a sand, without much change in porosity as the first bit of cement is added. Very different processes such as mechanical compaction and pressure solution can yield very similar trends (J. M. Florez, personal communication, 2003).

The depositional or sorting trends can be described by a series of *modified Hashin–Shtrikman lower bounds* (gray curves in Figure 1.9). The lowest gray curve is the suspension line, indicating a depth of zero (i.e., at the depositional surface). The additional gray curves above the suspension line indicate diagenetically older and older sediments. Each is a line of constant depth, but variable texture, sorting, and/or clay content. (Avseth *et al.*, 2000, also refer to these as constant cement lines.) Moving to the right on the gray sorting curves corresponds to cleaner, better sorted sands, while moving to the left corresponds to more clay-rich or more poorly sorted sands. Note that as we move from the right to the left along these trends, we are simply crossing Han's contours from clean sands to clay-rich sands.

Figure 1.11 shows some laboratory data examples. The data from Han (1986), which span a great range of depths and ages, show the steepest average trend, dominated by diagenesis. Data from the Troll (Blangy, 1992), Oseberg, and two North Sea fields "B" and "C" have flatter trends, dominated by *sedimentation-controlled* textural variations. The data from the North Sea "A" field are from a narrow depth range, but in this case the porosity is *diagenetically* controlled, related to varying amounts of chlorite, and this gives a steep slope, close to Han's. The second plot in Figure 1.11 shows

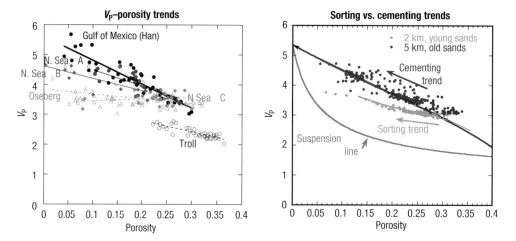

Figure 1.11 Left: Trends in V_P versus porosity, showing a broad range of behaviors. Steep trends (Han; N. Sea A) are dominated by diagenesis. Flatter trends are dominated by textural variations. Right: Another comparison of fields showing the difference between diagenetic trends and sedimentological trends.

another comparison of North Sea fields showing the difference between *diagenetic* trends and *depositional* trends. In both cases, the data are from fairly narrow depth ranges.

More detailed theory and applications of various rock physics models for characterization of sand-shale systems are presented in Chapter 2.

> Porosity has an enormous impact on P- and S-wave velocities.
> - Usually, an increase in porosity will result in a decrease of P- and S-wave velocities. Often the correlation is good, allowing porosity to be estimated from impedance.
> - For sandstones, clay causes scatter around the velocity–porosity trend, although data grouped by nearly constant clay sometimes yield systematic and somewhat parallel trends (Figure 1.8). In consolidated sandstones, clay tends to decrease velocity and increase V_P/V_S ratio. In unconsolidated sands, clay sometimes slightly stiffens the rock.
> - Variations in pore shape also cause variable velocity–porosity trends. This is usually modeled in terms of round vs. crack-like aspect ratio for pores. We now understand that deposition-controlled textural variations, such as sorting, lead to specific, similar variations in clastics. Increasing clay and poorer sorting act roughly in the direction of smaller aspect ratios.
> - Popular relations, like those of Han (1986), Wyllie *et al.* (1958), and Raymer *et al.* (1980), describe steep velocity–porosity trends (as in Figure 1.6), which, when correct, indicate conditions that are favorable for mapping porosity from velocity.

> These are only appropriate when porosity is controlled by diagenesis, often seen over great depth ranges. These relations can be misleading for understanding lateral variations of velocity within narrow depth ranges. They should certainly never be used for fluid substitution analysis.
> - We expect very shallow velocity–porosity trends when porosity varies texturally, because of sorting and clay content. These trends, when appropriate, indicate conditions where mapping porosity from velocity is difficult. However, these texturally controlled rocks tend to be elastically softer and have a larger sensitivity to pore fluids and pore pressure, and these characteristics are advantageous for 4D studies.

1.3 Fluid substitution analysis

This section focuses on *fluid substitution*, which is the rock physics problem of understanding and predicting how seismic velocity and impedance depend on pore fluids. At the heart of the fluid substitution problem are Gassmann's (1951) relations, which predict how the rock modulus changes with a change of pore fluids.

For the fluid substitution problem there are two fluid effects that must be considered: the change in rock bulk density, and the change in rock compressibility. The compressibility of a dry rock (reciprocal of the rock bulk modulus) can be expressed quite generally as the sum of the mineral compressibility and an extra compressibility due to the pore space:

$$\frac{1}{K_{\text{dry}}} = \frac{1}{K_{\text{mineral}}} + \frac{\phi}{K_\phi} \tag{1.7}$$

where ϕ is the porosity, K_{dry} is the dry rock bulk modulus, K_{mineral} is the mineral bulk modulus, and K_ϕ is the pore space stiffness defined by:

$$\frac{1}{K_\phi} = \frac{1}{v_{\text{pore}}} \frac{\partial v_{\text{pore}}}{\partial \sigma} \tag{1.8}$$

Here, v_{pore} is the pore volume, and σ is the increment of hydrostatic confining stress from the passing wave. Poorly consolidated rocks, rocks with microcracks, and rocks at low effective pressure are generally soft and compressible and have a small K_ϕ. Stiff rocks that are well cemented, lacking microcracks, or at high effective pressure have a large K_ϕ. In terms of the popular but idealized ellipsoidal crack models, low-aspect-ratio cracks have small K_ϕ and rounder large-aspect-ratio pores have large K_ϕ. In simple terms we can write approximately $K_\phi \approx \alpha K_{\text{mineral}}$ where α is aspect ratio. (This approximation is best at low porosity.)

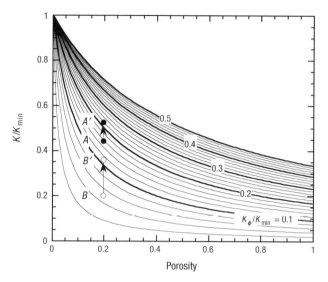

Figure 1.12 Normalized rock bulk modulus versus porosity, with contours of constant pore space stiffness. Points A and B correspond to two different dry rocks at the same porosity. Points A' and B' are the corresponding water-saturated values. The sensitivity to fluid changes is proportional to the contour spacing.

Similarly, the compressibility of a *saturated* rock can be expressed as

$$\frac{1}{K_{sat}} = \frac{1}{K_{mineral}} + \frac{\phi}{K_\phi + K_{fluid} K_{mineral}/(K_{mineral} - K_{fluid})} \quad (1.9)$$

or approximately as

$$\frac{1}{K_{sat}} \approx \frac{1}{K_{mineral}} + \frac{\phi}{K_\phi + K_{fluid}} \quad (1.10)$$

where K_{fluid} is the pore-fluid bulk modulus. Comparing equations (1.7) and (1.10), we can see that changing the pore fluid has the effect of changing the pore-space stiffness. From equation (1.10) we see also the well-known result that a stiff rock, with large pore-space stiffness K_ϕ, will have a small sensitivity to fluids, and a soft rock, with small K_ϕ, will have a larger sensitivity to fluids.

Figure 1.12 shows a plot of normalized rock bulk modulus $K/K_{mineral}$ versus porosity (where $K = K_{sat}$ or K_{dry}) computed for various values of normalized pore-space stiffness $K_\phi/K_{mineral}$. Since K_ϕ and K_{fluid} always appear added together (as in equation (1.10)), then fluid substitution can be thought of as computing the change $\Delta(K_\phi + K_{fluid}) = \Delta K_{fluid}$ and jumping the appropriate number of contours in the graph. For the contour interval in Figure 1.12, the difference between a dry rock and a water-saturated rock is three contours, anywhere in the plot. For the example shown, the starting point A was one of Han's (1986) dry sandstone data points, with effective dry rock bulk modulus $K_{dry}/K_{mineral} = 0.44$ and porosity $\phi = 0.20$. To saturate, we move up the amount $\Delta K_{fluid}/K_{mineral} = 0.06$, or three contours. The water-saturated modulus can be read off directly as $K_{sat}/K_{mineral} = 0.52$, point A'. The second example shown (points B–B') is for the same two pore fluids and the same porosity. However, the change in rock stiffness during fluid substitution is much larger.

Pitfall

Seismic sensitivity to pore fluids is not uniquely related to porosity.

Solution

Seismic fluid sensitivity is determined by a combination of porosity and pore-space stiffness. A softer rock will have a larger sensitivity to fluid substitution than a stiffer rock at the same porosity. On Figure 1.12, we can see that regions where the contours are far apart will have a large sensitivity to fluids, and regions where the contours are close will have a small sensitivity. Gassmann's relations simply and reliably describe these effects.

Equations (1.7) and (1.9) together are equivalent to Gassmann's (1951) relations. If we algebraically eliminate K_ϕ from equations (1.7) and (1.9) we can write one of the more familiar but less intuitive forms:

$$\frac{K_{sat}}{K_{mineral} - K_{sat}} = \frac{K_{dry}}{K_{mineral} - K_{dry}} + \frac{K_{fluid}}{\phi(K_{mineral} - K_{fluid})} \tag{1.11}$$

and the companion result

$$\mu_{sat} = \mu_{dry} \tag{1.12}$$

Gassmann's equations (1.11) and (1.12) predict that for an isotropic rock, the rock bulk modulus will change if the fluid changes, but the rock shear modulus will not.

These dry and saturated moduli, in turn, are related to P-wave velocity $V_P = \sqrt{(K + (4/3)\mu)/\rho}$ and S-wave velocity $V_S = \sqrt{\mu/\rho}$, where ρ is the bulk density given by

$$\rho = \phi \rho_{fluid} + (1 - \phi)\rho_{mineral} \tag{1.13}$$

In equations (1.7)–(1.11), ϕ is normally interpreted as the total porosity, although in shaly sands the proper choice of porosity is not clear. We sometimes find a better fit to field observations when effective porosity is used instead. The uncertainty stems from Gassmann's assumption that the rock is monomineralic, in which case the porosity is unambiguously all the space not occupied by mineral. Clay-rich sandstones actually violate the monomineralic assumption, so we end up forcing the Gassmann relations to apply by adapting the porosity and/or the effective mineral properties. Should the clay be considered part of the mineral frame? If so, does the bound water inside the clay communicate sufficiently with the other free pore fluids to satisfy Gassmann's assumption of equilibrated pore-fluid pressure, or should the bound water be considered part of the mineral frame? Alternatively, should the clay be considered part of the pore fluid? If so, then the functional Gassmann porosity is actually larger than the total porosity, but the pore fluid should be considered a muddy suspension containing clay particles.

1.3.1 The Gassmann fluid substitution recipe

The most common scenario is to begin with an initial set of velocities and densities, $V_P^{(1)}$, $V_S^{(1)}$, and $\rho^{(1)}$ corresponding to the rock with an initial set of fluids, which we call "fluid 1." These velocities often come from well logs, but might also be the result of an inversion or theoretical model. Then fluid substitution is performed as follows:

Step 1: Extract the dynamic bulk and shear moduli from $V_P^{(1)}$, $V_S^{(1)}$, and $\rho^{(1)}$:

$$K^{(1)} = \rho\left((V_P^{(1)})^2 - \frac{4}{3}(V_S^{(1)})^2\right)$$

$$\mu^{(1)} = \rho(V_S^{(1)})^2$$

Step 2: Apply Gassmann's relation, equation (1.11), to transform the bulk modulus:

$$\frac{K_{\text{sat}}^{(2)}}{K_{\text{mineral}} - K_{\text{sat}}^{(2)}} - \frac{K_{\text{fluid}}^{(2)}}{\phi(K_{\text{mineral}} - K_{\text{fluid}}^{(2)})} = \frac{K_{\text{sat}}^{(1)}}{K_{\text{mineral}} - K_{\text{sat}}^{(1)}} - \frac{K_{\text{fluid}}^{(1)}}{\phi(K_{\text{mineral}} - K_{\text{fluid}}^{(1)})}$$

where $K_{\text{sat}}^{(1)}$ and $K_{\text{sat}}^{(2)}$ are the rock bulk moduli saturated with fluid 1 and fluid 2, and $K_{\text{fluid}}^{(1)}$ and $K_{\text{fluid}}^{(2)}$ are the bulk moduli of the fluids themselves.

Step 3: Leave the shear modulus unchanged:

$$\mu_{\text{sat}}^{(2)} = \mu_{\text{sat}}^{(1)}$$

Step 4: Remember to correct the bulk density for the fluid change:

$$\rho^{(2)} = \rho^{(1)} + \phi(\rho_{\text{fluid}}^{(2)} - \rho_{\text{fluid}}^{(1)})$$

Step 5: Reassemble the velocities:

$$V_P^{(2)} = \sqrt{\left(K_{\text{sat}}^{(2)} + \frac{4}{3}\mu_{\text{sat}}^{(2)}\right)/\rho^{(2)}}$$

$$V_S^{(2)} = \sqrt{\mu_{\text{sat}}^{(2)}/\rho^{(2)}}$$

1.3.2 Pore fluid properties

When calculating fluid substitution, it is obviously critical to use appropriate fluid properties. To our knowledge, the Batzle and Wang (1992) empirical formulas are the state of the art. It is possible that some oil companies have internal proprietary data that are alternatives.

One unresolved question is the effect of gas saturation of brine. There is disagreement on how much this affects brine properties, and there is even disagreement on how much gas can be dissolved in brine.

Another fuzzy question is how to model gas condensate reservoirs. These are not mentioned specifically in the Batzle and Wang paper, although Batzle (personal communcation) says that the empirical formulas should extend adequately to both the gaseous and liquid phases in a condensate situation.

- The density and bulk modulus of most reservoir fluids increase as pore pressure increases.
- The density and bulk modulus of most reservoir fluids decrease as temperature increases.
- The Batzle–Wang formulas describe the empirical dependence of gas, oil, and brine properties on temperature, pressure, and composition.
- The Batzle–Wang bulk moduli are the **adiabatic** moduli, which we believe are appropriate for wave propagation.
- In contrast, standard PVT data are **isothermal**. Isothermal moduli can be ~20% too low for oil, and a factor of 2 too low for gas. For brine, the two do not differ much.

1.3.3 Cautions and limitations

A gas-saturated rock is not a "dry rock"

The "dry rock" or "dry frame" moduli that appear in Gassmann's relations and Biot's (1956) relations correspond to a rock containing an infinitely compressible pore fluid. When the seismic wave squeezes on a "dry rock" the pore-filling material offers no resistance. This is equivalent to what is sometimes called a "drained" experiment, in which the pore fluid can easily escape the rock and similarly offers no resistance when the rock is squeezed.

A gas is not infinitely compressible. For example, an ideal gas has isothermal bulk modulus $K_{ideal} = P_{pore}$, where P_{pore} is the gas pore pressure. Hence, an ideal gas at a reservoir pressure of 300 bar is 300 times stiffer (less compressible) than the same gas at atmospheric conditions. Conveniently, air at atmospheric conditions is sufficiently compressible that an air-filled rock with a pore pressure of 1 bar is an excellent approximation to the "dry rock." Hence, laboratory measurements of gas-saturated rocks with $P_{\text{pore}} = 1$ bar can be treated as the dry rock properties, except for extremely unconsolidated materials.

> **Pitfall**
>
> It is not correct simply to put gas-saturated rock properties in place of the "dry rock" or "dry frame" moduli in the Gassmann and Biot equations.

> **Solution**
>
> Treat gas as just another fluid when computing fluid substitution.

Low frequencies

Gassmann's relations are strictly valid only for low frequencies. They are derived under the assumption that wave-induced pore pressures throughout the pore space have time to equilibrate during a seismic period. The high-frequency wave-induced pressure gradients between cracks and pores that characterize the "squirt mechanism" (Mavko and Nur, 1975; O'Connell and Budiansky, 1977; Mavko and Jizba, 1991), for example, violate the assumptions of Gassmann's relations, and are the primary reason why Gassmann's relations usually do not work well for fluid effects in laboratory ultrasonic velocities. The inhibited flow of pore fluids between micro- and macroporosity can also violate Gassmann's assumptions.

What frequencies are low enough? This is difficult to answer precisely. The critical frequency is determined by the characteristic time for fluids to diffuse in and out of cracks and grain boundaries. One rough estimate can be written as O'Connell and Budiansky (1977) did:

$$f_{\text{squirt}} \approx \frac{K_{\text{mineral}} \alpha^3}{\eta} \qquad (1.14)$$

where α is the crack aspect ratio and η is the fluid viscosity. The crack aspect ratio, of course, is poorly known. Gassmann fluid substitution is expected to work well at seismic frequencies significantly lower than f_{squirt}.

Jones (1983) compiled scant laboratory data that suggest $f_{\text{squirt}} \approx 10^4$ Hz for water-saturated sandstones. Obviously, slight changes in the rock microstructure can drastically change this. We do expect that the frequency should scale with viscosity as shown in equation (1.14), so that higher viscosities will decrease f_{squirt}.

We believe that Gassmann's relations are usually appropriate for 3D surface seismic frequencies. Exceptions would include reservoirs with very heavy (high-viscosity) oils, and rocks with tight microporosity.

Logs, at frequencies of 1–20 kHz, unfortunately fall right in the expected transition range. Gassmann's relations sometimes work quite well at log frequencies, and sometimes not. Nevertheless, our recommendation is to use Gassmann's relations at logging and surface seismic frequencies unless there are specific reasons to the contrary. The recommended procedure for relating laboratory work to the field is to use dry ultrasonic velocities and saturate them theoretically using the Gassmann relations.

Isotropic rocks

Gassmann's relations are strictly valid only for isotropic rocks. Brown and Korringa (1975) have published an anisotropic form, but even in laboratory experiments, there is seldom a complete enough characterization of the anisotropy to apply the Brown and Korringa relations.

Real rocks are almost always at least slightly anisotropic, making Gassmann's relations inappropriate in the strictest sense. We usually apply fluid substitution analysis to V_P and V_S measured in a single direction and ignore the anisotropy. This can sometimes lead to overprediction and at other times underprediction of the fluid effects (Sava et al., 2000). Nevertheless, the best approach, given the limited data available in the field, is to use Gassmann's relations on measured V_P and V_S, even though the rocks are anisotropic.

Single mineralogy

Gassmann's relations, like many rock physics models, are derived assuming a homogeneous mineralogy, whose bulk modulus is K_{mineral}. The standard way to proceed when we have a mixed mineralogy is to use an "average mineral." A simple way to estimate the bulk modulus of the average mineral is to compute upper and lower bounds of the mixture of minerals, and take their average: $K_{\text{mineral}} \approx (K^{\text{HS}+} + K^{\text{HS}-})/2$.

A common approach for adapting Gassmann's relations to rocks with mixed mineralogy is to estimate an "average" mineral. For example, we might estimate the mineral bulk modulus as $K_{\text{mineral}} \approx (K^{\text{HS}+} + K^{\text{HS}-})/2$, where $K^{\text{HS}+}$ and $K^{\text{HS}-}$ are the Hashin–Shtrikman upper and lower bounds on bulk modulus for the mineral mix. Another approach is to simply ignore the mixed mineralogy and use the modulus of the dominant mineral, for example $K_{\text{mineral}} \approx K_{\text{quartz}}$ for a sand or $K_{\text{mineral}} \approx K_{\text{calcite}}$ for a carbonate. One way to understand the impact of these assumptions is by looking at Figure 1.12. In this figure, the Gassmann fluid sensitivity is proportional to the contour spacing. When the rock modulus is low relative to the mineral modulus, the rock is "soft" and the Gassmann relations predict a large sensitivity to pore-fluid changes; when the rock modulus is high relative to the mineral modulus, the rock is "stiff" and Gassmann predicts a small sensitivity to fluids. Hence, picking a mineral modulus that is too stiff (i.e., ignoring soft clay) will make the rock look too soft, and therefore predict a fluid sensitivity that is too large. Sengupta (2000) showed that the sensitivity of the Gassmann prediction to uncertainty in mineral modulus is small, except for low porosities.

Other approaches to generalizing Gassmann's relations to mixed mineralogy have been explored theoretically. For example, Brown and Korringa (1975) generalized the Gassmann problem to anisotropic rocks and to the case where the pore compressibility and sample compressibility are unequal – a possible consequence of mixed mineralogy. Berryman and Milton (1991) solved the problem of fluid substitution in a composite consisting of two porous media, each with its own mineral and dry frame bulk moduli. Mavko and Mukerji (1998b) presented a probabilistic formulation of Gassmann's relations to account for distributions of porosity and dry bulk moduli, arising from natural variability.

Mixed saturation

Gassmann's relations were originally derived to describe the change in rock modulus from one pure saturation to another – from dry to fully brine-saturated, from fully brine-saturated to fully oil-saturated, etc. Domenico (1976) suggested that mixed gas–oil–brine saturations can also be modeled with Gassmann's relations, if the mixture of phases is replaced by an effective fluid with bulk modulus $\overline{K}_\text{fluid}$ and density $\overline{\rho}_\text{fluid}$ given by

$$\frac{1}{\overline{K}_\text{fluid}} = \frac{S_\text{gas}}{K_\text{gas}} + \frac{S_\text{oil}}{K_\text{oil}} + \frac{S_\text{br}}{K_\text{br}} = \left\langle \frac{1}{K_\text{fluid}(x,y,z)} \right\rangle \tag{1.15}$$

$$\overline{\rho}_\text{fluid} = S_\text{gas}\rho_\text{gas} + S_\text{oil}\rho_\text{oil} + S_\text{br}\rho_\text{br} = \langle \rho_\text{fluid}(x,y,z) \rangle \tag{1.16}$$

where $S_\text{gas,oil,br}$, $K_\text{gas,oil,br}$, and $\rho_\text{gas,oil,br}$ are the saturations, bulk moduli, and densities of the gas, oil, and brine phases. The operator $\langle \cdot \rangle$ refers to a volume average and allows for more compact expressions, where $K_\text{fluid}(x, y, z)$ and $\rho_\text{fluid}(x, y, z)$ are the spatially varying pore-fluid modulus and density.

Substituting equation (1.15) into Gassmann's relation is the procedure most widely used today to model fluid effects on seismic velocity and impedance for low-frequency field applications.

A problem with mixed fluid phases is that velocities depend not only on saturations but also on the spatial distributions of the phases within the rock. Equation (1.15) is applicable only if the gas, oil, and brine phases are mixed uniformly at a very small scale, so that the different wave-induced increments of pore pressure in each phase have time to diffuse and equilibrate during a seismic period. Equation (1.15) is the Reuss (1929) average or "isostress" average, and it yields an appropriate equivalent fluid when all pore phases have the same wave-induced pore pressure. A simple dimensional analysis suggests that during a seismic period pore pressures can equilibrate over spatial scales smaller than a critical length $L_\text{c} \approx \sqrt{\kappa K_\text{fluid}/f\eta}$, where f is the seismic frequency, κ is the permeability, and η and K_fluid are the viscosity and bulk modulus of the most viscous fluid phase. We refer to this state of fine-scale, uniformly mixed fluids as "uniform saturation." Table 1.1 gives some estimates of the critical mixing scale L_c.

1.3 Fluid substitution analysis

Table 1.1 *Critical diffusion length or patch size for some values of permeability and seismic frequency*

Frequency (Hz)	Permeability (mD)	L_c (m)
100	1000	0.3
	100	0.1
10	1000	1.0
	100	0.3

Permeabilities are in milliDarcy.

In contrast, saturations that are heterogeneous over scales larger than $\sim L_c$ will have wave-induced pore pressure gradients that cannot equilibrate during the seismic period, and equation (1.15) will fail. We refer to this state as "patchy saturation." Patchy saturation can easily be caused by fingering of pore fluids and spatial variations in wettability, permeability, shaliness, etc. The work of Sengupta (2000) suggests that patchy saturation is most likely to occur when there is free gas in the system. Patchy saturation leads to higher velocities and impedances than when the same fluids are mixed at a fine scale. The rock modulus with patchy saturation can be approximated by Gassmann's relation, with the mixture of phases replaced by the Voigt average effective fluid (Mavko and Mukerji, 1998):

$$\overline{K}_{\text{fluid}} = S_{\text{gas}} K_{\text{gas}} + S_{\text{oil}} K_{\text{oil}} + S_{\text{br}} K_{\text{br}} \tag{1.17}$$

Equation 1.17 appears to be an upper bound, and data seldom fall on it, except at very small gas saturations.

Figure 1.13 shows low-frequency P- and S-wave velocities versus water saturation for Estaillades limestone, measured by Cadoret (1993), using the resonant bar technique, near 1 kHz. The closed circles show data measured during increasing water saturation via an imbibition process combined with pressurization and depressurization cycles designed to desolve trapped air. The imbibition data can be accurately described by replacing the air–water mix with the fine-scale mixing model, equation (1.15), and putting the average fluid modulus into Gassmann's equations.

The open circles (Figure 1.13) show data measured during drainage. At saturations greater than 80%, the V_P fall above the fine-scale uniform saturation line but below the patchy upper bound, indicating a heterogeneous or somewhat patchy fluid distribution. The V_S data fall again on the uniform fluid line, as expected, since patchy saturation is predicted to have no effect on V_S (Mavko and Mukerji, 1998). Cadoret (1993) used X-ray CAT (computerized axial tomography) scans to confirm that the imbibition process did indeed create saturations uniformly distributed at a fine, sub-millimeter scale, while the drainage process created saturation patches at a scale of several centimeters.

24 Introduction to rock physics

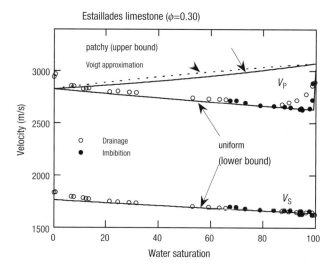

Figure 1.13 Low-frequency data from Cadoret (1993). Closed circles show data during imbibition and are in excellent agreement with the fine-scale effective fluid model (lower bound). Open circles show data during drainage, indicating heterogeneous or patchy fluid distributions for saturation greater than about 80%. The approximation (dashed line) using the Voigt average effective fluid does a good job of estimating the exact patchy upper bound shown by the solid line.

Brie and others (1995) presented an empirical fluid mixing equation, which spans the range of fine-scale to patchy mixing:

$$\overline{K}_{\text{fluid}} = (K_{\text{liquid}} - K_{\text{gas}})(1 - S_{\text{gas}})^e + K_{\text{gas}} \qquad (1.18)$$

where e is an empirical coefficient. When $e = 1$ equation (1.18) becomes the patchy upper bound, equation (1.17); when $e \to \infty$ equation (1.18) gives results resembling those of the fine-scale lower bound, equation (1.15). Values of $e \approx 3$ have been found empirically to give a better description of laboratory and simulated patchy behavior than the more extreme upper bound. In equation (1.18), K_{liquid} is the Reuss average mix of the oil and water moduli.

1.4 Pressure effects on velocity

There are at least four ways that pore pressure changes influence seismic signatures:
- Reversible elastic effects on the rock frame
- Permanent porosity loss from compaction and diagenesis
- Retardation of diagenesis from overpressure
- Pore fluid changes caused by pore pressure

1.4 Pressure effects on velocity

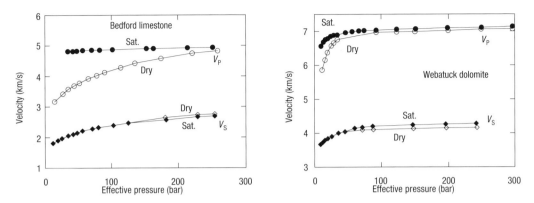

Figure 1.14 Seismic P- and S-wave velocities vs. *effective pressure* in two carbonates.

1.4.1 Reversible elastic effects in the rock frame

Seismic velocities in reservoir rocks almost always *increase* with effective pressure. Any bit of pore space tends to elastically soften a rock by weakening the structure of the otherwise rigid mineral material. This decrease in elastic moduli usually results in a decrease of the rock P- and S-wave velocities. *Effective pressure* (confining pressure – pore pressure) acts to stiffen the rock frame by mechanically eliminating some of this pore space – closing microcracks and stiffening grain contacts (Nur and Simmons, 1969; Nur, 1971; Sayers, 1988; Mavko *et al.*, 1995). This most compliant, crack-like part of the pore space, which can be manipulated with stress, accounts for many of the seismic properties that are interesting to us: the sensitivity to pore pressure and stress, the sensitivity to pore fluids and saturation, attenuation and dispersion.

Figure 1.14 shows examples of velocity vs. effective pressure measured on a limestone and a dolomite. In each case the pore pressure was kept fixed and the confining pressure was increased, resulting in frame stiffening which increased the velocities. In consolidated rocks this type of behavior is relatively elastic and reversible up to effective pressures of 30–40 MPa; i.e., reducing the effective stress *decreases* the velocities, just as increasing the effective stress *increases* the velocities (usually with only small hysteresis).

> **Caution**
>
> In soft, poorly consolidated sediments significant compaction can occur (see next section). This causes the velocity vs. effective pressure behavior to be inelastic and irreversible, with very large hysteresis.

Laboratory experiments have indicated that the reversible pressure effects illustrated in Figure 1.14 depend primarily on the *difference* between confining pressure and

26 Introduction to rock physics

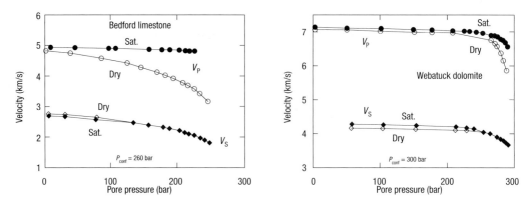

Figure 1.15 Seismic P- and S-wave velocities vs. pore pressure in two carbonates.

pore pressure. An increase of pore pressure tends approximately to cancel the effect of confining pressure, pushing open the cracks and grain boundaries, and hence decreasing the velocity. Figure 1.15 shows the data from Figure 1.14 replotted to illustrate how increasing *pore pressure* decreases velocities. The same effective pressures are spanned by varying pore pressure with confining pressure (P_{conf}) fixed.

> A trivial, but critical, point is that the sensitivity of velocity or impedance to pressure (the slopes of the curves in Figures 1.14 and 1.15) depends on what part of the curve we are working on. At low effective stresses, when cracks are open and grain boundaries are loose, there is large sensitivity to pressure; at high effective pressures, when the rock is stiff, we expect much smaller sensitivity. Related to this, there is a limit to the range of pressure change we can see. For the rocks in Figures 1.14 and 1.15, there is a large change in velocity for a pore pressure drop of 200 bar (20 MPa). After that, the cracks are closed and the frame is stiffened, so additional pressure drops might be difficult to detect (except for bubble-point saturation changes).

Numerous authors (Nur and Simmons, 1969; Nur, 1971; Mavko *et al.*, 1995; Sayers, 1988) have shown that these reversible elastic stress and pore pressure effects can be described using crack and grain contact models. Nevertheless, *our ability to predict the sensitivity to pressure from first principles is poor.* The current state of the art requires that we calibrate the pressure dependence of velocity with core measurements. Furthermore, when calibrating to core measurements, it is very important to use *dry* core data, to minimize many of the artifacts of high-frequency dispersion.

A convenient way to quantify the dependence, taken from the average of several samples, is to normalize the velocities for each sample by the high-pressure value, as shown for some clastics in Figure 1.16. This causes the curves to cluster at the high-pressure point. Then we fit an average trend through the cloud, as shown. The velocity

1.4 Pressure effects on velocity

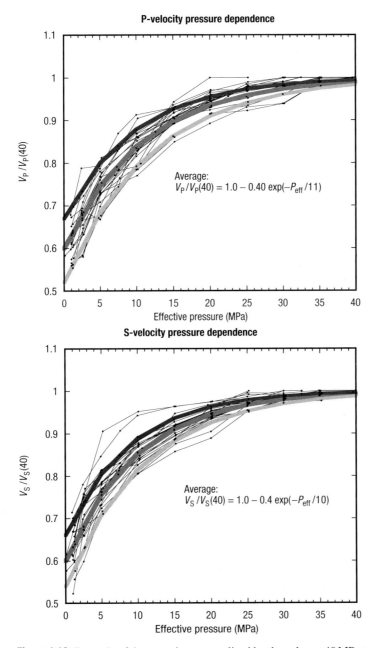

Figure 1.16 Example of dry core data, normalized by the value at 40 MPa to extract the average velocity–pressure function.

change between any two *effective pressures* P_{eff1} and P_{eff2} can be conveniently written as:

$$V_P(P_{\text{eff2}}) = V_P(P_{\text{eff1}}) \frac{1 - A_P \exp(P_{\text{eff2}}/P_{0P})}{1 - A_P \exp(P_{\text{eff1}}/P_{0P})};$$
$$V_S(P_{\text{eff2}}) = V_S(P_{\text{eff1}}) \frac{1 - A_S \exp(P_{\text{eff2}}/P_{0S})}{1 - A_S \exp(P_{\text{eff1}}/P_{0S})}$$
(1.19)

where A_P, A_S, P_{0P}, and P_{0S} are empirical constants. We write separate equations for V_P and V_S to emphasize that they can have different pressure sensitivities. For example, we often observe in the laboratory that dry-rock V_P/V_S increases with effective pressure.

> We know of no systematic relation between the parameters in a rock's pressure dependence, as in equation (1.19), and the rock type, age, or depth. Hence, site-specific calibration is recommended.

1.4.2 Permanent porosity loss from compaction, crushing, and diagenesis

Effective stress, if large enough, or held long enough, will help to reduce porosity permanently and inelastically. In the first few tens of meters of burial at the ocean floor, mechanical compaction (and possibly crushing) is the dominant mechanism of porosity reduction. At greater depths, all sediments suffer porosity reduction via pressure solution, which occurs at points of stress concentration. Stylolites are some of the most dramatic demonstrations of stress-enhanced dissolution in carbonate rocks. Seismic velocity varies inversely with porosity. Therefore, as stress leads to a permanent reduction of porosity, we generally expect a corresponding irreversible increase of velocity.

1.4.3 Retardation of diagenesis from overpressure

Earlier in this chapter, we discussed velocity–porosity relations. Some of these can be understood in terms of porosity loss and rock stiffening with time and depth of burial. A number of authors have discussed how anomalous overpressure development can act to retard the normal porosity loss with depth. In other words, overpressure helps to maintain porosity and keep velocity low. Hence, anomalously low velocities can be an indicator of overpressure (Fertl *et al.*, 1994; Kan and Sicking, 1994).

1.4.4 Pore fluid changes caused by pore pressure

Seismic velocities can depend strongly on the properties of the pore fluids. All pore fluids tend to increase in density and bulk modulus with increasing pore pressure. The pressure effect is largest for gases, somewhat less for oil, and smallest for brine.

1.4 Pressure effects on velocity

When calculating these fluid effects, it is obviously critical to use appropriate fluid properties. To our knowledge, the Batzle and Wang (1992) empirical formulas represent the best summary of published fluid data for use in seismic fluid substitution analysis. It is possible that some oil companies have internal proprietary data that are equally good alternatives.

Rock physics results regarding pore pressure
- The elastic frame effects are important for 4D seismic monitoring of man-made changes in pore pressure during reservoir production.
- Numerous authors have shown that these reversible elastic stress and pore pressure effects can be described using crack and grain contact models. Nevertheless, our ability to predict the sensitivity to pressure from first principles is poor. The current state of the art requires that we calibrate the pressure dependence of velocity with core measurements.
- Since the pressure dependence that we seek results from microcracks, we must be aware that at least part of the sensitivity of velocity to pressure that we observe in the lab is the result of damage to the core. Therefore, we believe that laboratory measurements should be interpreted as an upper bound on pressure sensitivity, compared with what we might see in the field.
- Permanent pore collapse during production has been studied extensively in rock mechanics, particularly to understand changes of reservoir pressure and permeability during production in chalks. However, we are not aware of much work quantifying the corresponding seismic velocity changes. These must be measured on cores for the reservoirs of interest.
- Overpressure tends to lower seismic velocities by retarding normal porosity loss, in the same sense as the reversible frame effects of pore pressure discussed above. Nevertheless, these result from entirely different mechanisms and the two should not be confused.
- A few authors claim to understand the relation between overpressure and porosity from "first principles." We believe that porosity–pressure relations can be developed that have great predictive value, but the most reliable of these will be empirical.
- The effect of pore pressure on fluids is opposite to the effect of pore pressure on the rock frame. Increased pore pressure tends to decrease rock velocity by softening the elastic rock frame, but it tends to increase rock velocity by stiffening the pore fluids. The net effect of whether velocity will increase or decrease will vary with each situation.

> **Pitfall**
>
> The decrease in the P-wave velocity with increasing pore pressure has been commonly used for overpressure detection. However, velocity does not uniquely indicate pore pressure, because it also depends, among other factors, on pore fluids, saturation, porosity, mineralogy, and texture of rock.

1.5 The special role of shear-wave information

This section focuses on the rock physics basis for use of shear-wave information in reservoir characterization and monitoring. Adding shear-wave information to P-wave information often allows us to better separate the seismic signatures of lithology, pore fluid type, and pore pressure. This is the fundamental reason why, for example, AVO and Elastic Impedance analysis have been successful for hydrocarbon detection and reservoir characterization. Shear data also provide a strategy for distinguishing between pressure and saturation *changes* in 4D seismic data. Shear data can provide the means for obtaining images in gassy sediments where P-waves are attenuated. Shear-wave splitting provides the most reliable seismic indicator of reservoir fractures.

One practical problem is that shear-wave information does not always help. Factors such as rock stiffness, fluid compressibility and density, target depth, signal-to-noise, acquisition and processing can limit the effectiveness of AVO. We will try to give some insights into these problems.

Another issue is the choice of shear-related attributes. Rock physics people tend to think in terms of the measured quantities V_P, V_S, and density, but one can derive equivalent combinations in terms of P and S impedances, acoustic and elastic impedances, the Lamé elastic constants λ and μ, R_0, G (AVO), etc. While these are mathematically equivalent, different combinations are more natural for different field data situations, and different combinations have different intrinsic sources of measurement error.

1.5.1 The problem of nonuniqueness of rock physics effects on V_P and V_S

Figure 1.17 shows ultrasonic velocity data for the water-saturated sandstones (Han, 1986) that we discussed earlier. Recall the general trend of decreasing velocity with increasing porosity. We know from Han's work that the scatter around the trend is well correlated with the clay content. More generally, increasing clay content or poorer sorting tend to decrease the porosity with a small change in velocity. The result is that within this data set, the combined variations in porosity and clay (lithology) account for almost a factor of 2 variation in V_P and more than a factor of 2 variation in V_S. A single measurement of V_P or V_S would do little to constrain the rock properties.

Figure 1.18 shows similar laboratory trends in velocity vs. porosity for a variety of typical limestones reported by Anselmetti and Eberli (1997), with some model trends superimposed. In this case, we can understand the scatter about the trends in terms of pore microstructure. Somewhat like the sorting effect in sandstones, we can think of this as a textural variation. In this case there is a factor of 3 variation in V_P resulting from variations in porosity and pore geometry.

1.5 The special role of shear-wave information

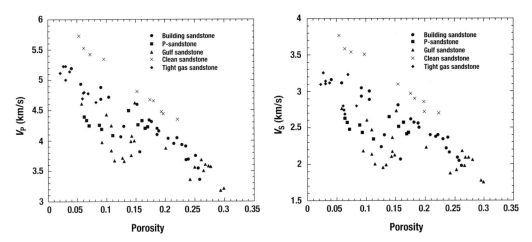

Figure 1.17 Velocity vs. porosity for water-saturated sandstones at 40 MPa. Data are ultrasonic measurements from Han (1986).

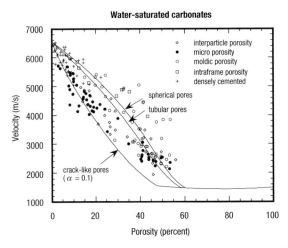

Figure 1.18 Comparison of carbonate data with classical, idealized pore-shape models. The aspect ratio of the idealized ellipsoidal pores is denoted by α.

Figure 1.19 again shows velocities from the sandstones in Figure 1.17. The figure on the left now includes a range of effective pressures: 5, 10, 20, 30, and 40 MPa. The figure on the right adds velocities for the corresponding gas-saturated case. Now we observe nearly a factor of 3 variation in P-wave velocity, resulting from a complicated mix of porosity, clay, effective pressure, and saturation. Clearly, attempting to map any one of these parameters from P-velocity alone would produce hopelessly nonunique answers.

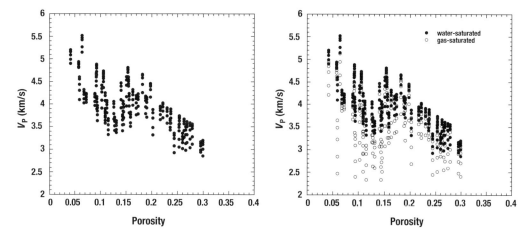

Figure 1.19 Left: Sandstone velocity vs. porosity data from the same sandstone samples as in Figure 1.17, but now with effective pressures at 5, 10, 20, 30, and 40 MPa, showing the additional scatter or ambiguity caused when both pressure and porosity are variable. Right: Additional variation due to gas vs. water saturation.

> The point we emphasize is that effects on seismic V_P and V_S from pore-fluid saturation, pore pressure, porosity, and shaliness can all be comparable in magnitude, and intermixed. Each of these can be an important reservoir parameter, but separating them is one of the fundamental sources of nonuniqueness in both 4D studies and reservoir characterization. Quite simply, there are many more interesting unknown rock and fluid properties than there are independent acoustic measurements. We will discuss in the next section the fundamental result that combining V_P and V_S allows some of the effects to be separated.

1.5.2 The rock physics magic of V_P combined with V_S

Figure 1.20 shows all of Han's water-saturated data from Figure 1.19, plotted as V_P vs. V_S. Blangy's (1992) water-saturated Troll data and Yin's (1992) water-saturated unconsolidated sand data are also included. We see that all of the data now fall along a remarkably simple and narrow trend, in spite of porosity ranging from 4% to 40%, clay content ranging from 0% to 50%, and effective pressure ranging from 5 to 50 MPa.

We saw earlier that porosity tends to decrease velocity. We see here (Figure 1.20) that porosity acts similarly enough on both V_P and V_S that the data stay tightly clustered within the same trend. We also saw that clay tends to lower velocity. Again, clay acts similarly enough on both V_P and V_S that the data stay tightly clustered within the same trend – and the same for effective pressure. The only thing common to the data in Figure 1.20 is that they are all water-saturated sands and sandstones.

Figure 1.21 shows the same data as in Figure 1.20, with gas-saturated rock velocity data superimposed. The gas- and water-saturated data fall along two well-separated trends.

1.5 The special role of shear-wave information

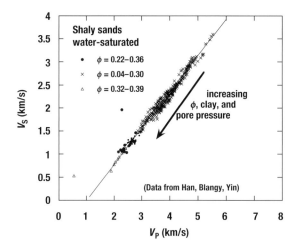

Figure 1.20 V_S vs. V_P for water-saturated sandstones, with porosities, ϕ, ranging from 4% to 40%, effective pressures 5–50 MPa, clay fraction 0–50%. Arrow shows direction of increasing porosity, clay, pore pressure. Data are from Han (1986), Blargy (1992) and Yin (1992).

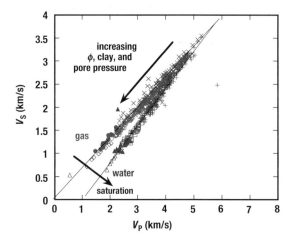

Figure 1.21 Plot of V_S vs. V_P for water-saturated and gas-saturated sandstones, with porosities of 4–40%, effective pressures 5–50 MPa, clay fraction 0–50%. Arrow shows direction of increasing porosity, clay, pore pressure. The trend of saturation is perpendicular to that for porosity, clay, pore pressure.

> The remarkable pattern in Figure 1.21 is at the heart of virtually all direct hydrocarbon detection methods. In spite of the many competing parameters that influence velocities, the nonfluid effects on V_P and V_S are similar. Variations in porosity, shaliness, and pore pressure move data up and down along the trends, while changes in fluid saturation move data from one trend to another. (Large changes in lithology,

> such as to carbonates, also move data to separate trends.) For reservoir monitoring, the key result is that changes in saturation and changes in pore pressure are nearly perpendicular in the (V_P, V_S) plane. Similar separation of pressure and saturation effects can be seen in other related attribute planes (R_0, G) (Landrø, 2001), (ρV_P, ρV_S), etc.

1.5.3 V_P–V_S relations

Relations between V_P and V_S are key to the determination of lithology from seismic or sonic log data, as well as for direct seismic identification of pore fluids using, for example, AVO analysis. Castagna *et al.* (1993) give an excellent review of the subject. There is a wide, and sometimes confusing, variety of published V_P–V_S relations and V_S prediction techniques, which at first appear to be quite distinct. However, most reduce to the same two simple steps:

(1) Establish empirical relations among V_P, V_S, and porosity ϕ for one reference pore fluid – most often water-saturated or dry.
(2) Use Gassmann's (1951) relations to map these empirical relations to other pore-fluid states.

Although some of the effective-medium models predict both P and S velocities assuming idealized pore geometries, the fact remains that the most reliable and most often used V_P–V_S relations are empirical fits to laboratory and/or log data. The most useful role of theoretical methods is in extending these empirical relations to different pore fluids or measurement frequencies – hence the two steps listed above.

We summarize here a few of the popular V_P–V_S relations, compared with lab and log data sets.

Limestones

Figure 1.22 shows laboratory ultrasonic V_P–V_S data for water-saturated limestones from Pickett (1963), Milholland *et al.* (1980), and Castagna *et al.* (1993), as compiled by Castagna *et al.* (1993). Superimposed, for comparison, are Pickett's (1963) empirical limestone relation, derived from laboratory core data:

$$V_S = V_P/1.9 \text{ (km/s)}$$

and a least-squares polynomial fit to the data derived by Castagna *et al.* (1993):

$$V_S = -0.05508 V_P^2 + 1.0168 V_P - 1.0305 \text{ (km/s)}$$

At higher velocities, Pickett's straight line fits the data better, although at lower velocities (higher porosities), the data deviate from a straight line and trend toward the water point, $V_P = 1.5$ km/s, $V_S = 0$. In fact, this limit is more accurately described as a suspension of grains in water at the critical porosity (see discussion below), where the grains lose contact and the shear velocity vanishes.

1.5 The special role of shear-wave information

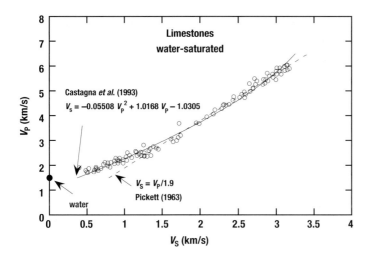

Figure 1.22 Plot of V_P vs. V_S data for water-saturated limestones with two empirical trends superimposed. Data compiled by Castagna *et al.* (1993).

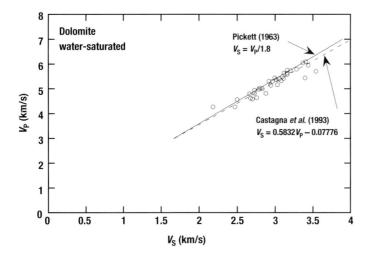

Figure 1.23 Plot of V_P vs. V_S data for water-saturated dolomites with two empirical trends superimposed. Data compiled by Castagna *et al.* (1993).

Dolomite

Figure 1.23 shows laboratory V_P–V_S data for water-saturated dolomites from Castagna *et al.* (1993). Superimposed, for comparison, are Pickett's (1963) dolomite (laboratory) relation:

$$V_S = V_P/1.8 \text{ (km/s)}$$

and a least-squares linear fit (Castagna *et al.*, 1993):

$$V_S = 0.5832 V_P - 0.0777 \text{ (km/s)}$$

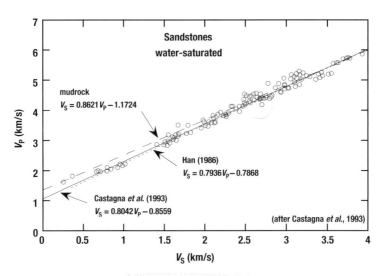

Figure 1.24 Plot of V_P vs. V_S data for water-saturated sandstones with three empirical trends superimposed. Data compiled by Castagna et al. (1993).

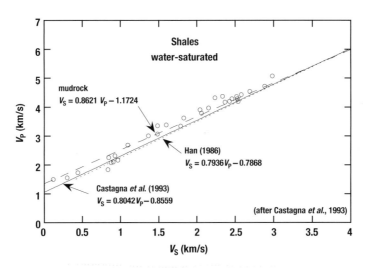

Figure 1.25 Plot of V_P vs. V_S data for water-saturated shales with three empirical trends superimposed. Data compiled by Castagna et al. (1993).

For the data shown, the two relations are essentially equivalent. The data range is too limited to speculate about behavior at much lower velocity (higher porosity).

Sandstones and shales

Figures 1.24 and 1.25 show laboratory V_P–V_S data for water-saturated sandstones and shales from Castagna et al. (1985, 1993) and Thomsen (1986) as compiled by

1.5 The special role of shear-wave information

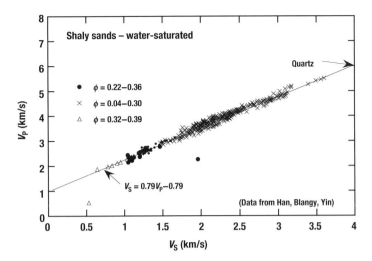

Figure 1.26 Plot of V_P vs. V_S data for water-saturated sandstones with three empirical trends superimposed. Data from Han (1986), Blangy (1992) and Yin (1992).

Castagna *et al.* (1993). Superimposed, for comparison, are a least-squares linear fit to these data offered by Castagna *et al.* (1993):

$$V_S = 0.8042 V_P - 0.8559 \text{ (km/s)}$$

together with the famous "mudrock line" of Castagna *et al.* (1985), which was derived from *in situ* data:

$$V_S = 0.8621 V_P - 1.1724 \text{ (km/s)}$$

and the empirical relation of Han (1986), based on laboratory ultrasonic data:

$$V_S = 0.7936 V_P - 0.7868 \text{ (km/s)}$$

Of these three relations, those by Han and by Castagna *et al.* are essentially the same and give the best overall fit to the sandstones in Figure 1.24. The mudrock line predicts systematically lower V_S, because it is best suited for the most shaly samples, as seen in Figure 1.25. Castagna *et al.* (1993) suggest that if the lithology is well known, then one might fine-tune these relations to slightly lower V_S/V_P for high shale content, and higher V_S/V_P in cleaner sands. When the lithology is not well constrained, then the Han and Castagna *et al.* lines give a reasonable average.

Figure 1.26 compares laboratory ultrasonic data for a larger set of water-saturated sands. The lowest-porosity samples ($\phi = 0.04$–0.30) are from a set of consolidated shaly Gulf Coast sandstones studied by Han (1986). The medium porosities ($\phi = 0.22$–0.36) are poorly consolidated North Sea samples studied by Blangy (1992). The very-high-porosity samples ($\phi = 0.32$–0.39) are unconsolidated clean Ottawa sand studied by Yin (1992). The samples span clay volume fractions from 0 to 55%, porosities from 0.22 to

Introduction to rock physics

Table 1.2 *Regression coefficients for the Greenberg–Castagna relations for V_S prediction*

Lithology	a_{i2}	a_{i1}	a_{i0}
Sandstone	0	0.80416	−0.85588
Limestone	−0.05508	1.01677	−1.03049
Dolomite	0	0.58321	−0.07775
Shale	0	0.76969	−0.86735

0.39, and confining pressures from 0 to 40 MPa. In spite of this, there is a remarkably systematic trend, represented well by Han's relation:

$$V_S = 0.79\, V_P - 0.79 \text{ (km/s)}$$

Greenberg and Castagna (1992) have given empirical relations for estimating V_S from V_P in multimineralic, brine-saturated rocks based on empirical, polynomial V_P–V_S relations in pure monomineralic lithologies. For each single lithology, they estimate $V_S = a_{i2}\, V_P^2 + a_{i1}\, V_P + a_{i0}$, where V_P is the water-saturated P-wave velocity, and V_S is the predicted water-saturated S-wave velocity. Both V_P and V_S are in km/s. Their regression coefficients a_{ij} for the individual lithologies are listed in Table 1.2.

The shear-wave velocity in brine-saturated composite lithologies is approximated by a simple average of the arithmetic and harmonic means of the constituent pure-lithology shear velocities:

$$V_S = \frac{1}{2}\left\{\left[\sum_{i=1}^{L} X_i \sum_{j=0}^{N_i} a_{ij} V_P^j\right] + \left[\sum_{i=1}^{L} X_i \left(\sum_{j=0}^{N_i} a_{ij} V_P^j\right)^{-1}\right]^{-1}\right\}$$

$$\sum_{i=1}^{L} X_i = 1$$

where

L = number of pure monomineralic lithologic constituents
X_i = volume fractions of lithologic constituents
a_{ij} = empirical regression coefficients
N_i = order of polynomial for constituent i
V_P^j = water-saturated P-wave velocity in the jth rock facies
V_S = S-wave velocities (km/s) in composite brine-saturated, multimineralic rock

Figure 1.27 shows the relations for monomineralic lithologies. Note that the above relation is for 100% brine-saturated rocks. To estimate V_S from measured V_P for other fluid saturations, Gassmann's equation has to be used in an iterative manner.

1.5 The special role of shear-wave information

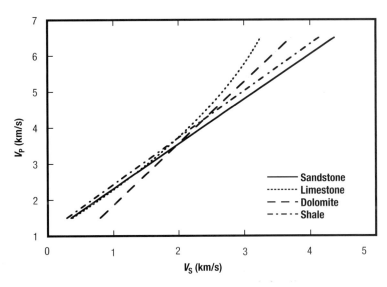

Figure 1.27 Relations between V_P and V_S obtained by Greenberg and Castagna (1992) for monomineralic lithologies.

1.5.4 Shear-related attributes

Our rock physics discussion has been generically stated in terms of P- and S-wave velocities (V_P, V_S). However, various field acquisition schemes naturally suggest other related attributes. We simply review a few definitions here.

The rock physics bottle-neck: only three key seismic parameters

Virtually all of rock physics applied to the seismic problem deals with just three fundamental pieces of information: P-wave velocity, S-wave velocity, and density. These are the only three seismic parameters that are typically measured in the laboratory, and these are generally the most we can ever hope for from field data (logs or seismic). An exception might be attenuation (equivalent to velocity dispersion), which is not very well understood or accurately measured. So this leaves us with at most 3 or 3.5 bits of seismic information on which to base our interpretations.

Similarly, the most a seismologist can ever hope to learn about interval properties from inverting seismic data is the same 3 or 3.5 bits of information at each location. Any seismic inversion process amounts to building a 3D Earth model, with V_P, V_S, and density assigned to every pixel. Synthetic seismograms are computed from the model and compared with observed seismic wiggles. The V_P, V_S, and density values are adjusted until a sufficiently good match is obtained. When the best match is poor, we consider using finer grids, or blame the modeling algorithm, processing artifacts, noise, etc. We never consider that more than V_P, V_S, and density are needed to specify

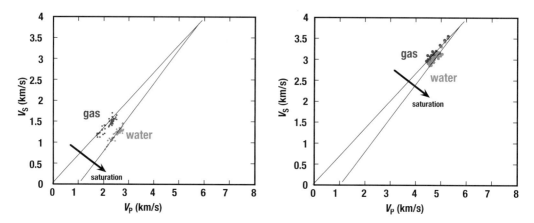

Figure 1.28 Estimating the best-case, error-free rock properties can be useful. On the left, soft, high-porosity rocks with good potential for detecting pressure and saturation changes. There is a lot of overlap in V_P only, but not in the 2D V_P–V_S plane. On the right, low-porosity stiff rocks offering little chance of detecting fluid and pressure effects – the problem is close to hopeless, suggesting that special shear acquisition would be a waste.

the rock properties, except for Q or attenuation, or velocity dispersion, or anisotropy. In fact, the wave equation doesn't require anything more. The main reason that many more attributes are sometimes measured is to reveal the *geometric arrangement* of rock types.

A related discussion is to show how to use calibration data to estimate quantitatively the rock physics uncertainty when interpreting data. For any given reservoir, we believe it is useful to quantify the "best-case" interpretation uncertainty that we would have if we could measure V_P, V_S, and density *error-free*. In this case, the interpretation accuracy will be limited by geologic parameters, such as mineralogy, pore stiffness, fluid contrasts, shaliness, etc. This is the "intrinsic resolvability" of the reservoir parameters. The value of quantifying the best-case uncertainty is that we will be able to identify and avoid hopeless interpretation problems right from the start. These will be the field problems where no amount of geophysical investment will allow accurate rock physics interpretation (Figure 1.28).

For most other situations, we can estimate how the uncertainty worsens compared with the best case, (1) when measurement errors are introduced, (2) when we drop from three parameters to two (e.g. V_P, V_S), or (3) when using alternative pairs of attributes (R_0, G, or ρV_P, ρV_S, etc.). We believe that this kind of analysis can be helpful in the decision-making process to find the most cost-effective use of shear-wave data.

Other shear-related attributes

The attribute pairs in the following short list are all algebraically derivable from V_P, V_S, and density, ρ or their contrasts.

1.5 The special role of shear-wave information

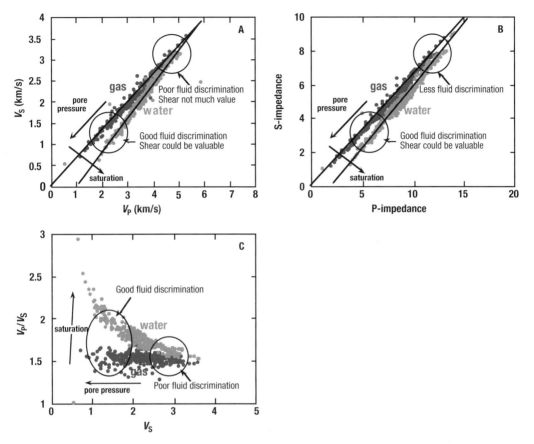

Figure 1.29 Saturation and pressure discrimination are very similar in different attribute domains. A. Plot of V_P vs. V_S sandstone data, showing the value of combining P- and S-wave data for separating lithology, pore pressure, and saturation (same as Figure 1.21). B. Same rock samples as in A, but plotted as P- and S-wave impedances. C. Same rock samples plotted as V_P/V_S vs. V_S.

V_P, V_S	P and S velocities
R_0, G	P–P AVO intercept and gradient
$\rho V_P, \rho V_S$	P and S impedances
AI, EI	acoustic impedance, elastic impedance (inverted from far offset stacks)
R_0, G_{PS}	P–S AVO gradient, with normal incidence P–P reflectivity
λ, μ	elastic Lamé coefficients

Figure 1.29 compares water- and gas-saturated data (same as in Figure 1.21) in three different domains: (V_P, V_S), $(\rho V_P, \rho V_S)$, and $(V_P/V_S, V_S)$. All three plots show a similar interpretability. Gas- vs. water-saturated rocks are well separated in all three domains when velocities are low (rocks are soft), and they are poorly separated when the velocities are high. Also, the trend for changes in pore pressure is essentially perpendicular to the trend for a change in saturation, in all three domains.

Information theory tells us that the intrinsic information in a data set does not change under coordinate transformation. Hence, interpretation of data plotted in the (V_P, V_S) domain should not be different than, for example, in the (V_P/V_S, V_S) domain. On the other hand, the problem is different in the (ρV_P, ρV_S) domain, because of the additional parameter, density.

The more important practical consideration, which this theoretical statement ignores, is the difference in measurement errors associated with the different domains. For example, P–P AVO attributes (R_0, G) have errors associated with noise, amplitude picking, phase changes with offset, velocity estimation, nonhyperbolic moveout, anisotropy, fitting a $\sin^2 \theta$ function to the amplitudes, etc. Values of V_P/V_S determined from comparing interval times on P-wave and converted shear-wave stacks have errors associated with incorrect moveout, migration difficulties, anisotropy, time-picking, correlation of the P and S events, etc. We suspect that image quality, signal-to-noise, and measurement uncertainty, in general, are the biggest practical differences among the different types of shear data. A critical part of acquisition decision-making is forward modeling to estimate these errors.

1.6 Rock physics "*What ifs?*": fluid and lithology substitution

One of the most powerful uses of rock physics is *extrapolation*. At a well – assuming that data quality is good – we pretty much know "the answer." Cuttings, cores, and logs tell us about the lithology, porosity, permeability, and fluids. And, assuming that there is a good tie between seismic and synthetics, we might even say that we understand the seismic data at the well.

The problem is, often, knowing what happens as we move away from the well. As we move laterally or vertically, what happens if the porosity changes, or if the lithology changes, or if the fluids change? How would the well-log V_P, V_S, and density change? How would the seismic data change?

This is the role of the rock physics "*What ifs?*" Using the various trends and transformations presented in this chapter, we can extrapolate to conditions that might exist away from the well, and then compute synthetics to explore how the seismic signatures might change. This is particularly useful when we wish to understand the seismic signatures of fluids and facies that are not represented in the well. For statistical methods, such as clustering analysis or neural networks, such extrapolations are critical for extending the training data.

The best known example is, "What if the pore fluids change?" This is the fluid substitution problem. In Plate 1.30 we show well logs penetrating a sandy North Sea turbidite sequence. Along with it (top right) are the corresponding normal-incidence synthetics, assuming a 50 Hz Ricker wavelet. The initial logs showed an average water saturation of about 10% in the thick sand, with light oil of 35 API, GOR (Gas Oil

Ratio) of 200 Sm^3/Sm^3. We apply a Gassmann fluid substitution, increasing the water saturation to 90%, with brine salinity of 30 000 parts per million (ppm). The predicted results are shown at the bottom of Plate 1.30. On the left, we see that replacing light oil with water increases the density and P-wave velocity in the sand. The impedance increases by about 8%. The synthetics show that the fluid substitution results in both amplitude changes and traveltime pullup (earlier arrivals).

A very different rock physics "*What if?*" is what we call "lithology substitution." Plate 1.31 again shows the logs through the sandy turbidite sequence as in Plate 1.30. Now we ask, "What if the porosity changes?" At the top of Plate 1.31, we predict a porosity reduction of 3% associated with an increase of cementing. We model the velocity change using the *cementing trend*, described by a modified upper Hashin–Shtrikman bound, as in Figures 1.9 and 1.10. We see that decreasing the porosity increases the density in the sand, and causes large increases in V_P and impedance. The synthetics show a large change in amplitude and traveltime.

At the bottom of Plate 1.31, we predict a porosity reduction of 3%, this time associated with a deterioration of sorting. We model the velocity change using the *sorting trend*, described by a lower Hashin–Shtrikman bound, as in Figures 1.9 and 1.10. We see that decreasing the porosity increases the density in the sand, and causes much smaller increases in V_P and impedance.

Finally, Plate 1.32 shows what happens if the pore pressure changes. In this case, the pore pressure declines by 5 MPa (effective pressure increasing from 10 MPa to 15 MPa). The result is virtually no effect on density, but fairly large increases in V_P and impedance.

1.7 All models are wrong . . . some are useful

Most rock physics models relevant to the scope of this book are aimed at describing relations between measurable seismic parameters and rock-fluid properties. While it is not our intention to review all models exhaustively, we find that many fall within three general classes: theoretical, empirical, and heuristic.

1.7.1 Theoretical models

The *theoretical models* are primarily continuum mechanics approximations of the elastic, viscoelastic, or poroelastic properties of rocks. Among the most famous are the poroelastic models of Biot (1956), who was among the first to formulate the coupled mechanical behavior of a porous rock embedded with a linear viscous fluid. The Biot equations reduce to the famous Gassmann (1951) relations at zero frequency; hence, we often refer to "Biot–Gassmann" fluid substitution. Biot (1962) generalized his formulation to include a viscoelastic frame, which was later pursued by Stoll and Bryan (1970). The "squirt model" of Mavko and Nur (1975) and Mavko and Jizba (1991) quantified a grain-scale fluid interaction, which contributed to the frame viscoelasticity. Dvorkin and Nur (1993) explicitly combined Biot and squirt mechanisms in their "Bisq" model.

Elastic models tend to be (1) inclusion models, (2) contact models, (3) computational models, (4) bounds, and (5) transformations.

(1) Inclusion models usually approximate the rock as an elastic solid containing cavities (inclusions), representing the pore space. Because the inclusion cavities are more compliant than solid mineral, they have the effect of reducing the overall elastic stiffness of the rock in either an isotropic or anisotropic way. The vast majority of inclusion models assume that the pore cavities are ellipsoidal or "penny-shaped" (Eshelby, 1957; Walsh, 1965; Kuster and Toksöz, 1974; O'Connell and Budiansky, 1974, 1977; Cheng, 1978, 1993; Hudson, 1980, 1981, 1990; Crampin, 1984; Johansen et al., 2002). Berryman (1980) generalized the self-consistent description so that *both* the pores and grains are considered to be ellipsoidal "inclusions" in the composite. Mavko and Nur (1978) and Mavko (1980) also considered inclusion cavities that were non-ellipsoidal in shape. Schoenberg (1983) and Pyrak-Nolte et al. (1990a,b) have considered inclusions in the form of infinite planes of slip or compliance, as models of fractures.

Inclusion models have contributed tremendous insights as elastic analogs of rock behavior. However, their limitation to idealized (and unrealistic) pore geometries has always made comparing the models to actual pore microgeometry difficult. For example, relating inclusion models to variations of rock texture resulting from different depositional or diagenetic processes is not feasible. Quite simply, if we observe a rock in thin section, scanning electron microscope (SEM) image, or outcrop, we really do not have a satisfactory way of choosing model parameters (such as inclusion density and aspect ratio) to describe what we see. Workers often "invert" for the model parameters that give a good fit to measured elastic properties, but the question always remains, "How well do these parameters represent real rock textures?"

(2) Contact models approximate the rock as a collection of separate grains, whose elastic properties are determined by the deformability and stiffness of their grain-to-grain contacts. Most of these (Walton, 1987; Digby, 1981; Norris and Johnson, 1997; Makse et al., 1999) are based on the Hertz–Mindlin (Mindlin, 1949) solution for the elastic behavior of two elastic spheres in contact. The key parameters determining the stiffness of the rock are the elastic moduli of the spherical grains and the area of grain contact, which results from the deformability of the grains under pressure. Dvorkin and Nur (1996) described the effect of adding small amounts of mineral cement at the contacts of spherical grains.

As with the inclusion models, the spherical contact models have served as useful elastic analogs of soft sediments, but they also suffer from their extremely idealized geometries. They are not easy to extend to realistic grain shapes, or distributions of grain size. Furthermore, the most rigorous part of the contact models is the formal description of a single grain-to-grain contact. To extrapolate this to a random packing of spheres makes sweeping assumptions about the number of contacts per grain, and the distribution of contact forces throughout the composite.

(3) Computational models are a fairly recent phenomenon. In these, the actual grain–pore microgeometry is determined by careful thin-section or CT-scan imaging. This geometry is represented by a grid, and the elastic, poroelastic, or viscoelastic behavior is computed by "brute force," using finite-element, finite-difference, or discrete-element modeling. Clear advantages of these models are the freedom from geometric idealizations, and the ability to elastically quantify features observed in thin section. A second, older class of computational contact models are the discrete-element models, which attempt to simulate the simultaneous interactions of many free-body grains in a soft sediment.

(4) Bounds, such as the Voigt–Reuss or Hashin–Shtrikman bounds presented in this chapter are, in our opinion, the "silent heroes" of rock models. They are extremely robust and relatively free of idealizations and approximations, other than representing the rock as an elastic composite. Originally, bounds were treated only as describing the limits of elastic behavior; some even considered them of only limited usefulness. However, as shown in Chapters 1 and 2, they have turned out to be valuable "mixing laws" that allow accurate interpolation of sorting and cementing trends, as well as being the rigorously correct equations to describe suspensions and fluid mixtures.

(5) Transformations include models such as the Gassmann (1951) relations for fluid substitution, which are relatively free of geometric assumptions. The Gassmann relations take measured V_P and V_S at one fluid state and predict the V_P and V_S at another fluid state. Berryman and Milton (1991) presented a geometry-independent scheme to predict fluid substitution in a composite of two porous media having separate mineral and dry-frame moduli. Mavko et al. (1995) derived a geometry-independent transformation to take hydrostatic velocity vs. pressure data and predict stress-induced anisotropy. Mavko and Jizba (1991) presented a transformation of measured dry velocity vs. pressure to predict velocity vs. frequency in fluid-saturated rocks.

1.7.2 Empirical models

Empirical models do not require much explanation. The approach is generally to assume some function form and then determine coefficients by calibrating a regression to data. Some of the best known are Han's (1986) regressions for velocity–porosity–clay behavior in sandstones, the Greenberg–Castagna (1992) relations for V_P–V_S, and the V_P–density relations of Gardner et al. (1974). A particularly popular form of empirical approach is to use neural networks as a way to determine nonlinear relations among the various parameters.

Empirical relations are sometimes disguised as theoretical. For example, the popular model of Xu and White (1995) for V_S prediction in shaly sands is based on the Kuster–Toksöz ellipsoidal inclusion formulation. One unknown aspect ratio is assigned to represent the compliant clay pore space, and a second unknown aspect ratio is assigned to represent the stiffer sand pore space. These aspect ratios are determined empirically by calibrating to training data. In other words, this is an empirical model, in which the function form of the regression is taken from an elastic model.

It is useful to remember that *all* empirical relations involve this two-step process of a modeling step to determine the functional form followed by a calibration step to determine the empirical coefficients. We sometimes forget that the common linear regression involves a very deliberate modeling step of deciding that there is a linear relation between the two variables.

1.7.3 Heuristic models

Heuristic models are what we might call "pseudo-theoretical." A heuristic model uses intuitive, though nonrigorous, means to argue why various parameters should be related in a certain way.

The best-known heuristic rock physics model is the Wyllie time average, relating velocity to porosity, $1/V = \phi/V_{\text{fluid}} + (1-\phi)/V_{\text{mineral}}$. At face value, it looks like there might be some physics involved. However, the time-average equation is equivalent to a straight-ray, zero-wavelength approximation, neither of which makes any sense when modeling wavelengths that are very long relative to grains and pores. We find that the Wyllie equation is sometimes a useful *heuristic* description of clean, consolidated, water-saturated rocks, but certainly not a theoretically justifiable one.

Other very useful heuristic models are the use of the Hashin–Shtrikman upper and lower bounds to describe cementing and sorting trends, as discussed in Chapters 1 and 2. Certainly the Hashin–Shtrikman curves are rigorous *bounds*, for mixtures of different phases. However, as we will discuss in Chapter 2, we often use the bounds as *interpolators* to connect the mineral moduli at zero porosity, with moduli of well-sorted end members at critical porosity. We give heuristic arguments justifying why an upper bound equation might be expected to describe cementing, which is the stiffest way to add mineral to a sand, and why a lower bound equation might be expected to describe sorting. But we are not able to derive these from first principles.

1.7.4 Our hybrid approach

We have had considerable rock physics success using a hybrid of theoretical, empirical, and heuristic models to describe clastic sediments. In this sense, we find ourselves thinking more as engineers than physicists – what Amos Nur likes to call "dirty science."

It started with Han's *empirical* discovery (Chapter 1) that velocity–porosity in shaly sands could be well described by a set of parallel contours of constant clay. Amos Nur noted that each of these contours had high- and low-porosity intercepts that had a clear physical interpretation: in the limit of zero porosity, any model should rigorously take on the properties of pure mineral, while in the limit of high porosity (the critical porosity), the rock should fall apart. When a rock is falling apart, it becomes a suspension, which is rigorously modeled with a lower bound equation. Eventually Han's contours were replaced by *modified upper bounds* (Chapters 1 and 2), partly because they fit the data better over a large range of porosities, and partly because we could heuristically defend them. We came to understand that these modified upper bounds described the *diagenetic* or *cementation trend* for sedimentary rocks.

We eventually found that a *modified lower bound* (Chapters 1 and 2) was an excellent description of the *sorting trend* in velocity–porosity. Again, this was more of an empirical observation, aesthetically symmetric to the modified upper bound, but not rigorously defendable. Jack Dvorkin introduced the friable sand model (Chapter 2), which uses a theoretical elastic contact model to describe clean, well-sorted sands, combined with a modified lower bound to interpolate these to lower porosity, poorly sorted sands.

In summary, the rock physics modeling approach presented throughout this book is the one that we have found works well. We avoid over-modeling with too much theory, frankly, because we have lost patience with model parameters that follow from assuming spheres and ellipsoids, rather than from geologic processes. We have also found it dangerous to become attached to meticulously derived theoretical models, which can never approach the complexity of nature. At the same time, we like to honor physical principles, because they make the models universal. As time goes on, it almost seems that we throw away more and more equations, and replace them with clever use of various bounds. Another important driver in our approach is our desire *to discover, understand, and quantify the elastic properties of geologic processes.*

We hope that the following chapters will not only illustrate this modeling approach, but also justify it.

2 Rock physics interpretation of texture, lithology and compaction

How does it feel
To be without a home
Like a complete unknown
Like a rolling stone?

Bob Dylan

2.1 Introduction

The main goal of conventional, *qualitative* seismic interpretation is to recognize and map geologic elements and/or stratigraphic patterns from seismic reflection data. Often hydrocarbon prospects are defined and drilled entirely on the basis of this qualitative information. Today, however, *quantitative* seismic interpretation techniques have become common tools for the oil industry in prospect evaluation and reservoir characterization. Most of these techniques, which are discussed in Chapter 4, seek to extract extra information about the subsurface rocks and their pore fluids from the reflection amplitudes. The seismic reflections are physically explained by contrasts in elastic properties, and rock physics models allow us to link seismic properties to geologic properties. Hence, the application of rock physics models can guide and improve on the qualitative interpretation. Figure 2.1 shows a schematic depiction of the relationship between geology, rock physics properties and seismic response.

In Chapter 1 we summarized how seismic properties are controlled by a wide range of different factors, including porosity, lithology, pore fluids and pressure. As of today, the application of rock physics in seismic interpretation has mainly been on prediction of porosity and discrimination of different fluid and pressure scenarios. Little work has been done on quantitative prediction of geologic parameters from seismic amplitudes, like sorting, cement volume, clay content, sand–shale ratio, and lithofacies. These factors have often just been lumped into porosity, which is often calculated from acoustic impedances derived from post-stack seismic inversion. In these cases, a linear relationship between impedance and porosity is assumed. However, as emphasized in Chapter 1 and indicated in Figure 2.1, different rock types or lithologies can

2.1 Introduction

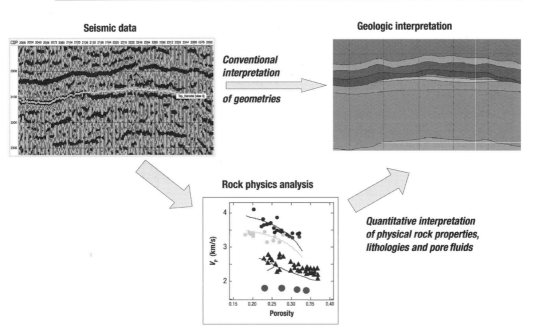

Figure 2.1 Conceptual illustration of the rock physics link between seismic data and geology. Rock physics helps to explain reflection signatures by quantifying the elastic properties of rocks and fluids.

have overlapping porosities but large differences in seismic velocities and impedances. Furthermore, during fluid substitution going from an observed brine saturation scenario to predicted hydrocarbon-saturated scenario, it is very common to assume the rock type and porosity to be constant, neglecting the possibility that lithology can change from the brine zone to the hydrocarbon zone. In this chapter, we link rock physics properties to various geologic parameters in siliciclastic environments, including sorting, cement volume, clay content, lithofacies, and compaction. We summarize important models, focusing on grain-contact models for sands, shaly sands and shales. These models allow us to perform lithology substitution, and solve important rock physics "*What if*" questions.

We also investigate how geologic trends in an area can be used to constrain rock physics models. Geologic trends can be split into two types: compactional and depositional. If we can predict the expected change in seismic response as a function of depositional environment or burial depth, we will increase our ability to predict hydrocarbons, especially in areas with little or no well-log information. Understanding the geologic constraints in an area of exploration reduces the range of expected variability in rock properties and hence reduces the uncertainties in seismic reservoir prediction. Figure 2.2 depicts this problem, where we have well-log control only in the shallow interval on the shelf edge. Before extending the exploration into more deeply buried

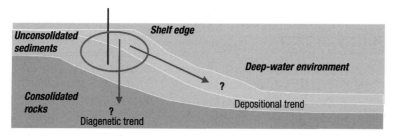

Figure 2.2 Rock physics properties change with depositional environment and burial depth. These geologic trends must be taken into account during hydrocarbon prediction from seismic data. Vertical line and ellipse depict the known area from well control in the shallow shelf edge. Arrows indicate deeper and more distal waters where we would like to predict the changes in seismic response.

zones, or to more distal deep-water environments, it is important to understand the rock physics trends in the area.

In this chapter, we first look at the effect of rock microstructure on the seismic properties of sands and shales, using well-log data from different wells in two North Sea oil fields. We apply the technology of *rock physics diagnostics* (Dvorkin and Nur, 1996), where we quantify various sedimentologic/diagenetic factors in terms of rock physics properties. This technology is applied by adjusting an effective-medium theoretical model curve to a trend in the velocity–porosity data, and then interpreting the rock microstructure as that used in the model. By superimposing such model curves on cross-plots in the velocity–porosity plane, we can sort (*diagnose*) data into clusters, characteristic of various facies and pore fluids in the reservoir.

These clusters, which are defined by characteristic sedimentologic and rock physics properties, are referred to as *seismic lithofacies*. Facies have a major control on reservoir geometries and porosity distributions. By relating lithofacies to rock physics properties one can improve the ability to use seismic amplitude information for interpretation of depositional systems and characterization of reservoir heterogeneities within these systems.

Next, we study the effect of burial depth on the seismic properties of sands and shales. Seismic amplitude contrasts will change with depth, because the contrasts between sands and shales are depth dependent. This is related to diagenesis and compaction, where the compaction histories of sands and shales are markedly different. Establishing local *rock physics depth trends* is important for constraining AVO classification when the target depth is variable.

Finally, we present the concept of *rock physics templates*, where we constrain rock physics models to local geologic parameters including texture, lithology and depth, and show how these can help us in better discrimination of hydrocarbons from well-log and seismic data.

> **Why relate rock physics to geology?**
>
> - Seismic exploration is increasingly focusing on searching for subtle traps. Therefore, seismic amplitude maps are becoming increasingly important in prospect evaluation and reservoir delineation. As shown in Plate 1.1, the amplitude patterns often reveal good insight into depositional patterns. However, a quantitative link between the geologic parameters and the rock physics properties is needed to make sure we understand the meaning of the seismic amplitudes.
> - If we understand the link between geological parameters and rock physics properties, we can avoid certain ambiguities in seismic interpretation, including fluid–lithology, sand–shale, and porosity–saturation ambiguities.
> - The link between rock physics and various geologic parameters, including cement volume, clay volume and degree of sorting, allows us to perform lithology substitution from observed rock types at a given well location to rock types assumed to be present nearby (cf. Chapter 1). Hence, we can do sensitivity analysis not only of fluid types, but also on the quality of the reservoir during quantitative seismic interpretation of a reservoir.
> - During statistical classification of different rock types, lithology substitution allows us to extend our training data set to include geologic scenarios not observed at any well location (see Chapter 3 and Chapter 5).

2.2 The link between rock physics properties and sedimentary microstructure: theory and models

If we wish to predict the seismic velocities of a rock, knowing only the porosity, the mineralogic composition, and the elastic moduli of the mineral constituents, we can at best predict the upper and lower bounds of seismic velocities (see Chapter 1). However, if we know the geometric details of how the mineral grains and pores are arranged relative to each other, we can predict more exact seismic properties. There are several models that account for the microstructure and texture of rocks, and these in principle allow us to go the other way: to predict the type of rock and microstructure from seismic velocities. The rock physics diagnostic technique was introduced by Dvorkin and Nur (1996) as a means to infer rock microstructure from velocity–porosity relations. This diagnostic is conducted by adjusting an effective-medium theoretical model curve to a trend in the data, assuming that the microstructure of the sediment matches that used in the model.

Below we present a collection of models that describe the velocity–porosity–pressure behavior of medium- to high-porosity sediments and rocks (low-porosity models are mentioned in Chapter 1). A very effective approach that we have found is to begin by defining the elastic properties of the "end members." At zero porosity, the rock must have the properties of mineral. At the high-porosity limit, the elastic properties are determined by elastic contact theory. Then, we interpolate between these two "end members" using either upper or lower Hashin–Shtrikman bounds. The upper bound explains the theoretical stiffest way to mix load-bearing grains and pore-filling material, while the lower bound explains the theoretical softest way to mix these. Hence, we have found that the upper bound is a good representation of contact cement, while the lower bound nicely describes the effect of sorting. It is found that rocks with very little contact cement (a few percent) are not well described by the Hashin–Shtrikman upper bound, because there is a large stiffening effect during the very initial porosity reduction as cement fills in the microcracks between the contacts. Then it is dangerous to interpolate between the high-porosity and zero-porosity end members. We therefore include a high-porosity contact-cement model that takes into account the initial cementation effect. In summary, we consider the following rock physics models, each to be used for different geologic scenarios:

- **The friable- (unconsolidated) sand model.** This describes the velocity–porosity behavior versus sorting at a specific effective pressure. The velocity for the well-sorted, high-porosity member (normally selected to be around 0.4) is determined by contact theory, and intermediate (poorly sorted) porosities are "interpolated" using a lower bound.
- **The contact-cement model.** This model describes the velocity–porosity behavior versus cement volume at high porosities. The contact cement fills the crack-like spaces near the grain contacts. This has the effect of very rapidly stiffening the rock with very little change in porosity. This cement tends to eliminate further sensitivity to effective pressure in the model. The high-porosity member is the critical porosity, which can vary as a function of sorting. For practical purposes, we assume this porosity to be equal or close to the well-sorted end member of the friable-sand model. More poorly sorted cemented sandstones are then modeled using the constant-cement model.
- **The constant-cement model.** This describes the velocity–porosity behavior versus sorting at a specific cement volume, normally corresponding to a specific depth. The high-porosity member is defined by first applying the contact-cement model and calculating the velocity–porosity for a well-sorted sandstone with a given cement volume. A lower bound interpolation between this well-sorted end member and zero porosity will then describe more poorly sorted sandstones with the constant-cement volume.
- **The increasing-cement model (modified Hashin–Shtrikman upper bound).** This model describes the velocity–porosity behavior versus cement volume for low to intermediate porosities. The high-porosity end member is determined by contact theory. The first 2–3% cement should be modeled with the contact-cement model.

Further increase in cement volume and decrease in porosity is described by an upper bound interpolation between the high-porosity end member and the mineral point. The theory and application of this model are shown in Chapter 1.

- **The constant-clay model.** This describes the velocity–porosity behavior for shales or sands with a constant clay–quartz ratio. The same equations are used as for the friable-sand model. However, the high-porosity end member will vary as a function of clay content. The mineral end member is defined by effective mineral moduli of quartz and clay. A lower bound interpolates between the two end members.
- **Dvorkin–Gutierrez shaly sand model.** This describes the velocity–porosity relation versus clay content for shaly sands, assuming the clay is pore-filling. The high-porosity end member is the same as for the friable-sand model, while the low-porosity end member is when the original sand porosity is completely filled with clay. However, this is not at zero porosity, since clay has intrinsic porosity.
- **Dvorkin–Gutierrez sandy/silty shale model.** This model describes the velocity–porosity behavior versus clay content for sandy/silty shales. The high-porosity end member is the pure shale, which can be calculated using the constant-clay model with 100% clay. The low-porosity end member is the same as the low-porosity end member for the Dvorkin–Gutierrez sandy shale model.
- **Yin–Marion shaly sand model.** This model explains the relationship between velocity–porosity–clay in unconsolidated sands. It assumes the clay to be a part of the pore fluid, hence there is no increase in shear stiffness with increasing clay content for shaly sands. Gassmann theory is used to calculate velocity–porosity relations for increasing clay content.
- **Yin–Marion silty shale model.** This model describes velocity–porosity behavior in shales with dispersed silt particles. The high-porosity member is the pure shale, while increasing silt content will reduce this porosity. The low-porosity end member will be when quartz grains start to touch each other, and the sediment becomes a shaly sand. Yin and Marion used the Reuss (lower) bound to interpolate between these end members.
- **Jizba's cemented shaly sand model.** Jizba's model expands on the Yin–Marion model and takes into account quartz cementation. The quartz cement volume is estimated as a function of clay content. Empirical equations are used to calculate velocity–porosity relations as a function of quartz volume and clay content.
- **The laminated sand–shale model.** This model calculates the effective velocity–porosity trends in thin-laminated sand–shales using the absolute lower elastic bound, which is the Reuss bound.

For each of these models we present elastic moduli and bulk density as a function of porosity. For some of the models, we present the dry moduli. The corresponding saturated elastic moduli are then calculated using Gassmann's formulas (equations (1.11) and (1.12)). However, for some of the models, the saturated elastic moduli are directly derived from the models. In any case, the P-wave and S-wave velocities (V_P and V_S) are calculated using the following equations:

$$V_P = \sqrt{\frac{K + 4/3\mu}{\rho}} \tag{2.1}$$

and

$$V_S = \sqrt{\frac{\mu}{\rho}} \tag{2.2}$$

where K and μ are either the dry or the saturated bulk and shear moduli, and ρ is corresponding dry or saturated bulk density.

Before we elaborate on each of these models, we want to emphasize that there exist several alternative models that try to explain the relationship between velocity and porosity in sedimentary rocks. A common approach, which differs from our use of contact theory combined with upper and lower bounds, is to model the rock as a solid with increasing numbers of pore inclusions. The shape of the inclusions, determined by the so-called aspect ratio, determines the stiffness of the rock at a given porosity. For instance, shaly sands will have softer pores and lower aspect ratios than clean sands with equal porosity. The inclusion models have become popular among some rock physics experts and practitioners, and are found to give good results in describing velocity–porosity trends for sands, shaly sands and sandy shales. Some excellent references on rock physics models using inclusion models include Berryman (1980 and 1995), Hornby *et al.* (1994), Xu and White (1995), Sams and Andrea (2001), Johansen *et al.* (2002), and Jakobsen *et al.* (2003).

2.2.1 Rock physics properties of clean sands

The friable-sand model

Dvorkin and Nur (1996) introduced two theoretical models for high-porosity sands. The friable-sand model, or the "unconsolidated line," describes how the velocity–porosity relation changes as the sorting deteriorates. The "well-sorted" end member is represented as a well-sorted packing of similar grains whose elastic properties are determined by the elasticity at the grain contacts. The "well-sorted" end member typically has a critical porosity ϕ_c around 40%. The friable-sand model represents poorly sorted sands as the "well-sorted" end member modified with additional smaller grains deposited in the pore space. These additional grains deteriorate sorting, decrease the porosity, and only slightly increase the stiffness of the rock (Figure 2.3).

The elastic moduli of the dry well-sorted end member at critical porosity are modeled as an elastic sphere pack subject to confining pressure. These moduli are given by the

2.2 Rock physics and sedimentary microstructure

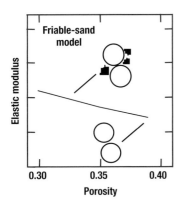

Figure 2.3 Schematic depiction of the friable-sand model and the corresponding sedimentologic variation. (Courtesy of Jack Dvorkin.)

Hertz–Mindlin theory (Mindlin, 1949) as follows:

$$K_{\text{HM}} = \left[\frac{n^2(1-\phi_c)^2\mu^2}{18\pi^2(1-\nu)^2}P\right]^{\frac{1}{3}} \qquad (2.3)$$

$$\mu_{\text{HM}} = \frac{5-4\nu}{5(2-\nu)}\left[\frac{3n^2(1-\phi_c)^2\mu^2}{2\pi^2(1-\nu)^2}P\right]^{\frac{1}{3}} \qquad (2.4)$$

where K_{HM} and μ_{HM} are the dry rock bulk and shear moduli, respectively, at critical porosity ϕ_c (i.e., depositional porosity); P is the effective pressure (i.e., the difference between the overburden pressure and the pore pressure); μ and ν are the shear modulus and Poisson's ratio of the solid phase; and n is the coordination number (the average number of contacts per grain).

The Poisson's ratio can be expressed in terms of the bulk (K) and shear (μ) moduli as follows:

$$\nu = \frac{3K-2\mu}{2(3K+\mu)} \qquad (2.5)$$

Effective pressure versus depth is obtained with the following formula:

$$P = g\int_0^Z (\rho_b - \rho_{\text{fl}})\,dz \qquad (2.6)$$

where g is the gravity constant, and ρ_b and ρ_{fl} are the bulk density and the fluid density, respectively, at a given depth, Z.

The coordination number, n, depends on porosity, as shown by Murphy (1982). The relationship between coordination number and porosity can be approximated by the following empirical equation:

$$n = 20 - 34\phi + 14\phi^2 \qquad (2.7)$$

Hence, for a porosity $\phi = 0.4$, $n = 8.6$.

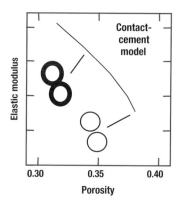

Figure 2.4 Schematic depiction of the contact-cement model and the corresponding diagenetic transformation. (Courtesy of Jack Dvorkin.)

The other end point in the friable-sand model is at zero porosity and has the bulk (K) and shear (μ) moduli of the mineral. Moduli of the poorly sorted sands with porosities between 0 and ϕ_c are "interpolated" between the mineral point and the well-sorted end member using the lower Hashin–Shtrikman (1963) bound (Chapter 1). One heuristic argument for this is that adding small grains passively in the pore space is the softest way to add mineral to the well-sorted sands; the lower bound equation is always the softest way to mix two phases. Another argument follows from Figure 2.3. Here we envision the poorly sorted sand as a few large grains enveloped by soft "shells" of fine-grained sand. This is the realization of a Hashin–Shtrikman bound, as discussed in Figure 1.4, Chapter 1.

At porosity ϕ the concentration of the pure solid phase (added to the sphere pack to decrease porosity) in the rock is $1 - \phi/\phi_c$ and that of the original sphere-pack phase is ϕ/ϕ_c. Then the bulk (K_{dry}) and shear (μ_{dry}) moduli of the dry friable sand mixture are:

$$K_{\text{dry}} = \left[\frac{\phi/\phi_c}{K_{\text{HM}} + 4\mu_{\text{HM}}/3} + \frac{1 - \phi/\phi_c}{K + 4\mu_{\text{HM}}/3} \right]^{-1} - \frac{4}{3}\mu_{\text{HM}} \qquad (2.8)$$

$$\mu_{\text{dry}} = \left[\frac{\phi/\phi_c}{\mu_{\text{HM}} + z} + \frac{1 - \phi/\phi_c}{\mu + z} \right]^{-1} - z \qquad (2.9)$$

where

$$z = \frac{\mu_{\text{HM}}}{6} \left(\frac{9K_{\text{HM}} + 8\mu_{\text{HM}}}{K_{\text{HM}} + 2\mu_{\text{HM}}} \right)$$

The saturated elastic moduli, K_{sat} and μ_{sat}, can now be calculated from Gassmann's equations (equations (1.11) and (1.12)).

2.2 Rock physics and sedimentary microstructure

Density is given by:

$$\rho_b = \phi \rho_{fl} + (1 - \phi)\rho_{min} \tag{2.10}$$

where ρ_{min} is the mineral density, which equals 2.65 g/cm³ for quartz, and ρ_{fl} is the fluid density, normally varying from around 1.0 g/cm³ to 1.15 g/cm³ for saline brine water. For dry rocks, the fluid density is zero.

> The friable-sand model represents velocity–porosity–sorting variation within a sand unit. For quartz-rich sands the sorting variation is due to smaller quartz grains filling into the pore space between larger quartz grains. However, deteriorating sorting is normally associated with increasing clay content, and if the clay content is relatively large (>20%) we are talking about a shaly sandstone (see Section 2.2.3).

The contact-cement model

During burial, sands are likely to become cemented sandstones. This cement may be diagenetic quartz, calcite, albite, or other minerals. Cementation has a more rigid stiffening effect, because grain contacts are "glued" together. The *contact-cement model* assumes that porosity reduces from the initial porosity of a sand pack because of the uniform deposition of cement layers on the surface of the grains (Figure 2.4). The contact cement dramatically increases the stiffness of the sand by reinforcing the grain contacts. In particular, the initial cementation effect will cause a large velocity increase with only a small decrease of porosity. The mathematical model is based on a rigorous contact-problem solution by Dvorkin *et al.* (1994).

In this model, the effective bulk (K_{dry}) and shear (μ_{dry}) moduli of dry rock are:

$$K_{dry} = n(1 - \phi_c)M_c S_n/6 \tag{2.11}$$

and

$$\mu_{dry} = 3K_{dry}/5 + 3n(1 - \phi_c)\mu_c S_\tau/20 \tag{2.12}$$

where ϕ_c is critical porosity; K_s and μ_s are the bulk and shear moduli of the grain material, respectively; K_c and μ_c are the bulk and shear moduli of the cement material, respectively; $M_c = K_c + 4\mu_c/3$ is the compressional modulus of the cement; and n is the coordination number, defined as average number of contacts per grain. The variables S_n and S_τ are:

$$S_n = A_n(\Lambda_n)\alpha^2 + B_n(\Lambda_n)\alpha + C_n(\Lambda_n), \quad A_n(\Lambda_n) = -0.024153\Lambda_n^{-1.3646}$$
$$B_n(\Lambda_n) = 0.20405\Lambda_n^{-0.89008}, \quad C_n(\Lambda_n) = 0.00024649\Lambda_n^{-1.9864}$$
$$S_\tau = A_\tau(\Lambda_\tau, \nu_s)\alpha^2 + B_\tau(\Lambda_\tau, \nu_s)\alpha + C_\tau(\Lambda_\tau, \nu_s)$$
$$A_\tau(\Lambda_\tau, \nu_s) = -10^{-2} \times \left(2.26\nu_s^2 + 2.07\nu_s + 2.3\right)\Lambda_\tau^{0.079\nu_s^2 + 0.1754\nu_s - 1.342}$$
$$B_\tau(\Lambda_\tau, \nu_s) = \left(0.0573\nu_s^2 + 0.0937\nu_s + 0.202\right)\Lambda_\tau^{0.0274\nu_s^2 + 0.0529\nu_s - 0.8765}$$

$$C_\tau(\Lambda_\tau, \nu_s) = 10^{-4} \times \left(9.654\nu_s^2 + 4.945\nu_s + 3.1\right)\Lambda_\tau^{0.01867\nu_s^2 + 0.4011\nu_s - 1.8186}$$

$$\Lambda_n = 2\mu_c(1-\nu_s)(1-\nu_c)/[\pi\mu_s(1-2\nu_c)], \quad \Lambda_\tau = \mu_c/(\pi\mu_s)$$

$$\alpha = [(2/3)(\phi_c - \phi)/(1-\phi_c)]^{0.5}$$

$$\nu_c = 0.5(K_c/\mu_c - 2/3)/(K_c/\mu_c + 1/3)$$

$$\nu_s = 0.5(K_s/\mu_s - 2/3)/(K_s/\mu_s + 1/3)$$

A detailed explanation of these equations and their derivation are given in Dvorkin and Nur (1996). Saturated elastic moduli are calculated using Gassmann's equations (equations (1.11) and (1.12)). Dry and saturated bulk densities are calculated using equation (2.10).

> The contact-cement model represents the initial stage of the "diagenetic trend" in the data. It is found to be applicable to high-porosity sands. During more severe cementation where the diagenetic cement is filling up the pore space, the contact theory breaks down and one should use Jizba's cement model (see Section 2.2.3) or the modified Hashin–Shtrikman upper bound (also referred to as the "increasing-cement model"; see Chapter 1).

The constant-cement model

The *constant-cement* model was introduced by Avseth et al. (2000), and assumes that sands of varying sorting (and therefore varying porosity) all have the same amount of contact cement. Porosity reduction is solely due to noncontact pore-filling material (e.g., deteriorating sorting). Mathematically, this model is a combination of the contact-cement model, where porosity reduces from the initial sand-pack porosity to porosity ϕ_b because of contact-cement deposition, and the friable-sand model where porosity reduces from ϕ_b because of the deposition of the solid phase away from the grain contacts (Figure 2.5). Considering a given reservoir, this is the most likely scenario, since the amount of cement is often related to depth, whereas sorting is related to lateral variations in flow energy during sediment deposition. In these cases we can refer to this model as a *constant-depth* model for clean sands. However, it is possible that cement can have a local source, and therefore cause a considerable lateral variation in velocity.

To use the constant-cement model, one must first adjust the well-sorted end-member porosity, ϕ_b, that corresponds to the point shown as an open circle in Figure 2.5. The dry-rock bulk and shear moduli at this porosity (K_b and μ_b, respectively) are calculated from the contact-cement model. Equations for the dry-rock bulk (K_{dry}) and shear (μ_{dry}) moduli at a smaller porosity ϕ are then interpolated with a lower bound:

$$K_{dry} = \left[\frac{\phi/\phi_b}{K_b + (4/3)\mu_b} + \frac{1-\phi/\phi_b}{K + (4/3)\mu_b}\right]^{-1} - \frac{4}{3}\mu_b \qquad (2.13)$$

2.2 Rock physics and sedimentary microstructure

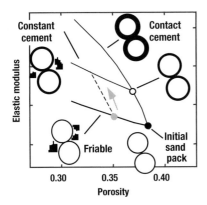

Figure 2.5 Schematic depiction of three effective-medium models for high-porosity sands in the plane of elastic modulus versus porosity, and corresponding diagenetic transformations. The elastic modulus may be compressional, bulk, or shear.

and

$$\mu_{\text{dry}} = \left[\frac{\phi/\phi_b}{\mu_b + z} + \frac{1 - \phi/\phi_b}{\mu + z} \right]^{-1} - z \qquad (2.14)$$

where

$$z = \frac{\mu_b}{6} \left(\frac{9K_b + 8\mu_b}{K_b + 2\mu_b} \right)$$

The effect of pore fluid can be accounted for by using Gassmann's (1951) equations. Dry and saturated bulk densities are calculated using equation (2.10).

Notice that it is possible to arrive at the constant-cement line by first moving along the friable-sand line and then adding contact cement to the rock (dashed line in Figure 2.5), which is consistent with diagenesis following deposition.

> For a reservoir at a given depth, where the sands are consolidated, the constant-cement model represents the most likely scenario. The amount of cement is often related to depth, whereas sorting is related to lateral variations in flow energy during sediment deposition. Making this assumption, we can refer to this model as a "constant-depth model" for clean sands. However, it is possible that cement can have a local source and therefore result in a considerable lateral variation in velocity. Also, if more poorly sorted sands means increased grain contacts, the amount of contact cement will be somewhat correlated with degree of sorting.

2.2.2 Rock physics properties of shales

In general, shales are mixtures of clay-sized particles, consisting primarily of clay minerals, and silt-sized particles, which are mostly quartz or in some cases feldspar. Krynine (1948) estimated the "average" shale to be about 50% silt, whereas Pettijohn (1975) and others suggested that shales average about 65% silt. Mineralogically, shales tend to be composed of about half clay minerals, while the other half is primarily quartz plus a few percent feldspar and calcite.

The constant-clay model for sandy shales

In shales, silt grains are suspended in the clay matrix. Furthermore, shales are normally not cemented. Therefore, shales with constant-clay content can be modeled by once again using the unconsolidated (friable-sand) line. The constant-clay lines for shales can be helpful for identification of different types of shales with respect to sand–shale ratio and/or silt content (Figure 2.6).

Using equations (2.3)–(2.9) we obtain K_{dry} and μ_{dry} for the shales, and using Gassmann's equations (equations (1.11) and (1.12)) we can calculate the water-saturated moduli. As input to equations (2.3)–(2.9), the critical porosity should be set to be very high for shales (60–70%) owing to the "card-stack" arrangements of clay platelets, and because the clay particles have internal porosity (i.e., bound water). The higher the clay content, the higher the critical porosity will be. Pure marine shale (with almost 100% clay) can have a depositional porosity of almost 90% (Rieke and Chilingarian, 1974). The next input parameter to consider is the mineral moduli. The mineral moduli of clays are highly variable and not very well known. In Table 2.1 we suggest several values that have been reported for clay minerals. However, if the shales are silty, it is necessary to calculate effective mineral moduli. We assume that silt grains are suspended in the clay matrix even at the zero-porosity end member. This results in soft effective mineral moduli, which can be estimated using the Reuss average equations (Chapter 1).

$$\frac{1}{K_{mixed}} = \frac{1-C}{K_{qz}} + \frac{C}{K_{clay}} \qquad (2.15)$$

and

$$\frac{1}{\mu_{mixed}} = \frac{1-C}{\mu_{qz}} + \frac{C}{\mu_{clay}} \qquad (2.16)$$

where C is the fraction of clay in the solid phase. The silt fraction is normally assumed to be composed of 100% quartz.

The densities for the constant-clay lines are calculated using equation (2.10). The mineral density of clays can be assumed to be close or equal to the density of quartz.

2.2 Rock physics and sedimentary microstructure

Table 2.1 *Elastic moduli of different clay minerals*

Clay mineral	K (GPa)	μ (GPa)	Author
Smectite	17.5	7.5	Brevik (1996)
Illite	39.4	11.7	Katahara (1996)
Kaolinite	1.5	1.4	Woeber et al. (1963)
	37.9	14.8	Katahara (1996)
	12	6	Vanorio et al. (2003)
Chlorite	95.3	11.4	Katahara (1996)

Dvorkin–Gutierrez silty shale model

We can also model the velocity–porosity trend of decreasing clay content (i.e., increasing silt content) for shale. The porosity of a shale as a function of clay content, assuming silt grains to be dispersed in the clay matrix, can be expressed as:

$$\phi = \phi_{sh} C \tag{2.17}$$

where ϕ_{sh} is the porosity of a clean shale, and C is the volume fraction of clay. Dvorkin and Gutierrez (2001, 2002) inserted equation (2.17) into the Hashin–Shtrikman lower bound, in order to express the velocity–porosity trend of shale as a function of clay content. The elastic moduli of the saturated silty shale are then expressed as:

$$K_{sat} = \left[\frac{C}{K_{sh} + (4/3)\mu_{sh}} + \frac{1-C}{K_{qz} + (4/3)\mu_{sh}} \right]^{-1} - \frac{4}{3}\mu_{sh} \tag{2.18}$$

$$\mu_{sat} = \left[\frac{C}{\mu_{sh} + Z_{sh}} + \frac{1-C}{\mu_{qz} + Z_{sh}} \right]^{-1} - Z_{sh} \tag{2.19}$$

where

$$Z_{sh} = \frac{\mu_{sh}}{6} \left(\frac{9K_{sh} + 8\mu_{sh}}{K_{sh} + 2\mu_{sh}} \right)$$

Here, K_{sh} and μ_{sh} are the saturated elastic moduli of pure shale, respectively. These could be derived from well-log measurements of V_P, V_S and density in a pure shale zone. By adding silt or sand particles, the clay content reduces, and the elastic moduli will stiffen. The variables K_{qz} and μ_{qz} are the mineral moduli of the silt grains, commonly assumed to consist of 100% quartz. The bulk density of shales with dispersed silt is given by:

$$\rho_b = (1-C)\rho_{qz} + C(1-\phi_{sh})\rho_{clay} + C\phi_{sh}\rho_{fl} \tag{2.20}$$

where ρ_{qz} is the density of the silt mineral (2.65 g/cm^3 for quartz) and ρ_{clay} is the density of the solid clay.

Yin–Marion silty shale model

Marion *et al.* (1992) calculated the elastic moduli of shales with dispersed quartz grains, using the Reuss bounds:

$$\frac{1}{K_{ssh}} = \frac{C}{K_{sh}} + \frac{1-C}{K_{qz}} \quad (2.21)$$

$$\frac{1}{\mu_{ssh}} = \frac{C}{\mu_{sh}} + \frac{1-C}{\mu_{qz}} \quad (2.22)$$

where K_{ssh} and μ_{ssh} are the bulk and shear moduli of the silty shale, respectively; K_{sh} and μ_{sh} are the elastic moduli of the pure shale (i.e., the moduli of the shale rock including the saturated porosity); and K_{qz} and μ_{qz} are the elastic mineral moduli of the silt grains. Densities are calculated using equation (2.20).

The models for silty shales presented here predict the isotropic velocities of shales. Most shales, however, have significant anisotropy. Vernik and Liu (1997) performed experimental measurements to quantify the anisotropy of shales, while Sams and Andrea (2001) and Jakobsen *et al.* (2003) have provided models to calculate anisotropic velocities of shales.

> **Pitfalls**
>
> - Shales can be difficult to model because they can be composed of various minerals with highly variable elastic properties. It is often impossible to determine the exact composition of shales because cores and thin sections are rarely obtained in shaly intervals.
> - The depositional porosity of shales is uncertain. In contrast to sands, shales do not have a well-defined upper critical porosity. Very clay-rich shales can have depositional porosities higher than 80%. However, abundance of silt will reduce the depositional porosity dramatically.
> - Clay minerals have intrinsic porosity and presence of bounding water in the mineral structure. During compaction, this water will be released and the shale mineralogy will change, as will the elastic properties of the shales.
> - Shales are always anisotropic! The shale models presented here predict isotropic velocities. The isotropic velocities are close to, but not equal to the vertical velocities of shales (e.g., Sams and Andrea, 2001).

2.2.3 Rock physics properties of shaly sands

The constant-clay model for shaly sands

By analogy with the friable-sand model, we use the modified Hashin–Shtrikman lower bound to model constant-clay lines for shaly sands. These lines can be used to define

2.2 Rock physics and sedimentary microstructure

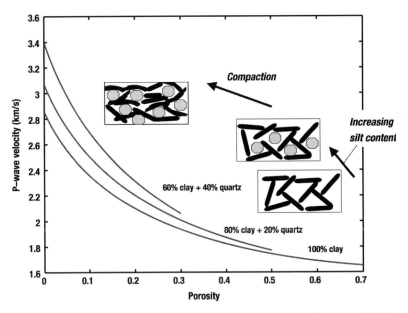

Figure 2.6 Lines of V_P versus porosity for shales with varying silt content. The lines are modeled using modified lower bound Hashin–Shtrikman combined with Hertz–Mindlin theory.

subfacies of sands where the facies changes are associated with varying clay content. Alternatively, we can use the more empirical V_P–porosity–clay trends of Han (1986) that are presented in Chapter 1, or the lithology lines of Vernik (1992).

In shaly sands, we can assume that the clay particles fill in the pores between the sand grains. Assuming shaly sands to be uncemented, we can apply the friable-sand model (i.e. Hertz–Mindlin plus modified lower bound Hashin–Shtrikman) to calculate constant-clay content lines for friable shaly sands. Using equations (2.3)–(2.9) we obtain K_{dry} and μ_{dry} for the mixed lithology, and using Gassmann's equation we can calculate the corresponding water-saturated moduli. As input to equations (2.3)–(2.9), the critical porosity will be lower for shaly sands than for sands (10–40%), depending on the clay content. The higher the clay content, the lower the critical porosity will be. The critical porosity will never reach zero since the clay particles have internal porosity. The next input parameters to consider are the mineral moduli. Since we are mixing quartz and clay particles, we need to calculate effective mineral moduli, just like we did for the silty shales. The mineral moduli represent the projections of the frame moduli, given by the modified lower Hashin–Shtrikman bound, at zero porosity. One can either use Voigt or Reuss averages (see Chapter 1) to calculate the mixed mineral's elastic moduli, or the average of these two. The moduli calculated from Voigt or Reuss should be approximately the same for most mineral mixtures. However, if we have very soft clays (e.g. smectite or illite) mixed with relatively stiff quartz, the difference between the two methods can be significant. We assume that the effective mineral moduli are given by the Voigt average equations, representing the stiffest possible alternative of the mixed quartz–clay mineral (cf. Chapter 1):

$$K_{mixed} = (1-C)K_{qz} + CK_{clay} \qquad (2.23)$$

and

$$\mu_{mixed} = (1-C)\mu_{qz} + C\mu_{clay} \qquad (2.24)$$

The densities along the constant-clay lines are calculated using equation (2.10).

Yin–Marion shaly sand model

As an alternative to constant-clay content lines, we can also model increasing clay content trends in the velocity–porosity plane. Marion (1990) introduced a topological model for sand–shale mixtures to predict the interdependence between velocity, porosity, and clay content. For shaly sands, it can be assumed that clay particles are strictly located within the sand pore space. Then, total porosity will decrease linearly with increasing clay content, C (Figure 2.7):

$$\phi = \phi_s - C(1-\phi_{sh}), \quad \text{for } C < \phi_s \qquad (2.25)$$

where ϕ_{sh} is porosity of pure shale, and ϕ_s is the clean sand porosity. Equation (2.25) holds until the sand porosity is completely filled up with clay, $C = \phi_s$. The total porosity at this point is the product of sand porosity and shale porosity,

$$\phi = \phi_s \phi_{sh}, \quad \text{for } C = \phi_s \qquad (2.26)$$

When clay content exceeds the sand porosity, the addition of clay will cause the sand grains to become disconnected, as we go from grain-supported to clay-supported sediments (i.e. shales). The total porosity will then be described by equation (2.17). The total porosity evolution of a sand–shale mixture, as a function of increasing clay content, is summarized in Figure 2.7.

Marion applied Gassmann's equations to calculate the velocities of shaly sand. As we know from Chapter 1, the Gassmann theory is used for fluid substitution of a porous rock. Similarily, we can use Gassmann to replace porous fluids with pore-filling clay (treating clay like a liquid). When clay content is less than the sand porosity, clay particles are assumed to be located within the pore space of the load-bearing sand. The clay will stiffen the pore-filling material, without affecting the frame properties of the sand. Therefore, increasing clay content will increase the stiffness and velocity of the sand–shale mixture as the elastic moduli of the pore-filling material (fluid and clay) increases. We can express the Gassmann's equations accordingly:

$$\frac{K_{sat}}{K_{qz} - K_{sat}} = \frac{K_{dry}}{K_{qz} - K_{dry}} + \frac{K_{pf}}{\phi_s(K_{qz} - K_{pf})} \qquad (2.27)$$

$$\mu_{sat} = \mu_{dry} \qquad (2.28)$$

2.2 Rock physics and sedimentary microstructure

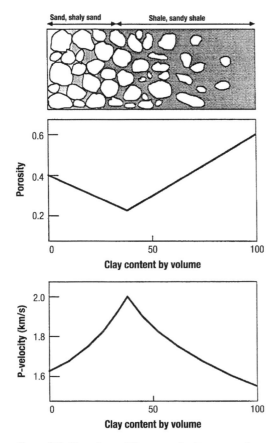

Figure 2.7 Porosity and P-wave velocity versus clay content for shaly sands and sandy shales. Note the porosity minimum and velocity maximum at the transition from grain-supported sediment to clay-supported sediment. (Adapted from Marion, 1990.)

where K_{dry} and K_{qz} are the bulk moduli of the solid frame and the frame-forming mineral, respectively, μ_{dry} is the shear modulus of the dry sand frame, and K_{pf} is the effective bulk modulus of the pore-filling material (fluid and clay). Marion assumed that like fluids, the pore-filling clay would not affect the shear modulus of the rock, so equation (2.28) is theoretically valid for increasing clay content. This assumption was supported by laboratory measurements (Yin, 1992).

Density of the sand–shale mixture where porosity is reduced by pore-filling clay can be calculated using the following formula:

$$\rho = (1 - \phi_s)\rho_{qz} + C(1 - \phi_{sh})\rho_{clay} + (\phi_s - C(1 - \phi_{sh}))\rho_{fl} \tag{2.29}$$

where ρ_{qz}, ρ_{clay}, and ρ_{fl} are the density of sand grains (quartz), clay mineral and saturating fluid, respectively.

The impact of increasing clay content in a sand–shale mixture on velocity–porosity relationship is depicted in Figure 2.8. From the measured data we observe that when

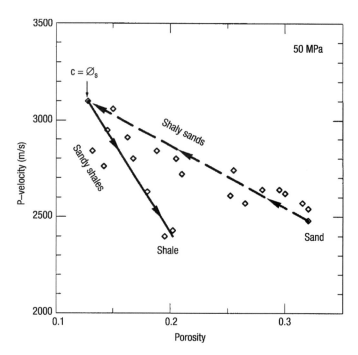

Figure 2.8 P-wave velocity versus porosity for unconsolidated sands and shales at constant effective pressure (50 MPa). A clear V-shaped trend is observed with increasing clay content, where velocity reaches a maximum and porosity a minimum when the clay content equals the sand porosity. (Adapted from Marion et al., 1992.)

clay content increases, porosity decreases and velocity increases till a given point, the critical clay content. This point represents the transition from shaly sands to sandy shales. After this point, porosity increases with increasing clay content, and velocity decreases.

The Dvorkin–Gutierrez shaly sand model

Instead of using Gassmann theory, we can use the lower bound Hashin–Shtrikman to calculate velocity–porosity trends for sands with increasing clay content:

$$K_{sat} = \left[\frac{1 - C/\phi_{ss}}{K_{ss} + (4/3)\mu_{ss}} + \frac{C/\phi_{ss}}{K_{cc} + (4/3)\mu_{ss}} \right]^{-1} - \frac{4}{3}\mu_{ss} \qquad (2.30)$$

$$\mu_{sat} = \left[\frac{1 - C/\phi_{ss}}{\mu_{ss} + Z_{ss}} + \frac{C/\phi_{ss}}{\mu_{cc} + Z_{ss}} \right]^{-1} - Z_{ss} \qquad (2.31)$$

$$Z_{ss} = \frac{\mu_{ss}}{6} \frac{9K_{ss} + 8\mu_{ss}}{K_{ss} + 2\mu_{ss}}$$

where K_{cc} and μ_{cc} are K_{sat} and μ_{sat} as calculated from the sandy shale model at critical clay content (using equations (2.18) and (2.19)), and K_{ss} and μ_{ss} are K_{sat} and μ_{sat} as calculated from any clean sandstone model (see Section 2.2.1). Plate 2.9 shows an

2.2 Rock physics and sedimentary microstructure

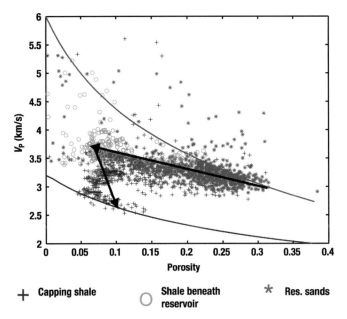

Figure 2.10 Example of V_P versus porosity for a Norwegian Sea reservoir zone. There are obvious transitions from clean sands to shaly sands at the base of the reservoir, and from shale via silty shale to shaly sand at the top of the reservoir. The upper curve is the clean sand line (i.e. constant-cement fraction line assuming 2% quartz cement). The lower curve is the shale line (i.e. constant-clay line assuming 80% clay and 20% quartz).

example of using the Dvorkin–Gutierrez models for a complete sandy shale to shaly sand sequence. The modeled lines show trends similar to those observed by Marion *et al.* (1992) in Figure 2.8.

Figure 2.10 shows velocity–porosity data from a Norwegian Sea oil field, where reservoir velocities are brine measurements. We clearly observe the shaly sand and silty shale trend as a characteristic V-shape. The seismic response of this reservoir will be very different at the top compared to the base. The cap-rock is a silty shale with predominantly lower velocities than the reservoir, while the rock beneath the reservoir is relatively stiff compared with the reservoir. Note that the reservoir unit itself spans a large range in terms of sorting and increasing clay content. A clean sand model (i.e., the constant-cement line with 2% cement) and a shale model (i.e., the constant-clay line with 80% clay) nicely describe the cleanest reservoir sands and the most clay-rich shales, respectively.

Jizba's cemented shaly sand model

Jizba (1991) established a simple model to predict velocity of cemented sands. First, she simulated the cementation process (i.e., volume of quartz, V_q) as a function of time:

$$V_q(t + \Delta t) = V_q(t) + V_q(\Delta t) \tag{2.32}$$

The quartz that precipitates during the cementation process is proportional to the amount of fluid that flows through the pore space within a given volume of rock (V). Using Darcy's law (see Jizba, 1991), the following expression is obtained:

$$V_q(\Delta t) = E_q \frac{\Delta t}{V} \left(\frac{kA}{v} \frac{dP}{dx} \right) \qquad (2.33)$$

where k is permeability, v is fluid viscosity, A is the cross-sectional area of the rock volume V under consideration, and dP/dx is the pressure gradient. E_q is the excess concentration of silica in solution that deposits, or the cementation efficiency. Permeability is computed using the Kozeny–Carman equation.

In the next step, velocity as a function of increasing quartz cement is calculated. Regression coefficients of linear velocity–porosity–clay relations in consolidated sands are used to describe the effect of quartz cementation on bulk and shear frame moduli. The quartz cementation results in increased frame bulk and shear moduli, as well as increased bulk density. Given initial values of dry elastic moduli for unconsolidated sands, one can calculate the perturbed moduli due to the cementation using:

$$K_{\text{fr}}(t + \Delta t) = K_{\text{fr}}(t) + 73.5 V_q(\Delta t) \qquad (2.34)$$

$$\mu_{\text{fr}}(t + \Delta t) = \mu_{\text{fr}}(t) + 87.5 V_q(\Delta t) \qquad (2.35)$$

$$\rho(t + \Delta t) = \rho \Delta t + V_q(\Delta t) \rho_q \qquad (2.36)$$

where the coefficients in equations (2.34) and (2.35) (in GPa) represent the empirical increase in stiffness of the dry frame with increase in quartz cement (in this case derived from Han's (1986) data set). The perturbed dry-frame moduli are used to compute the saturated moduli, using either Gassmann theory or the bounding average method (Marion and Nur, 1991). Note that for these empirical equations, the units of the elastic moduli must be in GPa.

Using equations (2.34) to (2.36), Jizba obtained the relationship between P-wave and S-wave velocities and clay content at various degrees of diagenesis. It was found that velocity increases dramatically from the initial unconsolidated model at low clay concentrations and increases less in shaly sandstones (Figure 2.11). The cleaner sandstones will be more extensively cemented, because the presence of clay inhibits quartz cementation.

Plate 2.12 shows a velocity versus gamma-ray cross-plot with log data from offshore Brazil. We clearly observe two trends: one is the increasing velocities with increasing clay content. The other is a more dramatic increase in velocities for clean sands becoming increasingly more cemented. The trends in this plot are in good agreement with Jizba's modeling results (Figure 2.11).

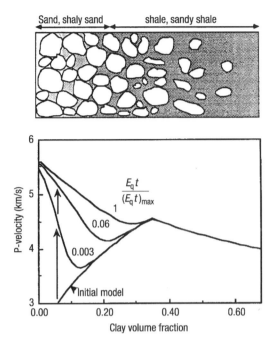

Figure 2.11 Jizba's modification of the Marion–Yin model as a function of quartz cement volume. Curves represent increasing degree of diagenesis as indicated by the normalized product of excess silica concentration and elapsed time, $E_q t$.

> NB Jizba's cement model yields linear increase in velocity with increasing cement volume. This is a good assumption for medium- to low-porosity sands, when the cement fills the macroporosity. However, the velocity increase is much larger during initial cementing of sands than later in the cement process. The contact-cement model represents a more realistic prediction of velocities during initial cementation of sands. In contrast, this model fails for low-porosity sandstone (porosity $\lesssim 0.2$), since it is based on grain contact theory.
>
> Jizba's model, however, requires a lot of geologic input during the modeling, including clay surface area and tortuosity, in addition to mineral moduli. An alternative model to predict velocity–porosity–clay trends in low-porosity sandstones is the increasing-cement model (i.e., the upper Hashin–Shtrikman bound). Nevertheless, Jizba's modeling serves as a useful demonstration of how quartz cementation affects the velocity–porosity–clay relationship of sandstones.

The laminated sand–shale model

In sandstone units clay particles can be distributed as laminas. The Marion–Yin model assumes that clay particles are located as pore-filling material, and is therefore not valid for clay-laminated sands. Dvorkin and Gutierrez (2001, 2002) presented a model for velocity–porosity–clay relationship in clay-laminated sands. Firstly, the total porosity is given by the weighted average of the sand and shale porosities:

$$\phi = C\phi_{sh} + (1 - C)\phi_{ss} \qquad (2.37)$$

where C is clay content, ϕ_{sh} is the porosity of the shale, and ϕ_{ss} is the porosity of the sand.

The bulk density is given by:

$$\rho_b = (1 - C)[(1 - \phi_{ss})\rho_{qz} + \phi_{ss}\rho_{fl}] + C[(1 - \phi_{sh})\rho_{clay} + \phi_{sh}\rho_{fl}] \qquad (2.38)$$

where ρ_{qz} is the density of the sand mineral, ρ_{clay} is the density of the solid clay mineral, and ρ_{fl} is the fluid density.

The effective properties of clay-laminated sands can be drastically reduced compared with clean sand, owing to the weakening effect of clay laminas. The clay particles are relatively soft compared with the sand grains, and given that the clay laminas are arranged perpendicular to the direction of wave propagation (i.e., transverse isotropic rock), the dry-frame elastic moduli will follow the lower bound Reuss average equations (Dvorkin and Gutierrez, 2001, 2002):

$$\frac{1}{M_{mix}} = \frac{1 - C}{M_{qz}} + \frac{C}{M_{clay}} \qquad (2.39)$$

and

$$\frac{1}{\mu_{mix}} = \frac{1 - C}{\mu_{qz}} + \frac{C}{\mu_{clay}} \qquad (2.40)$$

where $M = K + (4/3)\mu$, and C is clay content and varies between 0 and 1.

An example of clay-laminated shaly sand is shown in Figure 2.13. The cross-plot of P-wave velocity versus porosity shows well-log data from an offshore Angola turbidite field. We observe a gradual spread from clean sands, via shaly sands, to shales. No V-shape trends are seen, as we observed for shaly sands with pore-filling clay.

2.3 Example: rock physics interpretation of microstructure in North Sea turbidite systems

We apply the rock physics models above to diagnose rocks of Paleocene age in the North Sea. We use data from two wells, Well 1 and Well 2, located in two different oil fields in the Southern Viking Graben, North Sea. Well 1 is located in the Glitne field, while Well 2 is from the Grane field located about 100 km northeast of the Glitne field. The Paleocene interval is comprised of mostly pelagic/hemipelagic shales and turbidite sandstones, but volcanic tuff, marl, and limestone are also present. The Paleocene

2.3 Example: Interpretation of microstructure

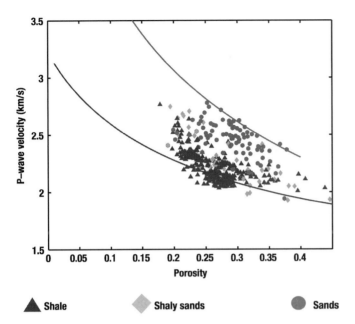

Figure 2.13 Example of velocity–porosity data from an offshore Angola turbidite field. There is a gradual spread from clean sands via shaly sands to shales. In this case, we do not observe the V-shape as seen in Figure 2.10, but a porosity-independent transition with increasing clay content. Hence, we interpret the shaly sands to be laminated sand–shale units.

sands encountered in both the Glitne and Grane fields are referred to as the Heimdal Formation, and hence represent the same stratigraphic level, yet separate turbidite systems.

2.3.1 Diagnosing Paleocene turbidite sands

Diagnosing microstructure from well-log data

The gamma-ray and P-wave velocity log curves for the two wells under examination are shown in Figure 2.14. In Well 1, we observe a gradual variation of clay content between very clean sand and shale. Only a relatively thin (10 m) sand interval (gray bar in Figure 2.14A) is identified as a practically clay-free reservoir sand. In Well 2, unlike in Well 1, a thick oil-saturated sand interval (gray bar in Figure 2.14C) is marked by extremely low and constant gamma-ray readings (about 55 GAPI) and high velocity (about 3 km/s). This sand layer is surrounded by shale packages whose gamma-ray readings and velocity strongly contrast with those of the pay zone sand. As mentioned, the clean sand zones in both wells represent the same stratigraphic unit, although located at different depths and in separate oil fields.

The velocity difference between the pay zones in the wells under examination is emphasized in Figure 2.15, where the P-wave velocity is plotted versus porosity.

72 Rock physics interpretation of texture, lithology and compaction

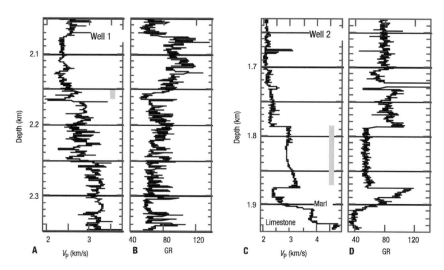

Figure 2.14 Gamma-ray (GR) and P-wave velocity curves for Wells 1 and 2. The pay zones are marked by gray vertical bars.

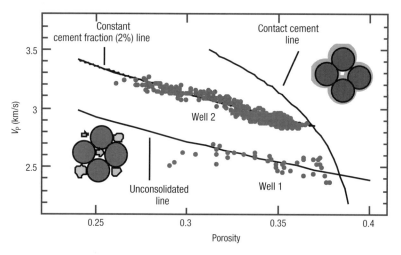

Figure 2.15 P-wave velocity versus porosity for the pay zones in Wells 1 and 2, with model curves superimposed. Porosity is calculated from bulk density.

In the same porosity range, with similar gamma-ray count, and close oil saturation, the velocity in Well 2 exceeds the velocity in Well 1 by about 500 m/s.

In order to understand the reason behind the observed velocity difference in the two wells, we superimpose the model lines on the velocity–porosity cross-plot in Figure 2.15. The three curves come from the contact-cement, constant-cement, and friable-sand models. The solid is assumed to be pure quartz; the porosity of the initial sand pack is 39%, and the initial-cement porosity ϕ_b is 37% (the latter corresponds to contact cement occupying about 2% of the pore space of the initial sand pack).

2.3 Example: Interpretation of microstructure

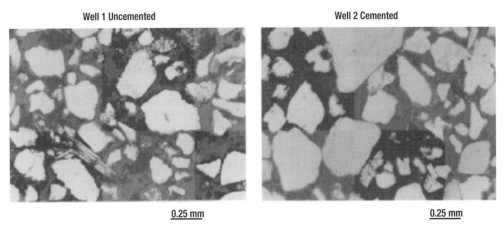

Figure 2.16 Thin sections of two selected samples from the reservoir zones of Well 1 (left), taken at 2154.0 m, and Well 2 (right), taken at 1800.25 m.

The rock diagnostic shown in Figure 2.15 implies that the sands in Well 2 have small initial contact cementation. The porosity decrease from the initial-cement porosity is likely to be due to deteriorating sorting (smaller grains fall in the pore space between larger grains and have a large effect on the porosity). The pay zone sands in Well 1 appear to lack any contact cementation, with porosity reducing from the initial sand-pack porosity because of deteriorating sorting.

Confirming the sandstone diagnostics from thin-section and SEM analysis

Quartz cementation

Thin sections of samples from both reservoir zones are shown in Figure 2.16. The porosity of both samples is about 35%, and they are predominantly composed of quartz. No contact cementation is apparent in either of the images. The Well 1 image (on the left), unlike the image from Well 2 (on the right), shows clay coating (black) around quartz grains.

The presence of contact cement in Well 2 reveals itself in scanning electronic microscope (SEM) image in Figure 2.17. Not detectable in the back-scatter light, it shows as a dark rim around a light grain in cathodoluminescent light. Energy dispersive spectroscopy (EDS) analysis, and X-ray analysis, confirm that both the grain and cement are pure quartz (Figure 2.18). The polygonal crystal shapes in the upper left corner in Figure 2.17 are also typical for overgrowth cementation. These shapes are observed throughout the reservoir zone in Well 2 (Figure 2.19). No cement rims or polygonal crystal shapes have been found in the sand interval from Well 1.

EDS analysis of the clay coating (Figure 2.20) shows the presence of pyrite (FeS) which is commonly associated with organic matter in sedimentary rocks (Boggs, 1987).

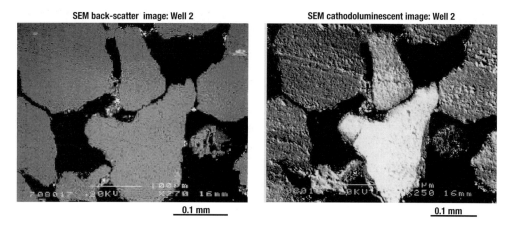

Figure 2.17 SEM images of a Well 2 sample in back-scatter light (left) and cathodoluminescent light (right).

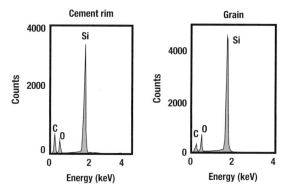

Figure 2.18 EDS spectrogram of cement rim and grain observed in the SEM image in Figure 2.17, confirming that both the grain and the cement are quartz, SiO_2. The carbon peak (C) is from the preparation of the sample.

The high Si peak is related to interference between coating and the quartz grain. The peaks of Al, Si, and K may reflect remnants of dissolved K-feldspar, or illitized kaolinite. We also identify mineral signatures typical for mixed smectite–illite (Al, Si, and K, with traces of Mg and Cl). The clay and organic matter that coat the sand grains may explain why the sands in Well 1 are not cemented, as clay coating tends to inhibit quartz cementation.

Thus, the thin-section analysis confirms the result of our mathematical rock diagnostic. Consistent with this conclusion is also the fact that the cores extracted from Well 1 appeared as piles of loose sand, while those from Well 2 supported external stress. This structural integrity of the samples from Well 2 is apparently due to the binding effect of contact quartz cement.

2.3 Example: Interpretation of microstructure

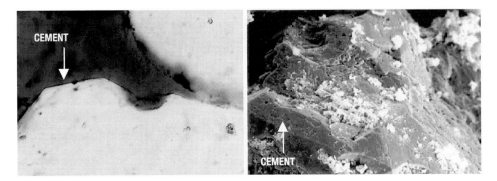

Figure 2.19 A thin section (left) and an SEM image (right) of grains with crystal cement shapes from different depths (1800.25 m and 1818.0 m, respectively) in the reservoir zone in Well 2.

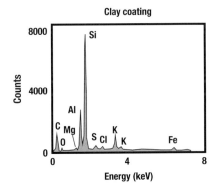

Figure 2.20 EDS spectrogram of clay coating observed in thin-section image in Figure 2.16 (Well 1), showing the presence of pyrite (FeS), indicative of organic matter. The aluminum (Al), potassium (K), and magnesium (Mg), together with silicon (Si) and chlorine (Cl), are indicative of mixed smectite–illite clays and/or illitized kaolinite.

Sorting variation

According to the diagnostics, the clean sands in Figure 2.15 have decreasing porosity with deteriorating sorting. We did thin-section analysis throughout a sand interval representing the same reservoir sands as in Well 2, but in a different well where an extensive coverage of thin sections was available. Figure 2.21 shows V_P and density-porosity in Well 3. We observe almost mirror-shaped patterns in the V_P and density-porosity logs. Thin-section analysis shows that clay and cement content is consistently close to zero. Hence, the porosity and velocity changes in this sand unit should be attributed to rock texture and grain-size variation (i.e., sorting). We carried out quantitative grain-size analysis of 12 thin sections, and Figure 2.22 shows the histograms of mean grain diameter within four of the thin sections analyzed. We observe a marked change in the character. The two upper histograms (at 1785.1 and 1790.1 m) show a much smaller spread in grain size than the two lower histograms, where much larger grains are present. The

Rock physics interpretation of texture, lithology and compaction

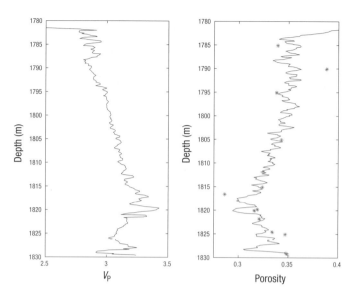

Figure 2.21 P-wave velocity and density-porosity versus depth. Note how the porosity trend is almost a perfect mirror-image of the velocity trend. Star symbols (*) represent helium porosities at the locations where thin sections are analyzed.

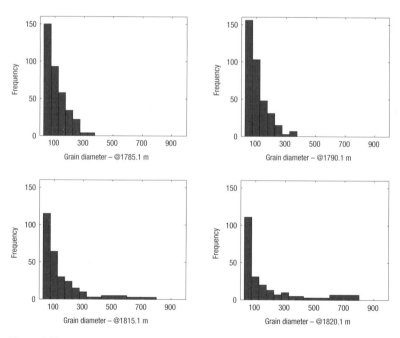

Figure 2.22 Histograms of grain-size distribution from different depth locations throughout the sand unit.

2.3 Example: Interpretation of microstructure

Figure 2.23 Thin-section images taken at depths 1785.1 m (upper left), 1790.1 m (upper right), 1815.1 m (lower left) and 1820.1 m (lower right).

corresponding thin sections are shown in Figure 2.23; we can see that the two upper pictures have a more evenly sized grain population, whereas the two lower pictures show larger variance in grain size, and indeed some larger grains. Furthermore, we observe that the porosity is lower and the grains more closely packed in the two lower pictures. Thus thin-section analysis confirms that the degree of sorting varies within the studied sand interval.

The grain-size measurements are conducted using an image analysis computer program. The next step is to study the relationship between derived sorting parameters from the quantitative thin-section analysis and rock physics properties from well-log (sonic velocities and density-porosities) and core measurements (helium porosities). Sorting is difficult to quantify. However, a reasonable parameter of sorting (S) is defined by the standard deviation of grain size (σ) normalized to the mean grain size (M):

$$S = \sigma/M \tag{2.41}$$

When this value is relatively large, the sands are relatively poorly sorted; when it is low the sands are relatively well sorted. Figure 2.24 shows how sorting affects

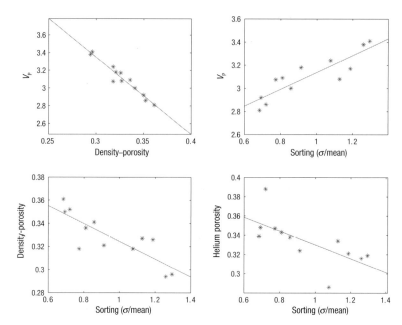

Figure 2.24 There is a very good correlation between velocity and porosity within the sand unit at the depths where thin sections have been studied (upper left). The derived sorting factor shows a very good correlation to V_P (upper right) and porosity (lower).

velocity–porosity relations in the studied sand unit. There is a nice correlation between V_P and density-porosity at the depth locations where the thin sections have been taken (upper left). Next, we observe a nice correlation between V_P and sorting (S), where velocity linearly increases when the sands become more poorly sorted. Hence, sorting is also nicely correlated to density-porosity (lower left), and we observe a marked decrease in porosity as the sands become more poorly sorted. Helium porosities versus sorting show the same trend.

Thus, the thin-section analysis confirms our hypothesis that the porosity decrease in the clean sands and sandstones of the Heimdal Formation is due to deteriorating sorting. However, the results in Figure 2.24 should be considered as a qualitative confirmation that deteriorating sorting increases velocities in saturated rocks. Bear in mind, too, that in these sands the smaller grains filling in the pores between larger grains are also quartz grains. More commonly, deteriorating sorting is associated with increasing clay content.

2.3.2 Diagnosing Paleocene shaly sands and shales

Next, we do rock diagnostics of shaly sands and shales. The goal is to better understand the composition of the shales, in particular the silt (i.e., quartz) content. Direct information in terms of core or thin sections is lacking in the Paleocene shaly intervals, and this is why diagnostics become extra important for these rocks. Petrographic or

2.3 Example: Interpretation of microstructure

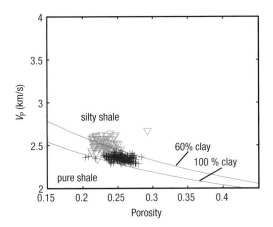

Figure 2.25 P-wave velocity versus porosity for two types of shale superimposed on rock physics models. We have included the unconsolidated shale line and silty shale line where the silt content is 40%.

mineralogic variations within the shales may cause internal seismic signatures, and it is important to understand the acoustic properties of shales and how we can distinguish these from other lithologies, in particular sands and sandstones.

Consider the data of the Sele Formation shales, located at about 2100 m depth in Well 1 (Figure 2.14 and Plate 2.31). This unit has been interpreted to be shales, on the basis of high gamma-ray values as well as mud-loggers' cuttings observations. By plotting the velocity–porosity values of this unit together with the diagnostic shale model line, we find that there is a good correlation between the shale data and the shale line (Figure 2.25). However, the shales (shown as crosses) are plotting slightly above the model line. This probably reflects the fact that the shales are not 100% clay-rich.

Consider another shaly interval in Well 1 (Figure 2.14 and Plate 2.31), this time the Lista Formation shales located just above the Heimdal reservoir (i.e., the cap-rock unit), at a depth of about 2140 m. These shales have slightly lower gamma-ray values than the Sele Formation shales. The question that arises is whether this is due to higher silt content. This question is confirmed using rock physics diagnostics. These shaly sands are plotted as open triangles in Figure 2.25. Using equations (2.3)–(2.9), we are able to quantify the silt content of the silty shales to be an average of 40%.

Now consider the zone below the reservoir sands in Well 1 (Figure 2.14 and Plate 2.31), ranging from about 2165 to 2200 m depth. This interval has core and thin-section information, just like the clean reservoir sands above. Thin-section analysis reveals two different lithofacies within this interval. The upper zone (2165–2180 m) is relatively clean sands, but with plane lamination of clay. Figure 2.26 shows thin-section images from these sands. The grains are slightly smaller than the clean sands in Figure 2.16, and the intergranular pore space has higher clay content. The lower zone (2183–2200 m) comprises thin-bedded sand and shales. The thin-bedded sands, shown in thin section (Figure 2.26, right) are even more shaly and show even smaller grains

Rock physics interpretation of texture, lithology and compaction

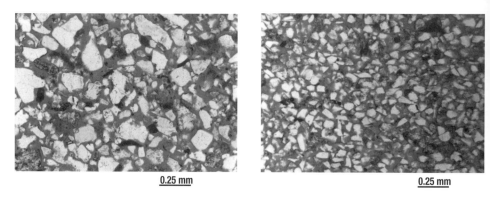

Figure 2.26 Thin-section images of the shaly sands encountered in Well 1. The left picture is taken from depth 2168 m, within a thick-bedded sand unit. Clay content in this shale is, according to XRD analysis, about 11% (Martinsen *et al.*, 1996). The picture to the right is taken from depth 2183 m, within a thin sand bed of an interbedded sand–shale unit. The sand grains are smaller and clay content higher, 17%. The scales of the images are the same as for the thin sections in Figure 2.16.

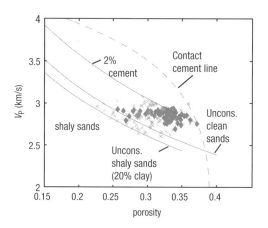

Figure 2.27 P-wave velocity versus porosity for shaly sands superimposed on rock physics models. We have included the unconsolidated clean sand line, the 2% cement line, the contact cement line and the unconsolidated shaly sand line where clay content is 20%. The two different types of shaly sands correspond to the thin sections in Figure 2.26 (diamonds here represent the thick-bedded shaly sands shown in Figure 2.26 (left), while the crosses here represent the shaly sands within the thin-bedded turbidites shown in Figure 2.26 (right)).

and more pore-filling clay than the overlying thick-bedded zone (Figure 2.26, left). Figure 2.27 shows the two zones of shaly sands, the plane-laminated shaly sands (diamonds) and thin-bedded sand–shales (crosses), cross-plotted in the velocity–porosity plane together with rock physics diagnostic models. The data points fall pretty much between the unconsolidated shaly sand line and the 2% cement fraction line. The unconsolidated shaly sand line is modeled assuming 20% clay in the matrix, and hence it has lower effective mineral moduli than the unconsolidated clean sand line. The "flat" projection of the data trend, from the contact cemented line to the uncemented shaly sand line, probably reflects that as clay content increases the cement content gradually drops (cf. Jizba's modeling results, Section 2.2.3). The velocity stays fairly constant because of the pore-filling effect of clay particles that counteracts the effect of decreasing cement volume.

The data also include the laminas or interbeds of shale, so the scatter in the V_P–porosity plane can be attributed to this interbedding. However, the laminas in the thick-bedded shaly sands are very thin (a few centimeters or millimeters) and not likely to be resolvable by the well logs. The thin-bedded sand–shales, however, can have a significant scatter due to the binary lithology composition.

2.4 Relating rock physics to lithofacies and depositional environments

One of the fundamental aspects of this chapter is to establish a link between rock physics and sedimentology. More specifically, we want to relate lithofacies to rock physics properties. This will improve the ability to use seismic amplitude information for interpretation of depositional systems, as facies have a major control on depositional geometries and porosity distributions. Facies furthermore occur in predictable patterns in terms of lateral and vertical distribution and can also be linked to sedimentary processes. They therefore represent an important parameter in seismic exploration and reservoir characterization.

Traditionally, seismic facies have been interpreted at a large scale from seismic traveltimes, that is from geometric patterns made out of the reflections. This has been a purely visual and qualitative methodology where pre-defined "seismic facies" have been interpreted from the seismic data (e.g., Weimer and Link, 1991). The first use of seismic amplitudes to interpret depositional facies was by Brown et al. (1981), who recognized river channels from seismic amplitude maps. Their work was followed up by other authors who imaged facies from seismic amplitude maps, most successfully in fluvial systems where channel facies have been easily recognized (Rijks and Jauffred, 1991; Brown, 1992; Enachescu, 1993; Ryseth et al., 1998). A few authors have studied the correlation between seismic amplitudes and lithology by seismic forward modeling (Varsek, 1985; Zeng et al., 1996). Zeng et al. (1996) linked lithofacies to rock physics properties and conducted a facies-guided seismic modeling study of a micro-tidal shore-zone depositional system. Furthermore, several authors have used seismic inversion to estimate lithology and reservoir properties from pre-stack seismic data (Loertzer and Berkhout, 1992; Buland et al., 1996).

Conventional seismic interpretation may be very uncertain in complex depositional environments (Tyler and Finley, 1991). Figure 2.28 shows how reservoirs in various depositional systems have been produced in conventional development. In the most

Rock physics interpretation of texture, lithology and compaction

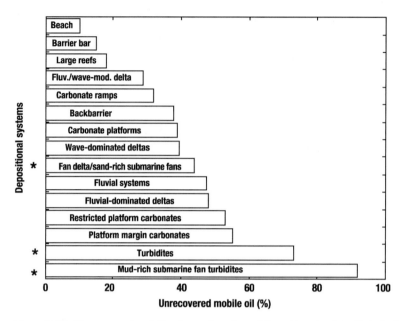

Figure 2.28 Unrecovered mobile oil as a function of depositional origin. Relatively simple, homogeneous reservoir systems (e.g., beach systems) are produced effectively, whereas much oil in highly compartmentalized systems (e.g., turbidites) is often left behind in conventional development. The latter reservoirs are particular targets of improved oil recovery (Tyler and Finley, 1991). (* indicates deep-water clastic systems.)

complex depositional environments, turbidite systems, more than 70% of the mobile oil is commonly left behind, because of the heterogeneous nature of these reservoirs.

Since conventional geophysical methods of interpreting and characterizing reservoirs from seismic data do not suffice in these complex systems, there is a need to use more quantitative seismic techniques to reveal reservoir units from seismic amplitude data. In this section we show how we can relate lithofacies and rock physics properties. In Chapter 5 we show several case studies where we apply this link to characterize reservoirs from seismic amplitude data.

2.4.1 Depositional systems and facies associations

Facies analysis and classification has been an important procedure among petroleum geologists for decades. A facies is defined as a rock unit with distinctive lithologic features, including composition, grain size, bedding characteristics, and sedimentary structures. Facies furthermore occur in predictable patterns of lateral and vertical distribution and can also be linked to sedimentary processes and depositional environments. When Walther (1893–4) formulated what is today known as the *Walther's law of facies*, a new concept was introduced that had great impact on the way geologists analyzed the stratigraphic record. Walther stated: "It is a basic statement of far-reaching significance that only those facies and facies areas can be superimposed primarily which can be observed beside each other at the present time." Careful application of Walther's law suggests that in a vertical sequence, a conformable transition from one facies to another implies that the two facies can also be found laterally adjacent to each other (Middleton, 1973). Prograding and retrograding depositional systems can explain these facies associations. Hence, if we have Walther's law in mind, we can understand lateral stratigraphic relationships by analyzing vertical well-log sections. This means that by linking facies to rock physics properties we can understand lateral patterns in seismic amplitude maps from rock physics analysis of vertical wells.

2.4.2 Seismic lithofacies

A seismic lithofacies is a seismic-scale sedimentary unit which is characterized by its lithology (sand, silt, and clay), bedding configuration (massive, interbedded, or chaotic), petrography (grain size, clay location, and cementation) and seismic properties (P-wave velocity, S-wave velocity, and density).

By introducing seismic lithofacies that represent seismic-scale sedimentary units, we try to improve our lateral facies prediction, as we can link facies observed in vertical well logs to seismic attribute maps. Facies have a major control on reservoir geometries and porosity distributions, so by relating lithofacies to rock physics properties one can improve the ability to use seismic amplitude information for reservoir prediction and characterization in these systems. Even better, the seismic lithofacies classified from well logs can serve as a calibration of statistical populations, each of which we can assume to have stationarity in the seismic parameters. These can serve as constraints in the seismic reservoir characterization (see Chapter 5).

2.5 Example: seismic lithofacies in a North Sea turbidite system

2.5.1 Seismic lithofacies description

A descriptive facies scheme is suggested in order to determine facies objectively from well logs, cores, and thin sections (Table 2.2 and Figure 2.29). Our scheme comprises six major facies (I–VI) that are geologically characterized by a specific grain size, clay content, and bedding configuration. Facies I represent gravels and conglomerates, Facies II are thick-bedded sandstones, Facies III are interbedded sands and shales where the individual bed is below seismic resolution (i.e., thinner than approximately 10 m), Facies IV are shales with a significant silt content (i.e., more than approximately 30%), while Facies V are relatively pure shales. This scheme is general and aims to include all possible siliciclastic lithofacies that can be encountered in deep-water clastic systems. Figure 2.29 must not be confused with any given depositional sequence (it is not meant to illustrate a coarsening or thickening upward sequence). It is a schematic illustration of lithofacies that can occur at a seismic scale.

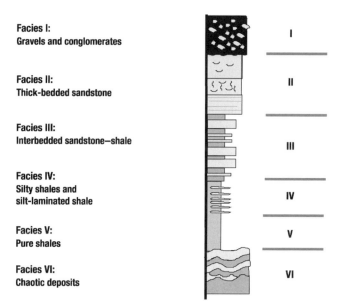

Figure 2.29 Seismic lithofacies in deep-water clastic systems. Geologic description.

In the Glitne turbidite system studied in this example and case studies 1 and 2 in Chapter 5, we only recognize Facies II, III, IV, and V. Facies II–V represent a gradual transition from clean sandstone to pure shale, whereas sand–shale ratio can vary considerably for Facies I and VI. In the Grane turbidite system, studied in case studies 3 and 5 in Chapter 5, we recognize other facies including tuff and carbonates, but these are not encountered within the target zone in the Glitne field.

Three subfacies of Facies II are recognized and honor seismically important petrographic variations within the thick-bedded sand facies. These subfacies are determined from core, thin-section and SEM analyses, and include cemented clean sands (Facies IIa), uncemented or friable clean sands (Facies IIb), and plane-laminated sands (Facies IIc). Thick-bedded shaly sands (Facies IId) are included as a possible facies to be encountered in deep-water clastic environments. These could be slurry-flow deposits as defined by Lowe *et al.* (1995), or sandy debris-flows as defined by Shanmugam *et al.* (1995). This type of facies, however, is not encountered in the area of study.

There is a gradual increase in clay content as we go from Facies IIa to IId, and the cleanest sandstones (IIa) are slightly cemented.

2.5.2 Facies associations in turbidite systems (classical submarine fans)

Our seismic lithofacies can be linked to depositional sub-environments and sedimentary processes within a deep-water clastic system. Walker (1978) suggested an idealistic depositional model which gives a simplified but good picture of how we expect sedimentary facies to be distributed in a "classical" submarine fan system. An even

2.5 Example: Seismic lithofacies

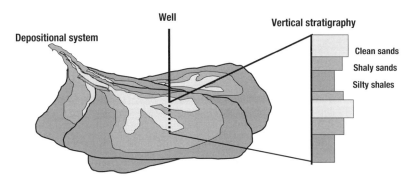

Figure 2.30 Walker's (1978) conceptual model for facies associations on a submarine fan.

more simplified and schematic version of Walker's model is depicted in Figure 2.30. The upper fan is characterized by channel-fill turbidite conglomerates, debris-flow, or slump deposits (Facies I and VI), but can also be characterized by starved shale units (Facies V). The turbidity currents on the upper fan are usually transported through a single deep channel depositing conglomerates and thick-bedded sands (Facies I and II). This feeder-channel is usually confined by stable overbanks. The overbank deposits are finer-grained, thin-bedded turbidites (Facies III). In the mid-fan and lower fan areas, a lot of the coarse-grained material is transported radially via channels and is deposited as thick elongated sand sheets, or as sandy lobes that spread out at the end of the channels. Fine-grained material is transported along the channels and then laterally as overbank deposits. The sand–shale ratio is therefore high within the channels (Facies II) and in the proximal parts of the lobes, but relatively low in interchannel areas and in the more distal fan environments (Facies III and IV). Outside the fan system, there will be mainly deposition of hemipelagic and pelagic shales (Facies IV and V).

> Turbidite systems can be very complex. Furthermore, these depositional systems can be very different from one basin to another, depending on hinterland characteristics, feeder-system (river or canyon), shore-to-shelf length, dominant grain size, and slope steepness. In particular, sand-dominated turbidite systems show very different depositional characteristics than mud-dominated turbidite systems. The Grane field in case studies 3 and 5, Chapter 5, represents a very sand-rich turbidite system, whereas the Glitne field in case studies 1 and 2, Chapter 5, represents a mixed sand/mud-rich turbidite system. The offshore Angola turbidite system in case study 4, Chapter 5, is more mud-dominated. The expected facies and sand–shale ratios will be very different for the different types of turbidite systems. Reading and Richards (1994) presented an excellent overview of 12 different types of turbidite systems and their facies characteristics as a function of various controlling factors.

Table 2.2 *Geological description of seismic lithofacies in North Sea deep-water clastic systems*

Facies	Geological description of facies and subfacies	Gamma-ray log motif
I Gravels and conglomerates	Gravels, conglomerates, and pebbly sands. Sand-rich or mud-rich debris flow deposits.	Complex. Can be blocky if "clean"
II Thick-bedded sandstone	IIa: Very clean, well-sorted, massive sandstones with small amounts of quartz overgrowths. Water-escape structures are common. Clay content less than 10%.	Usually blocky and smooth
	IIb: Clean, massive sandstones with clay coatings. Water-escape structures are prominent. Pore-filling clay content slightly higher than in Facies IIa (approximately 10–15%).	Bell and funnel shapes can occur
	IIc: Plane-laminated sandstone. Higher pore-filling clay content (10–20%) and grain size in general smaller (fine- to medium-grained) than in Facies IIa and IIb.	Low, but increasing GR values, from IIa–IIc
	IId: Shaly sandstone (clay content between 20–40%).	Intermediate in IId
III Interbedded sandstone–shale	Interbedded sand–shale couplets, where sand units are relatively thin-bedded compared with Facies II types of sand (i.e., below seismic resolution).	Serrated Intermediate GR values
IV Silty shales	Silty shales and thin-laminated silt–shale couplets. (In rock physics often referred to as "sandy" shales.)	Serrated High GR values
V Pure shales	Pure shales, often seen as thick, massive shale units.	Serrated/smooth Very high GR values
VI Chaotic deposits	Syn-depositional deformation units, slide blocks, slump deposits, injection sands, shale diapirs, etc.	Serrated/complex

2.5.3 Seismic lithofacies identification from well logs

We select a type-well for identification of seismic lithofacies from well-log data (Plate 2.31; this is the same well as Well 1 in Section 2.3). Primarily, we have used the gamma-ray log to determine the different facies, as it is a good clay indicator in the quartz-rich sediments of the North Sea. Facies II will usually show blocky log motifs and

2.5 Example: Seismic lithofacies

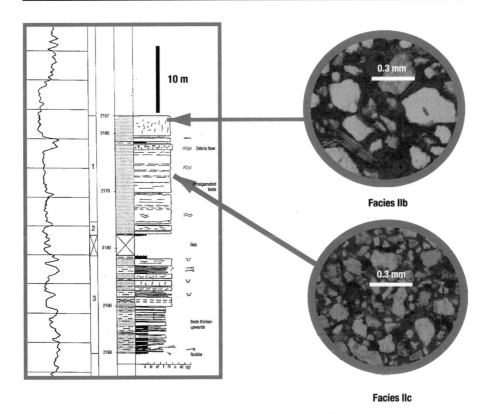

Figure 2.32 Subfacies of Facies II are defined by petrographic differences determined from thin sections and cores. (Core description is courtesy of Norsk Hydro.)

low gamma-ray values. Fining upwards or coarsening upwards trends may occur, but are not typically recognized on gamma-ray logs in deep-water clastic systems, as clay content tends to be sorted equally from the fine-grained sands to the coarse-grained sands (Rider, 1996). Facies III will show a serrated log pattern, and the overall gamma-ray values will be higher than for Facies II. However, individual sand beds within a Facies III unit may show gamma-ray values as low as Facies II sands. Facies IV shows a less serrated pattern, but higher gamma-ray values. Facies V can show serrated gamma-ray values, but ideally it should be smooth, with very high gamma-ray values. Facies I and VI, not observed in Plate 2.31, will normally show a complex pattern because of random arrangement of quartz and clays.

Density and sonic logs have also been used to ensure that each facies occurs as significant clusters in terms of rock physics properties. Rock physics analysis can furthermore be used diagnostically to determine lithofacies and to define training data when direct core and thin-section data are not available. The subfacies IIb and IIc have been determined from core, thin-section, and SEM analyses (Figure 2.32), whereas IIa, representing a zone where no cores were taken, has been diagnosed as cemented

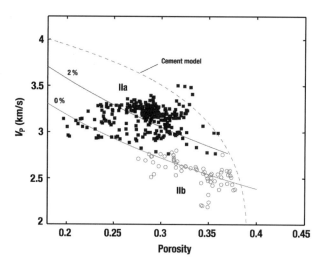

Figure 2.33 Rock physics diagnostics of two sandstone intervals in the type-well, indicating an unconsolidated zone (Facies IIb, open circles) and a cemented zone (Facies IIa, filled squares). The unconsolidated sands have been confirmed by core observations (Figure 2.32). Presence of cemented Heimdal Formation sands has been confirmed in Section 2.3.

thick-bedded sands using rock physics theory (Dvorkin and Nur, 1996). As confirmed in Section 2.3, the Heimdal Formation comprises both friable sands and cemented sandstones. Figure 2.33 shows the interval between 2252 and 2280 m in the type-well plotted in terms of velocity versus porosity, superimposed on the contact-cement model, the constant-cement fraction model (2% quartz cement), and the unconsolidated line. We diagnose the zone as cemented sands (~2%). Also included in this plot is the zone defined as Facies IIb, which we know from core and thin sections to be uncemented sands. These sands fit perfectly with the unconsolidated line. The cementation in Facies IIa is volumetrically not very significant, but in terms of elastic properties it has important impact. The seismic velocities and impedances are relatively high because of the stiffening effect of initial cementation (cf. Section 2.2).

2.5.4 Rock physics analysis of seismic lithofacies

Figure 2.34 shows the different seismic lithofacies plotted as P-wave velocity versus gamma ray (left), and density versus gamma ray (right). We observe an overturned V-shape, and an ambiguity exists between Facies IIb and IV/V. Cemented sands (IIa) and laminated sands (IIc) as well as interbedded sand–shales have relatively high velocities. The sand–shale ambiguity is not observed in density versus gamma ray. Here we see a more linear trend where density increases with increasing gamma-ray values (i.e., clay content) as we go from clean sands (Facies IIa and IIb) to silty shales (Facies IV). However, we observe that silty shales have higher densities than pure shales. The

2.5 Example: Seismic lithofacies

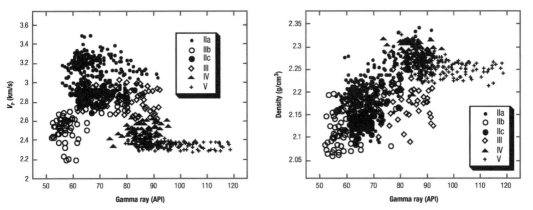

Figure 2.34 P-wave velocity versus gamma ray (left) and density versus gamma ray (right), for different seismic lithofacies in training data (i.e. Well 2). Note the ambiguity in P-wave velocity between Facies IIb and IV/V.

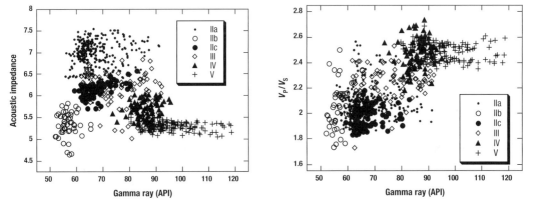

Figure 2.35 Acoustic impedance versus gamma ray (left) and V_P/V_S ratio versus gamma ray (right) in type-well.

sand–shale ambiguity observed in terms of velocity is also observed in acoustic impedance, which is the product of V_P and density (Figure 2.35; left). The overturned V-shape we observe can be explained physically: for grain-supported sediments, increasing clay content tends to reduce porosity (i.e. increase density) and therefore stiffen the rock. However, for clay-supported sediments, porosity will increase with increasing clay content because of the intrinsic porosity of clay, and the rock framework will weaken. Hence, velocity will reach a peak when clay content is approximately 40%. This effect was described by Marion *et al.* (1992) based on laboratory measurements of sand–shale mixtures (see Section 2.2.3 and Figure 2.7). Among others, Zeng *et al.* (1996) observed an ambiguity between seismic properties of clean sands and pure shales in an oil field, studying the acoustic impedance of Tertiary sediments in the Powderhorn Field, Texas.

The shear-wave sonic log provides us with shear-wave velocity (V_S). Figure 2.35 (right) shows the V_P/V_S ratio versus gamma-ray value. Here we observe that Facies IIb can be distinguished from shales (Facies IV and V), as the V_P/V_S ratio increases with increasing shaliness. Higher V_P/V_S ratios in shales than sands are expected, since the shear strength in shales tends to be relatively low compared with sands, owing to the platy shapes of clay particles.

2.6 Rock physics depth trends

Velocity–depth trends are important in seismic exploration and borehole drilling for several reasons. Commonly, these have been used for detection of overpressure zones from seismic velocity data (using traveltime inversion), indicated by negative velocity anomalies (e.g., Herring, 1973; Japsen, 1998; Dutta *et al.*, 2002a, 2002b). These are important to detect since they can cause hazardous blowouts during drilling. Also, velocity–depth trends can be used for calculation of interval velocities and depth conversion of seismic time horizons (e.g., Carter, 1989; Al-Chalabi, 1997). In areas where few wells are drilled, one often needs to assume a velocity trend based on an interpreted geologic depth-column. The trends for sands and shales can also be used to study the expected seismic signatures of sand–shale interfaces as a function of depth, and to identify anomalous lithologies (e.g., limestones) or diagenetic zones (e.g., cementation). Similarly, overcompacted zones related to uplift can be recognized, and erosion-thickness (i.e., missing overburden) can be estimated (e.g., Bulat and Stoker, 1987; Japsen, 1993; Al-Chalabi and Rosenkranz, 2002). Finally, expected brine-saturated velocity–depth trends can be applied to detect seismic velocity anomalies related to hydrocarbons (e.g., Avseth *et al.*, 2003).

Several authors have studied the impact of depth and compaction on porosity of sands and shales (e.g., Magara, 1980; Ramm and Bjørlykke, 1994; Lander and Walderhaug, 1999). Independently, empirical velocity–depth baselines have been applied to study burial anomalies. However, few authors have used more rigorous rock physics theory to study the effect of depth on seismic velocities. Japsen *et al.* (2001) investigated the relations between rock physics models and normal velocity–depth trends for different lithologies, with examples from the North Sea and the Gulf of Mexico. He presented baselines for sandstone and for shale that are based on a modified Voigt trend, and on a constrained relation between transit time and depth, respectively.

In this section, we use existing empirical porosity–depth trends for sands and shales, as input to rock physics models of V_P, V_S, and density. We use Hertz–Mindlin theory (Mindlin, 1949) to calculate the velocity–depth trends for unconsolidated sands and shales, whereas Dvorkin–Nur's contact-cement model (Dvorkin and Nur, 1996) is used

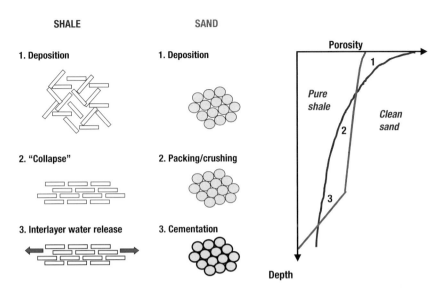

Figure 2.36 Schematic illustration of porosity–depth trends for sands and shales. Both the sand and shale trends can vary significantly because of composition, texture, pore fluids, temperature and pressure gradients. Hence, no attempt is made to assign absolute scales. However, there are a few rules of thumb. (1) The depositional porosity of shales is normally higher than that of sands. (2) The porosity gradient with depth is steeper for shales than for sands during mechanical compaction (i.e., at shallow depths). (3) The porosity gradient with depth will be steeper for sands than for shales during chemical compaction (i.e., quartz cementation of sands normally occurs at greater burial depth, beyond 2–3 km).

for cemented sands. The modeling results provide estimates of the parameters needed to calculate expected seismic response with depth for sand–shale interfaces. Hence, the depth trends allow us to study the ability to discriminate between pore fluids and lithologies at different depths.

> **Shale versus sand compaction**
>
> At deposition, shales tend to have relatively high porosities compared with sands. Sands will have depositional porosities of about 0.4, while shales can have depositional porosities of more than 0.8 (Figure 2.36). Shaly sands and heterolithics (i.e., mixed sands and shales) can have even lower depositional porosity than 0.4, as clay particles will fill the pore space of the sand frame.
>
> Shales tend to compact more easily than sands, causing a cross-over of the porosity–depth trends of sands and shales. At greater depths, different diagenetic processes occur. Sands lose porosity mainly through cementation, while bound water is released and intrinsic clay porosity is reduced in shales. Secondary porosity may occur in sands because of dissolution of mineral grains. Hence, porosity–depth trends can become very complex at great depths.

2.6.1 Compaction of sands and shales

During burial, the porosity of sediments changes dramatically through diagenesis. Diagenesis represents all mechanical and chemical alterations of a rock after deposition. Diagenetic processes change with burial depth, time (age), and/or temperature. The processes most damaging to porosity and permeability in sands and shales during early burial are packing change and ductile grain deformation (Surdam et al., 1989). In the North Sea, this mechanical compaction dominates the diagenetic reduction of porosity during burial from 0 to 2.5–3 km (Ramm et al., 1992).

During progressive compaction of sandstone, the number of grain contacts and the area of contacts between grains increase. If the grains are spherically shaped and there are no ductile grains present, the intergranular volume of a sandstone may be reduced towards 26% (closest packing of spheres). However, resistance to reorientation by angularities prevents compaction. Sandstones with more ductile components such as clay matrix, phyllitic rock fragments, and mica undergo a more severe loss of porosity by mechanical compaction (Surdam et al., 1989).

The rate of porosity decrease for sands and shales is more rapid at shallow depths and slows at greater depth of burial (Magara, 1980). Rubey and Hubbert (1959) proposed an exponential function to describe the porosity reduction with depth as follows:

$$\phi = \phi_0 e^{-cZ} \tag{2.42}$$

where ϕ is the porosity at depth Z, ϕ_0 is the depositional porosity (i.e., critical porosity) at the surface ($Z = 0$), and c is a constant of dimension (length^{-1}). Ramm and Bjørlykke (1994) suggested a clay-dependent exponential regression model for porosity versus depth of sands, valid only for mechanical compaction:

$$\phi = A e^{-(\alpha + \beta C_\mathrm{I})Z} \tag{2.43}$$

where A, α, and β are regression coefficients (see Figure 2.37). Coefficient A is related to the initial porosity at zero burial depth, α is a framework grain stability factor for clean sandstones ($C_\mathrm{I} = 0$) and β is a factor describing the sensitivity towards increasing clay index (C_I). The clay index is defined as the volume content of total clays (V_Cl) relative to the total volume content of stable framework grains, where we assume grains are quartz (V_qz):

$$C_\mathrm{I} = \frac{V_\mathrm{Cl}}{V_\mathrm{qz}} \tag{2.44}$$

However, chemical compaction also affects the porosity of rocks. In particular, quartz cementation is of great importance in quartz-rich sands, and drastically affects porosity, permeability, and seismic properties. It may occur during shallow burial, associated with meteoric flow precipitation (Dutton and Diggs, 1990), but is more common at deeper diagenetic levels associated with pressure solution. In quartz-rich sandstones, pressure

2.6 Rock physics depth trends

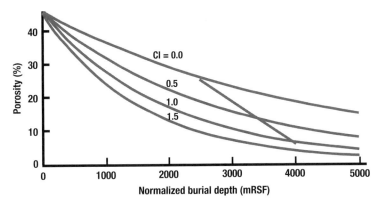

Figure 2.37 Sand and shale porosity models (equations (2.43) and (2.45)) with depth. During shallow burial, porosity change is mainly due to mechanical compaction (curved lines, equation (2.43)), and the porosity decreases with increasing clay content (i.e., increasing ductility). At a certain depth level, clean sands lose porosity mainly via pressure solution and quartz cementation (straight line, equation (2.45)). (Modified from Ramm and Bjørlykke, 1994.) Depth is in meters relative to sea floor (mRSF).

solution and related quartz cementation is probably the process most devastating to porosity during deep burial (Surdam *et al.*, 1989). Sandstones in continuously subsiding sedimentary basins, such as in the North Sea and the Gulf Coast, are mainly subject to mechanical compaction and tend to have poorly developed quartz cement down to a depth of 2.5–3.0 km. Chemical compaction in terms of pressure solution and quartz cementation will normally dominate below this depth level (Bjørlykke and Egeberg, 1993; Ramm and Bjørlykke, 1994). Ramm and Bjørlykke suggested that clean sandstones lose porosity mostly via pressure solution and quartz cementation approximated by the following formula:

$$\phi = \phi_D - k(Z - Z_D) \tag{2.45}$$

where ϕ_D is the porosity at depth Z_D where diagenetic cement initiates. The constant k is the rate at which the cement volume increases with depth.

Presence of clay in sandstones normally inhibits quartz cementation. Consequently, chemical compaction related to quartz cementation is most significant in clean sandstones. Quartz cementation is furthermore inhibited by the early migration of hydrocarbons, and/or overpressure (Dutton and Diggs, 1990).

Diagenesis of shales is restricted to mechanical compaction during shallow burial (less than ∼80 °C). However, a stable clay fabric tends to develop in the early stages of burial, and remains unchanged during the subsequent burial history. Hence, clay fabrics are relatively independent of depth, and pure shales tend to obtain a nearly constant porosity trend versus depth (Sintubin, 1994). Chemical processes in shales, including transformation of smectite to illite and liberation of organic acids from organic matter, begin at an intermediate diagenetic level (80–140 °C).

Baldwin and Butler (1985) introduced alternative compaction curves for sands and shales. Instead of deriving empirical porosity–depth trends, they derived solidity–depth trends, where solidity (N) is defined as the volume of solid grains as a percent of total volume of sediment, which is the complement of porosity. During compaction, the thickness of solid grains is constant; therefore the relation between solidity and sediment thickness reduction is linear, whereas the relation between porosity and sediment thickness reduction is nonlinear (Shinn and Robbin, 1983). Hence, the equations for solidity versus depth are normally simpler than the equations for porosity versus depth. Also, the resulting power-law equations for solidity versus depth were found to fit data for siliciclastic sediments better than the exponential equations for porosity versus depth. For normal shales they found the following power-law relationship:

$$Z = 6.02 N^{6.35} \tag{2.46}$$

where Z is the burial depth in km.

Similarly, they found that thick (>200 m), undercompacted shales followed the Dickinson equation (Dickinson, 1953) derived from Tertiary shales of the Gulf Coast. In terms of solidity, the resulting power-law version of the Dickinson equation is:

$$Z = 15 N^8 \tag{2.47}$$

Finally, they found that the Sclater and Christie (1980) sandstone curve for North Sea sediments should work well for rather mature, moderately cemented sandstones. The equation for their curve is exponential, and solved for burial depth it has the following logarithmic form:

$$Z = 3.7 \ln[0.49/(1 - N)] \tag{2.48}$$

2.6.2 Rock physics properties as a function of compaction

In order to understand the expected seismic response of a siliciclastic reservoir, at any given depth, it is of key interest to know the expected contrast in elastic properties between shales and sands as a function of depth. However, rock physics depth trends can be very complicated, depending on mineralogy, lithology, diagenesis, pore pressure, effective stress and fluid properties. In areas with good well coverage, one can establish empirical rock physics depth trends for different lithologies from statistical regressions to well-log data (V_P, V_S, and density). However, we want to stress the importance of modeling depth trends. Rock physics models allow for extrapolation of observed trends to depositional settings and depth ranges that are not covered by well-log data. This is often the case in an early exploration stage. Furthermore, modeled depth trends help us to better understand observed depth trends, and to detect anomalous zones that do not

2.6 Rock physics depth trends

follow the expected depth trends, whether these are pressure anomalies, unexpected lithologies, or abrupt diagenetic events.

In general, seismic velocities and densities of siliciclastic sedimentary rocks will increase with depth because of compaction and porosity reduction. However, the rock physics depth trends (i.e., V_P, V_S, and density versus depth) corresponding to the porosity–depth trends mentioned in the previous section can be rather complex because of the competing effects of porosity, pressure, mineralogy, texture, and pore fluids. In fact, we may observe more than one cross-over in velocity–depth trends of sands and shales. Rock physics models can be very useful in better understanding these depth trends. However, the models have to be calibrated to local geology before they can be used for further prediction of hydrocarbons and lithology. Geologic constraints include expected lithofacies and facies associations, sand and shale mineralogy (to determine effective elastic moduli and densities for the solid phase), fluid properties (oil density, GOR, gas gravity, brine salinity), as well as information about pressure and temperature gradients. For unconsolidated rocks we apply Hertz–Mindlin contact theory (see Section 2.2) to calculate elastic moduli of unconsolidated sediments as a function of porosity and pressure. Based on the elastic moduli, we calculate V_P and V_S versus depth. Density (ρ) is calculated directly from the porosity trends. From these parameters we can calculate acoustic impedance and V_P/V_S ratios versus depth. We can calculate depth trends for clean sands, shaly sands and shales using Hertz–Mindlin theory.

Using Hertz–Mindlin for unconsolidated sediments

It has been found that Hertz–Mindlin contact theory overpredicts velocities in unconsolidated sands, especially at low pressures and for dry samples. One possible explanation for this could be shear slip and/or rotation of grains around contact points which is not accounted for by the Hertz–Mindlin contact theory. However, we assume this to be a second-order effect for saturated sediments *in situ* at burial depths of several hundred meters to a few kilometers.

One can debate the applicability of Hertz–Mindlin theory for shales. The theory behind the model assumes spherical grains, but clay-rich shales certainly contain mainly platy grains and particles. However, when calculating the effective bulk and shear moduli of a dry sphere pack using equations (2.3) and (2.4), the coordination number to some degree takes into account the shape of the grains. At deposition, a shale tends to have very high porosity and very low coordination number, while a compacted, "collapsed" shale will have relatively low porosity and high coordination number.

Although the Hertz–Mindlin theory is not a completely rigorous model, we find it to work fairly well for shales as well as for sands.

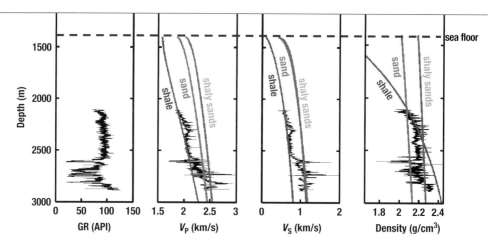

Figure 2.38 Examples of rock physics depth trends estimated for a well offshore Angola. This well is the same as the one used in case study 4 in Chapter 5. We observe a nice match between the calculated depth trends for different lithologies and the well-log data. The gamma-ray log to the left is a good clay indicator in the area, where extreme lows are representative of clean sands. Some of the swings in the logs which deviate sharply from the trends correspond to hydrocarbon zones (low V_P and density values) and a few cemented zones (high V_P and V_S values).

Rock physics depth trends for cemented sandstone can be calculated using the Dvorkin–Nur cement model (see Section 2.2). However, in addition to the porosity, we would then need to know or assume the amount of cement as a function of depth before calculating the elastic properties. To do this requires reliable information from geologists about the expected cement volume at a given depth.

Figure 2.38 shows an example of rock physics depth trends calculated for an offshore Angola well. This well is the same as the one used in case study 4 in Chapter 5. We observe a nice match between the calculated velocity–depth trends for different lithologies and the well-log data. The corresponding depth trends in acoustic impedance and V_P/V_S are shown in case study 4 in Chapter 5, and these are used to constrain the AVO classification in the area.

2.7 Example: rock physics depth trends and anomalies in a North Sea field

In this section we want to apply the physical models given in the previous section, to predict expected depth trends in seismic properties in a North Sea field. The empirical porosity–depth trends need to be calibrated to local observations, and this can be done in two ways: either by calibrating the equations to an inverted density log, or if core observations and helium porosity measurements are available, using these to calibrate the porosity–depth trends. On the basis of the porosity calibration, we can model expected depth trends of velocity, impedances and V_P/V_S ratio. For other areas, the

2.7 Example: Depth trends and anomalies

porosity trends need to be calibrated to the local conditions, and the porosity formulas below should not be applied.

In this example we also use local observations to constrain the transition from mechanical compaction to cementation. This transition can vary considerably from one area to another, but using only the mechanical compaction models can reveal where this transition zone happens. Reliable information about cement volume from geologic observations or geochemical modeling may be used to constrain the velocity–depth trend modeling for cemented rocks.

2.7.1 Modeling and calibration of porosity–depth trends

We apply the empirical porosity–depth models of Ramm and Bjørlykke (1994), given in Section 2.6, equation (2.43), to North Sea well-log data from the Glitne field. We calibrate the formulas to measured porosities at target level, where helium porosities of sands are available. We assume the surface depositional porosity (i.e., critical porosity) of sands to be 0.45, consistent with Ramm and Bjørlykke's regression. The regression is calibrated with clean Heimdal sands at 2150–2160 m depth, where the average porosity is 0.34. Hence we obtain the following regression formula for relatively clean sands ($C_I = 0.1$):

$$\phi = 45 e^{-(0.10 + (0.27 \times 0.1))Z} \tag{2.49}$$

We do the same procedure for shales. However, we select a higher critical porosity for shales than for sands, $A = 60$ versus $A = 45$, since we know that shales normally have higher depositional porosities related to the card-stack arrangements of clay particles (Rieke and Chilingarian, 1974). Lacking cores and helium-porosity measurements, we calibrate with well-log derived porosities, which are, on average, 0.28 for Lista Formation shales in the interval 2140–2154 m. These shales have a clay content of 60–70% and a quartz content (i.e., silt particles) of about 30–40% (see Section 2.3.2). Hence, we choose $C_I = 2$, and the regression formula for this shale becomes

$$\phi = 60 e^{-(0.01 + (0.22 \times 2.0))Z} \tag{2.50}$$

For modeling the porosity of the cemented sands, we use the linear cementation model of Ramm and Bjørlykke's (equation (2.45)), and calibrate it to cemented sands observed at 2250–2270 m depth in the North Sea. These sands have an average porosity of 0.28. The resulting formula is assumed to be valid below 2.2 km depth; hence the formula becomes

$$\phi = 30 - 13(Z - 2.2) \tag{2.51}$$

The porosity at 2.2 km depth is set to 30% (0.3) in order to be calibrated to the cemented sands at 2250–2270 m depth. The trends calculated from the formulas (equations (2.49), (2.50), and (2.51)) are shown in Figure 2.39 (left).

Rock physics interpretation of texture, lithology and compaction

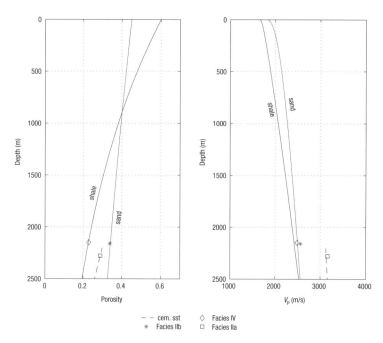

Figure 2.39 Porosity and P-wave velocity versus depth for unconsolidated sands and shales as modeled by Hertz–Mindlin theory, and for cemented sandstone as modeled by the Dvorkin–Nur cementation model. The models have been calibrated with North Sea observations in the porosity space, and the modeled velocities match the observations almost perfectly.

2.7.2 Modeling velocity–depth trends

Next we input the porosities calculated above into the theoretical velocity models outlined in the previous section. The porosities are shown in Figure 2.39 (left) and the corresponding velocities in Figure 2.39 (right). The formula for coordination number in equation (2.7) has been combined with equations (2.49) to (2.51) to obtain coordination number versus depth for sands and shales, respectively (see Figure 2.40). As input for the contact-cement model, the critical porosity value is set to 0.34 at 2.2 km depth, which is the value of unconsolidated sands at that depth. The calibrated porosity value at that depth, according to the regression formula in equation (2.48), is 0.3, which means that the cement volume at that depth is 4%.

Based on the elastic moduli, we calculate V_P and V_S versus depth. Density (ρ) is calculated directly from the porosity trends. P-wave velocity V_P is shown in Figure 2.39 (right), whereas acoustic impedance and V_P/V_S ratios versus depth are shown in Figure 2.41.

There is an excellent fit between the predicted P-wave velocities and the observed mean values at about 2200 m, for all the different rock types (Figure 2.39). The unconsolidated sands have slightly higher velocities than the shales. The cemented sands, however, show much higher velocities. The acoustic impedances also show an excellent

2.7 Example: Depth trends and anomalies

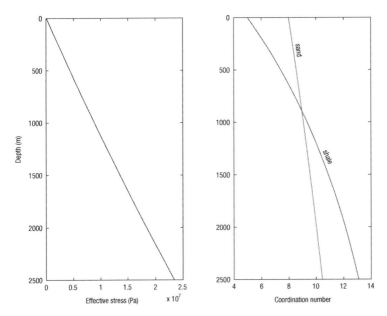

Figure 2.40 Effective stress and coordination number as a function of depth.

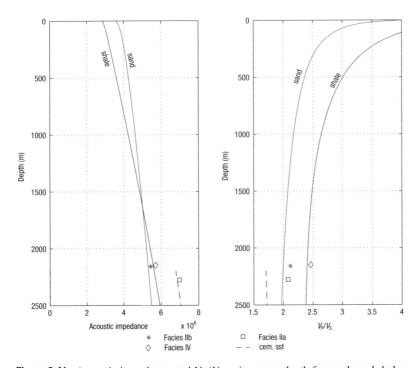

Figure 2.41 Acoustic impedance and V_P/V_S ratio versus depth for sands and shales.

match (Figure 2.41, left). The unconsolidated sands have slightly lower impedance than the shales, whereas the cemented sands have much higher impedance values. Moreover, there is a cross-over in impedance with depth between unconsolidated sands and shales occurring at about 1600 m depth. For V_P/V_S, the match between the models and observed data is not perfect (Figure 2.41, right). However, the Hertz–Mindlin models for sands and shales show a nice match with the observed values. The V_P/V_S ratio is consistently lower for sands than for shales. The contact-cement model, however, underpredicts the V_P/V_S of cemented sands.

Plate 2.42 shows the porosity and sonic velocity versus depth for relatively clean sandstones (red) and relatively pure shales (blue) from a different well, adjacent to the one from the previous section. The data are measured in the interval between the sea floor and rocks of Intra-Paleocene age (Tertiary). We have superimposed the Hertz–Mindlin models for unconsolidated sands and shales. Although there is significant discrepancy in the porosities, the predicted velocities show a remarkably good fit with the data.

Below 2 km depth, the sands have much higher velocities than the shales immediately above. This dramatic increase in velocity with depth for sands of the Heimdal Formation does not agree with purely mechanical compaction and corresponding porosity reduction with depth. Purely mechanical compaction would be represented by a gradually increasing velocity and a decreasing velocity gradient with depth, according to the Hertz–Mindlin model. The "jump" in velocity observed for the Heimdal sands, however, can be explained by chemical compaction. These sands most likely have slight quartz cementation, which produces a velocity increase through a stiffening effect on the grain contacts. The onset of cementation is interpreted to occur at about 2 km depth. We assume that mechanical compaction dominates above 2 km, while quartz cementation dominates below this depth.

The overall velocity contrasts between sands and shales in the mechanical compaction zone (0–2 km) are relatively weak. We do not observe a strong contrast in velocity between sands and shales until the sands enter the chemical compaction zone. The transition from mechanical to chemical compaction therefore represents a very significant seismic boundary in the North Sea.

By cross-plotting velocity versus porosity for various sand and shale intervals at the different depth levels observed in Plate 2.42, we can better evaluate what happens in terms of diagenesis and compaction of the rocks (Plate 2.43). We analyze velocity–porosity data for sand intervals and their overlying shales at three different depth intervals. The sandstone intervals include the Utsira sands (Sst 800) where data range from 820 to 830 m, the Frigg sands (Sst 1500) from 1500 to 1600 m, and the Heimdal sands (Sst 2200). The shale intervals include Shale 650 ranging from 620 to 670 m, Shale 1400 from 1400 to 1500 m, and Shale 2100 spanning the interval from 2120 to 2150 m. The various sand intervals create separate data clusters in an *echelon* pattern, with overlapping, decreasing porosity values, but discrete jumps and great separation in velocity. These velocity jumps can be attributed to the increased depth, causing an increase in effective pressure and more severe mechanical and chemical compaction, as discussed above. The individual sand clusters at given depths show relatively constant or slowly increasing velocity with decreasing porosity. According to rock physics diagnostics, presented in Section 2.2, this variation is related to deteriorating sorting and/or increased clay content in the sands.

The paths assumed to represent increasing clay from clean sands to pure shales are superimposed. The expected V-shaped paths are observed at all depths. However, at the shallowest level, the V-shaped path is highly compressed. The sands probably overlap with the shales, because this sand cluster includes both clean and shaly sands, while the shales are both pure and sandy. The whole clay spectrum is therefore represented in the data. With increasing depth, the V-shaped path is less compressed. This is consistent with Yin's (1992) laboratory observations, where he conducted pressure-dependent studies of the velocity–porosity relations in sand–shale mixtures. At the greatest depth, there is a large drop in velocity from the shaly sands to the pure shales. This gap is related to the effect of going from grain-supported to clay-supported sediments (Marion, 1990), but the gap is amplified because the sands are slightly cemented.

Figures 2.44 and 2.45 show an example of combined diagenetic modeling and rock physics modeling (Helseth *et al.*, 2004). The cement volume as a function of depth is modeled following the methodology of Lander and Walderhaug (1999), whereas the corresponding velocities are modeled using the clean sandstone models presented in Section 2.2. In this example, three different depths are selected (Figure 2.44), and we can see the expected velocity–porosity locations at these various depths (Figure 2.45).

2.8 Rock physics templates: a tool for lithology and fluid prediction

In this section we describe how we can combine the depositional and diagenetic trend models presented in this chapter with Gassmann fluid substitution, and make charts or templates of rock physics models for prediction of lithology and hydrocarbons. We refer to these locally constrained charts as *Rock Physics Templates* (RPTs), and this technology was first presented by Ødegaard and Avseth (2003). Furthermore, we expand on the rock physics diagnostics presented earlier in this chapter as we create RPTs of seismic parameters, in our case acoustic impedance versus V_P/V_S ratios. This will allow us to perform rock physics analysis not only of well-log data, but also of seismic data (e.g. elastic inversion results). Moreover, we show how important it is to constrain the rock physics models to local geology.

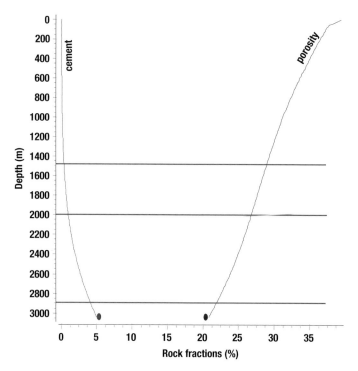

Figure 2.44 Quartz cement volume (left curve) and corresponding porosity trend (right curve) versus depth (Helseth *et al.*, 2004). The modeling is based on the method of Lander and Walderhaug (1999), and assumes temperature and pressure gradients representative for mid Norwegian Sea. These models can be converted to velocity versus depth using the rock physics models for sands in Section 2.2. In this example, three depths are selected where corresponding velocity versus porosity values are calculated (see Figure 2.45).

The motivation behind RPTs is to apply the models presented in this chapter (or other alternative rock physics models) and calculate a compilation of relevant RPTs for different basins and areas. Then the ideal interpretation workflow becomes a fairly simple two-step procedure:

- Select the appropriate RPT for the area and depth under investigation. Use well-log data to verify the validity of the selected RPT(s). (If no appropriate RPT exists, one should update the input parameters in the rock physics models, honoring the local geologic observations, and create new RPTs for the area under investigation.)
- Use the selected and verified RPT(s) to interpret elastic inversion results (see Chapter 5).

RPT interpretation of well-log data may also be an important stand-alone exercise, for interpretation and quality control of well-log data, and in order to assess seismic detectability of different fluid and lithology scenarios. Examples of this are included in Section 2.8.2.

2.8 Rock physics templates

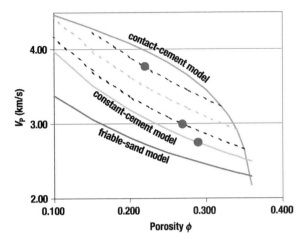

Figure 2.45 Velocity–porosity values (circles) corresponding to the three different depths selected in Figure 2.44 (Helseth *et al.*, 2004). Also included are the contact-cement model and various constant-cement models (light gray line, and dashed lines), as well as the friable-sand model. The modeled data points are based on combined diagenetic and rock physics models, and we observe that both cementation and mechanical packing (which will cause similar effects to sorting) increased with depth.

> **Intuitive cross-plots for cross-disciplinary applications**
>
> By creating intuitive RPTs, a broad range of geo-practitioners can use these as a toolbox for efficient lithology and pore-fluid interpretation of well-log data and elastic inversion results. The RPTs can be applied by petrophysicists who do formation evaluation, seismic interpreters who analyze inversion results, or rock physicists who evaluate seismic detectability from well logs.
>
> Moreover, the RPTs serve as a nice venue for different disciplines to meet for round-table discussions, as the plots show trends in seismic, petrophysical and geological parameters all together.

2.8.1 Rock physics models constrained by local geology

The first step in the modeling of RPTs is to calculate velocity–porosity trends for the expected lithologies, for various burial depths, using the models presented earlier in this chapter. To briefly summarize, we apply Hertz–Mindlin contact theory (Mindlin, 1949) to calculate the pressure dependency at the high-porosity end member. The other end point is at zero porosity and has the bulk and shear moduli of the solid mineral. These two points in the porosity–moduli plane are connected with a curve given by modified Hashin–Shtrikman (Hashin and Shtrikman, 1963) bounds (bulk and shear moduli) for the mixture of two phases: the original porous phase and the added solid phase. Porosity reduction related to packing and sorting, where smaller grains enter the pore space between larger grains, is modeled by the lower bound. For cemented rocks, we can either apply Dvorkin–Nur's cement model or the Hashin–Shtrikman upper bound model.

The next step is to calculate the elastic bulk moduli of brine- and hydrocarbon-saturated rocks. The dry rock properties calculated from the combined Hertz–Mindlin and Hashin–Shtrikman models are used as input into Gassmann's equations to calculate the saturated rock properties, assuming uniform saturation. From these we can calculate the V_P, V_S and density of brine- or gas-saturated rocks, and finally the acoustic impedance (AI) and V_P/V_S ratio. Estimates of AI and V_P/V_S are among the typical outputs from elastic inversion of seismic data, and this is the main reason for presenting the rock physics templates as cross-plots of V_P/V_S ratio versus AI. An example of an RPT is shown in Figure 2.46. It includes a background shale trend, a brine-sand trend, and curves for increasing gas saturation as a function of sand porosity.

> **Possible modifications to the RPTs**
>
> In the templates above we have calculated AI and V_P/V_S as a function of lithology, porosity and gas saturation. Alternatively, one could calculate similar templates for other parameters, such as AI versus EI (elastic impedance), AI versus SI (shear impedance), or λ (Lamé's parameter) versus μ (shear modulus). Also, one could add other dimensions to the templates, like attenuation and anisotropy. Finally, one could make templates with effective rock physics models to account for scale effects.

The RPTs are site- (basin-) specific and honor local geologic factors. Geologic constraints on rock physics models include lithology, mineralogy, burial depth, diagenesis, pressure, and temperature. All these factors must be considered when generating RPTs for a given basin. In particular, it is essential to include only the expected lithologies for the area under investigation when generating the rock physics templates. A siliciclastic system will comprise different lithofacies from a carbonate system. In this book we show examples for siliciclastic environments, so we expect the following lithologies: shale, shaly sands, and clean sands. But even for a siliciclastic system, the mineralogy can be highly variable. The sands can be either quartz-rich (arenite) or feldspar-rich (arkose). Quartz and feldspar have very different elastic properties, and this must be considered in the rock physics modeling. Other minerals may also be of significance. Shales are dominated by clay minerals such as smectite, illite, kaolinite or chlorite. Silty particles of quartz and feldspar are also very common in shales. Mavko *et al.* (1998) list elastic properties of common minerals. The sands modeled in the RPT in Figure 2.46 represent clean, quartz-rich sands (arenite), while the shales are assumed to be smectite-rich.

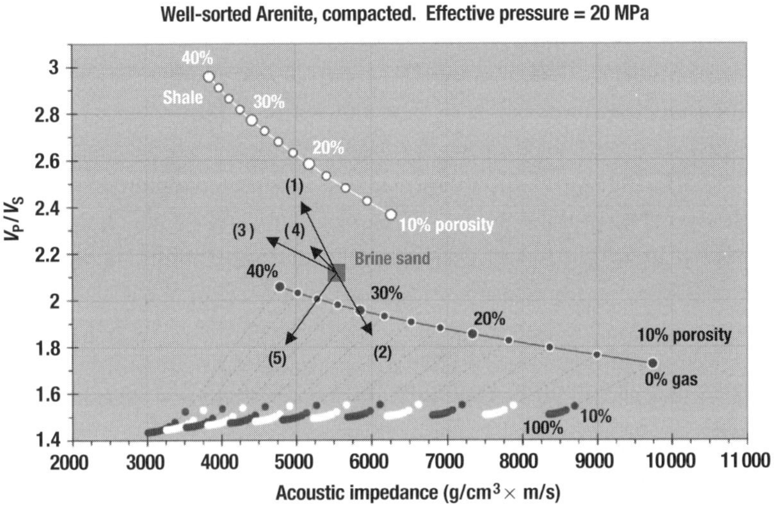

2.8 Rock physics templates

Figure 2.46 A rock physics template (RPT) presented as cross-plots of V_P/V_S versus AI includes rock physics models locally constrained by depth (i.e., pressure), mineralogy, critical porosity, and fluid properties. The template includes porosity trends for different lithologies, and increasing gas saturation for sands (assuming uniform saturation). The black arrows show various geologic trends (conceptually): (1) increasing shaliness, (2) increasing cement volume, (3) increasing porosity, (4) decreasing effective pressure, and (5) increasing gas saturation.

The water depth and the burial depth determine the effective pressure, pore pressure and lithostatic pressure. The pore pressure is important for the calculation of fluid properties, and to determine the effective stress on the grain contacts of the rock frame carrying the overburden. Porosity reduction associated with rock compaction and diagenesis are directly related to burial depth. At great depths quartz-rich sands tend to be quartz-cemented whereas smectite-rich shales will go through illitization and the release of bound water. In Figure 2.46 the effective pressure is 20 MPa. If the pore pressure is hydrostatic, this represents approximately 2 km burial depth.

In the modeling of the RPTs we also need to know the acoustic properties of mud filtrate, formation water and hydrocarbons in the area of investigation. Required input parameters include temperature, pressure, brine salinity, gas gravity, oil reference density (API), and oil GOR. In areas where hydrocarbons have yet to be encountered, gas gravity can be assumed (normally between 0.6 and 0.8). However, oil API is more uncertain. Also, the seismic response of oil can be difficult to distinguish from that of brine. Thus, in the templates presented in this chapter we consider only gas- and brine-saturated sands. One should, however, expect oil to show similar values to those of low gas saturation in cross-plots of AI vs. V_P/V_S ratio. Regarding saturation distribution, we have assumed uniform distribution in the modeling of the templates, which gives the famous effect where residual amounts of gas will cause almost the same seismic properties as commercial amounts of gas. However, a patchy distribution of gas would have shown a more linear change in seismic properties with increasing gas saturation.

Pitfalls

During modeling and application of RPTs, one should address the most likely geological scenario to be modeled. However, alternative pitfall scenarios should also be considered. Silica ooze and opal-A to opal-CT transitions, volcanic tuff, salt intrusions, calcite cement, and shallow overpressure all represent potential pitfalls that are typically not included in the models. Nonetheless, the RPTs may help discriminate some of these anomalies from hydrocarbon-related anomalies.

2.8.2 RPT analysis of well-log data: examples from Norwegian Sea and offshore West Africa

The well-log example presented in Plate 2.47 is from offshore mid-Norway. It shows acoustic impedance and V_P/V_S logs for a 100-m interval, and the corresponding V_P/V_S vs. AI cross-plot. The logs are color-coded based on the four populations defined in the cross-plot domain, and the cross-plot points are color-coded using the gamma-ray log (not shown).

Four populations can easily be identified in the log cross-plot domain, and a separate lithology or pore fluid can be attributed to each of the four populations based on additional log information: two different shales, gas sand and brine sand. The cross-plot interpretation would have been much more difficult without the additional log information. However, if we can use rock physics models to interpret these clusters, we can do more than interpret and quality-control observed well-log data. We can also use the templates to find the expected seismic properties for lithologies and fluid scenarios not encountered by any well. Moreover, having validated the templates with well-log data in an area, we can go forward and use the templates for enhanced qualitative interpretation of elastic inversion results (see case study 5 in Chapter 5).

Let us again consider the well-log data in Plate 2.47. Plate 2.48 is basically the same cross-plot as in Plate 2.47, superimposed onto the appropriate RPT. It includes porosity trends for different lithologies: the upper white dotted line represents pure shale while the black dotted line represents clean compacted brine-filled quartz sand. Increasing gas saturations are included for the clean sands. Note that the two "shale" populations fall exactly on the shale trend, and assuming that the trends are valid for the area, the obvious interpretation is that these two populations represent shales with different total porosities. The "brine-sand" population sits just above the theoretical brine-sand trend, and again assuming that the trends are valid for the area, the interpretation is that the brine sand is slightly shaly. The "gas-sand" population falls well below the brine-sand trend and roughly along the dotted lines indicating the effects of increasing gas saturation. For the "gas-sand" population it is possible to estimate the corresponding clean brine-sand porosities, but little can be inferred about the shaliness of the gas sand (this is a fundamental limitation in seismic data analysis).

Plate 2.49 shows examples of RPTs for different types of sands. The upper plot represents unconsolidated clean sand, whereas the lower one represents cemented sandstone. For the unconsolidated sands we observe that the V_P/V_S ratio drops dramatically with just a little increase in gas saturation, whereas the acoustic impedance drops more moderately. For the cemented sandstones, which have a stiffer rock frame, there is a much smaller fluid sensitivity. The V_P/V_S ratio shows a very small decrease with increasing gas saturation. However, the acoustic impedance still shows a marked decrease due to the density effect of the gas.

Log data for two different wells from offshore West Africa have been superimposed onto the templates in Plate 2.49. The logs represent the same Oligocene interval, but different burial depths; one well penetrated the Oligocene with approximately 1200 m overburden (top RPT in Plate 2.49), the other with approximately 2400 m overburden (bottom RPT). Hence, the sands and shales in the second well have been compacted more than in the first well, and the sands in the second well are cemented whereas the sands in the first well are unconsolidated. Quartz cementation tends to occur at temperatures over about 80 °C, corresponding to burial depths of about 1.5–2 km. The sands in the first well (top RPT in Plate 2.49) show a much bigger fluid response than in the second well (bottom RPT), even though the sands are from the same stratigraphic level! This illustrates the value of the RPTs. Even for the same basin, and the same stratigraphic level, different rock physics models will apply for different burial depths. Moreover, the RPTs indicate that the expected seismic response of hydrocarbon-saturated sands will be different at the two wells. For the shallow sands in the first well, we expect an AVO class II for oil-saturated sands, whereas the gas sands will be class III. For the deeper, cemented sands in the second well, the oil sands will be predominantly AVO class I, whereas gas sands will be AVO class II. Hence, the AVO response of hydrocarbons will be different at the two locations because of local diagenetic changes. This example of RPT analysis confirms how rock physics depth trends must be taken into account during AVO analysis, as shown in Chapter 4 and in case study 4, Chapter 5.

2.9 Discussion

Rock physics models can be used to predict or diagnose petrographic changes from velocity–porosity relations. By separating into different clusters of data, with characteristic sedimentary features and rock physics properties, we can use these as training data in a classification procedure (Chapter 3), and ultimately predict these various clusters from seismic data (Chapters 4 and 5). But why do we use physical models to diagnose the rocks? Why do we not just use thin-section and core information and correlate the various intervals with corresponding seismic properties, without caring much about the physical relations? First of all, thin-section and core data are not always available, and especially in non-sand lithologies, this direct information is rarely available. Even in sandy intervals where such information is available, we will not automatically find out what petrographic factor will be most important seismically, until we investigate the rock physics properties. The Heimdal sands are clearly a good example of this: the friable and cemented thick-bedded sands look very similar in core and thin section, but small amounts of quartz cement occurring in some of the sands make a big difference in seismic response. The rock physics diagnostics helped us to separate into these two different rock types.

The rock physics diagnostics used in this chapter suffer from two important factors. The first one is ambiguity in the velocity–porosity plane. A data point in this plane does not necessarily have a unique diagnostic result. For sands, there are ambiguities between clay content and sorting. Both these factors have similar paths in the velocity–porosity plane. There also seem to be ambiguities between different lithologies. Marl, tuffaceous muds and silty shales show great overlaps. One way to solve the ambiguities would be to use shear-wave information. In this chapter we have primarily used P-wave velocities. However, shear-wave information is often not available. Furthermore, the rock physics models for shear-wave velocity are known to vary widely from real data (Dvorkin and Nur, 1996), indicating that current models are not completely valid. One solution to this problem is to diagnose clusters of data using multivariate statistical methods as an alternative to physical rock diagnostics. Examples are shown in Chapter 3.

The second factor that may cause the rock physics diagnostics to fail is the issue of resolution. The well-log data can show effective values of small-scale heterogeneities, while the rock physics models used to diagnose the rocks in this chapter assume homogeneous rock types. This could be corrected by using effective rock physics models. The thin-bedded shaly sands showed a large scatter, and some of this scatter could be due to variation in sand–shale ratio and individual layer thicknesses. Ultrasonic measurements on cores could eliminate this problem, but since the core samples in the studied area are poorly consolidated or friable, velocity measurements in the laboratory would not give correct *in situ* values.

The seismic lithofacies defined in this study create a link between rock physics and sedimentology, which can be used in seismic reservoir characterization (see Chapter 5). They can also serve as "building blocks" in seismic modeling. In both cases, the use of seismic lithofacies becomes a predictive tool, as different seismic lithofacies will be associated with each other. Seismic lithofacies will therefore make it easier to interpret seismic amplitudes in terms of sedimentological features and depositional trends. Ultimately, the potential of seismic lithofacies is to improve reservoir characterization from 3D seismic data.

When rock physics is used in reservoir characterization, it is important to separate depth-related changes and constant-depth variations in seismic properties. Usually, hydrocarbon reservoirs are located within a small depth interval with little depth-dependent variation in the seismic properties. Depth-related factors include mechanical and chemical compaction and related porosity reduction. However, variations in these may occur at a given depth level. In the North Sea case considered here, we observed that the transition from purely mechanical compaction to chemical compaction (i.e., quartz cementation) occurred within the Paleocene interval of interest, and this is a very significant seismic horizon. As a consequence, sands are observed to change laterally from friable sands to cemented sandstones, giving completely different seismic signatures.

The RPT analysis presented in this chapter brings together the concept of rock physics diagnostics with rock physics depth trends, for improved prediction of hydrocarbons at any given depth and any given depositional environment. This RPT analysis represents the first step in a methodology where the second step is to apply RPTs for interpretation of elastic inversion results. The application to quantitative seismic interpretation is presented in Chapter 4, with case examples in Chapter 5. Nevertheless, as illustrated in this chapter, RPT analysis of well-log data may also be an important stand-alone exercise. It can be used for petrophysical interpretation (i.e., formation evaluation) and quality control of well-log data, as well as assessment of seismic detectability of lithologies and fluids.

However, the templates must be used with care, and the reliability of the information extracted depends on the quality of the input data and the model assumptions. RPTs may not always be 100% valid, but can in most cases be used for enhanced qualitative interpretation of well-log and seismic data. Moreover, one should be aware of potential scale effects distorting the similarities between well-log data and seismic data. Nevertheless, the rock physics templates provide a very useful interpretation tool that can improve communication between geologists and geophysicists and can help reduce risk in seismic exploration and prospect evaluation.

2.10 Conclusions

- The seismic properties of sedimentary rocks are highly dependent on geologic factors including burial compaction, diagenesis, rock texture, lithology, and clay content.
- Clay content, cement volume, degree of sorting, and lithology can be identified via rock physics diagnostics based on well-log data.
- Rock physics diagnostics are important to incorporate into seismic interpretation. Otherwise, the seismic response differences may be misinterpreted as fluid or porosity changes, and result in erroneous prediction of hydrocarbons.
- Rock physics diagnostics can be used as a tool to identify characteristic clusters of data (facies) that can serve as training data in classification procedures (Chapter 3).
- Seismic lithofacies are seismic-scale sedimentary units with characteristic rock physics properties. Geologically, these facies are defined by clay content, grain size, and bedding configuration. The potential benefits of seismic lithofacies are better

understanding of seismic signatures and consequently improved reservoir characterization in complex depositional systems.
- Hertz–Mindlin theory can be applied to predict expected velocity–depth trends for unconsolidated sands and shales. Dvorkin–Nur's contact-cement theory can be applied to cemented sands.
- Deviations from expected velocity–depth trends can be related to overpressure, gas, diagenesis, lithology, uplift, etc. One can apply the velocity–depth models to detect seismic anomalies related to these effects.
- Quartz cementation of sands tends to initiate at a certain burial depth. Even a small percentage of contact diagenetic cement strongly affects the elastic properties of sands, resulting in a drastic difference between the seismic response of slightly cemented and friable reservoirs.
- Rock physics templates (RPTs) are basin-specific rock physics models constrained by local geologic trends. A compilation of RPTs provides geoscientists with an easy-to-use "tool-box" for lithology and pore-fluid interpretation of sonic log data and elastic inversion results.
- The RPT technology has a broad range of applications, ranging from analysis and quality control of well-log data, to interpretation and quality control of elastic inversion results.

3 Statistical rock physics: Combining rock physics, information theory, and statistics to reduce uncertainty

Any physical theory is a kind of guesswork. There are good guesses and bad guesses. The language of probability allows us to speak quantitatively about some situation which may be highly variable, but which does have some consistent average behavior.
...
Our most precise description of nature must be in terms of probabilities. *Richard Phillips Feynman*

3.1 Introduction

This chapter introduces the concepts of *statistical rock physics* for seismic reservoir characterization. We will see how we can quantify uncertainties in reservoir exploration and management by combining rock physics models with statistical pattern recognition techniques to interpret seismic attributes. Plate 3.1 shows an example of results from a statistical rock physics study. Seismic impedances from near and far-offset inversions were interpreted using well logs and rock physics to estimate the probabilities of oil sands. The figure shows the iso-probability surface for 75% probability of oil-sand occurrence. Statistical rock physics is also useful for identifying additional information that may help to reduce the interpretation uncertainties. Seismic imaging brings indirect, but nevertheless spatially extensive information about reservoir properties that are not available from well data alone. Rock physics allows us to establish the links between seismic response and reservoir properties, and to extend the available data to generate training data for the classification system. Classification and estimation methods based on computational statistical techniques such as nonparametric Bayesian classification, bootstrap, and neural networks help to quantitatively measure the interpretation uncertainty and the misclassification risk at each spatial location. With geostatistical stochastic simulations, geologically reasonable spatial correlation and small-scale variability are added; these are hard to identify with seismic information alone because of the limits of resolution. Combining deterministic physical models with statistical techniques helps us to develop new methods for interpretation and estimation of reservoir rock properties from seismic data. These formulations identify not only the most likely interpretation of our seismic data but also the uncertainty of the interpretation, and serve as a guide for quantitative decision analysis.

3.2 Why quantify uncertainty?

It is well known that subsurface heterogeneity delineation is a key factor in reliable reservoir characterization. Heterogeneities contribute to interpretation uncertainty. These heterogeneities occur at various scales, and can include variations in lithology, pore fluids, clay content, porosity, pressure and temperature. Some of the methods used in seismic reservoir characterization are purely statistical, based on multivariate techniques (e.g. Fournier, 1989). Others are deterministic, based on physical models derived from elasticity theory as well as laboratory observations. Each group of techniques can have some degree of success depending on the particular study. The optimum strategy is to combine the best of each method to generate results much more powerful than would be possible from purely statistical or purely deterministic techniques alone. Examples where combined methodologies have been used in case studies include, among others, Lucet and Mavko (1991), Doyen and Guidish (1992), Avseth *et al.* (1998, 2001a, 2001b), Mukerji *et al.* (1998a, 2001), and Eidsvik *et al.* (2004); see also the case studies in Chapter 5.

Subsurface property estimation from remote geophysical measurements is always subject to uncertainty, because of many inevitable difficulties and ambiguities in data acquisition, processing, and interpretation (see Chapter 4). How can we express quantitatively the information content and uncertainty in rock property estimation from seismic data? Indeed, why quantify uncertainty at all? Most interpretation techniques give us some optimal estimate of the quantity of interest. Obtaining the uncertainty of that estimate usually requires further work and hence comes at an extra cost. So what extra benefits do we get thereby? Why care about quantifying uncertainty?

> Uses of uncertainty
> - Assessing risk
> - Integrating data from different sources
> - Estimating value of additional data

One fundamental reason stems from our accountability as responsible scientists. We know that models are approximate, data have errors, and rock properties are variable. So it is always appropriate to report error bars along with the interpretation results. Error bars lend credibility. A more practical reason for understanding uncertainty is for *risk analyses* and optimal decision-making. Quantifying uncertainty helps us to estimate our risks better, and possibly take steps to protect ourselves from those risks. Uncertainty estimates are useful also in data integration, and in estimating the value of additional information for reducing the uncertainty. Complex interpretational processes such as reservoir characterization usually require *integration of data* from different sources and of different types. Understanding the uncertainties associated with the different data sets helps us to assign proper weights (and discard unreliable data) before we combine them together in the interpretational model. Additional data (for example S-wave data in addition to P-wave seismic data) may help to clear away ambiguities and reduce uncertainty – but not always. Estimating the *value of additional information* requires quantitative estimates of uncertainty.

3.2.1 Uncertainty, probability distributions and the flaw of averages

Having understood some of the applications of estimating uncertainty, let us now look at the tools required for quantitatively expressing uncertainty. Probabilities, random variables, and probability distributions are the basic building blocks of uncertainty. Classically the concept of probability involves long run frequencies. When we say that the probability of heads in a fair coin toss is 1/2, we mean that if we were to toss the coin repeatedly in a long series of independent tosses, heads would occur half of the time. This frequency concept of probability may be applicable in some situations, but often it does not suffice. For example, in the problem of trying to estimate the net-to-gross (NG) of a reservoir, what does it mean to say that the probability that $0.6 < NG < 0.7$ is 1/2? Here we do not have repeated identical trials. The net-to-gross is an uncertain number that we are trying to estimate. It is either between 0.6 and 0.7 or it is not. It cannot be half of the time within the interval and half of the time outside. Subjective probability theory allows us to deal with probabilities when the frequency viewpoint does not apply. In the subjective view probability is the degree of belief about a statement on the basis of given evidence. Some subjective probabilists argue that the frequency concept never applies because it is impossible to have a long sequence of identical repetitions of any event, except in textbooks. There continues to be disagreement and posturing among statisticians about the various philosophical viewpoints of probability. The interested reader can find discussions of the differing philosophical perspectives in Berger (1985), and Kyburg and Smokler (1980). Berger's (1985) book is also an excellent source for Bayesian analysis and statistical decision theory.

Uncertainty arises through our imperfect knowledge. We model uncertainty mathematically by random variables. Random variables are uncertain numbers. They can be continuous variables (e.g. uncertain porosity) or categorical variables (e.g. uncertain shale/sand lithology). Statistical probability density functions (pdfs) and cumulative distribution functions (cdfs) give us one way to describe quantitatively the state of our knowledge – or lack of knowledge – about the random variable. Categorical random variables are described by their probability mass function (pmf). If a random variable X

has a pdf $f(x)$, then specifying $f(x)$ allows us to compute probabilities associated with X:

$$P(a < X < b) = \int_a^b f(x)\,dx$$

The cdf is obtained by integrating the pdf. These distribution functions are the shapes that describe the uncertain quantity. The moments of the pdf, such as the mean and standard deviation, tell us about the central tendency and spread of the random variable, but a complete description of the random variable is given by the full pdf. The mean and standard deviation are the height and weight, but the pdf is the complete DNA. How do we estimate the unknown pdf from observed data? One estimate is the parametric approach where we assume that the pdf is a known function (e.g. Gaussian) and estimate the parameters (e.g. mean and variance) of the function from the data. A nonparametric approach makes less rigid assumptions about the functional form of the pdf. Silverman's monograph (Silverman, 1986) is the standard reference for nonparametric density estimation. The oldest and simplest density estimate is the histogram (Figure 3.2). The choice of bin widths and number of bins can have quite an impact on the histogram. Nevertheless, histograms are very useful for exploring data variability. Kernel estimators can improve upon histograms. Kernel-based pdf estimates amount to a smoothing or interpolation of the observations by a kernel or window function. The pdf estimate is a superposition of n window functions centered at the n data points. The window functions themselves must be nonnegative and integrate to 1 to ensure that the estimate is a legitimate pdf. Examples of window functions are the triangular kernel, the Gaussian kernel, or the Epanechnikov kernel (Silverman, 1986). The shape of the kernel does not matter too much. Any legitimate kernel function can be chosen based on considerations such as ease of computation. The size of the kernel and the amount of smoothing does have significant impact on the estimated pdf. If the window is too large, the estimated pdf will be too smooth and have too little resolution. If the window is too small, the estimate will have too much variability. Figure 3.3 shows histograms and smoothed kernel-based pdf estimates for porosity observed in a well log. The appropriate choices of the smoothing parameter will depend on the purpose of the density estimate. For many applications it is desirable to choose the smoothing subjectively by plotting several curves with different amounts of smoothing. Several pdf plots smoothed by different amounts can give more insight about the data than a single automatic curve. For visual presentations, the curve should be somewhat undersmoothed. The eye can very easily do further smoothing but it is hard to unsmooth "by eye." When estimating pdfs for statistical classification purposes, the smoothing parameter can be selected by cross-validation using a training data set and a test set. The smoothing that gives the least classification error on the test set is selected.

3.2 Why quantify uncertainty?

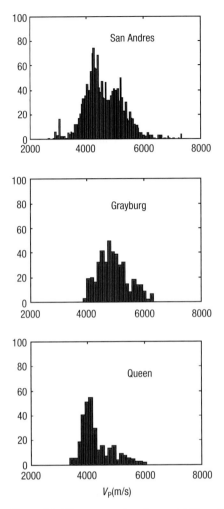

Figure 3.2 Histograms showing variability of P-wave velocity for three different formations in Texas. The histogram for the San Andres formation shows a bimodal behavior. The Queen formation has a skewed V_P histogram with a long tail. Data are from sonic logs.

The relations between multiple random variables (e.g. rock properties and seismic signatures), including their inherent uncertainty, can be described by joint pdfs. Nonparametric bivariate pdfs may be estimated as simple 2D histograms or by 2D kernel estimates analogous to the 1D pdf estimation. Plate 3.4 shows an example of a trivariate nonparametric joint density estimate for V_P, V_S, and rock bulk density. The power of nonparametric pdf estimates lies in their generality. We make no assumptions about their functional form but let the data speak. But computation and number of samples required for good estimation grow exponentially with dimensionality. The curse of dimensionality limits practical nonparametric pdf estimations in high dimensions (Duda et al., 2001).

Statistical rock physics

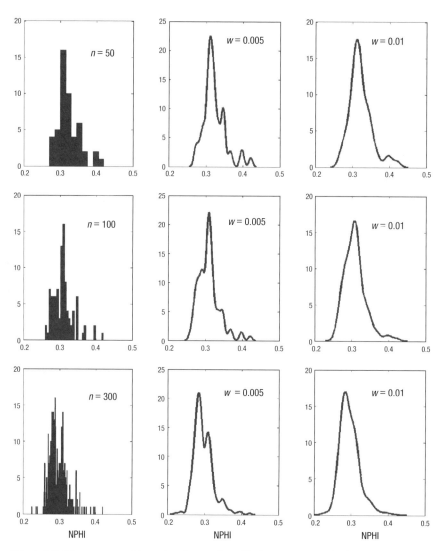

Figure 3.3 Histograms and kernel estimates for pdf of neutron porosity from well data. Increasing number of data points, n, from top to bottom, and for two different kernel bandwidths, w, for a Gaussian kernel. Larger bandwidth gives a smoother pdf estimate.

> Pdfs and random variables help us to quantify the distributions and variability of target reservoir properties. Estimates of the variability are data-dependent and model-dependent, and subject to our prior knowledge. All estimates of uncertainty are subjective.

The pdfs must be estimated from prior knowledge or available training data. The training set often has to be extended or enhanced using rock physics models to derive pdfs for situations not sampled in the original training data. For example, a well may encounter only brine sands, thus yielding no data for oil sands. The well data give us the pdf for brine-sand properties such as P-wave velocity or impedance. The pdf for oil-sand P-wave velocity then has to be derived by transforming the brine-sand pdf, using Gassmann's fluid substitution model. When the pdfs can be expressed analytically, it may be possible to carry out the mathematical transformations analytically to derive the new distributions. An example of analytically derived pdfs for hydrocarbon indicators using the congenial Gaussian distribution is presented in Mavko and Mukerji (1998b). Monte Carlo simulation is another, more powerful way of deriving transformed pdfs, without having to assume analytically tractable forms for the pdfs. Simple techniques for univariate and multivariate Monte Carlo simulations are described later, in Section 3.5.

> The concept of *derived distributions* is very important in statistical rock physics. Derived distributions, by combining deterministic rock physics models with the observed statistical variability, allow us to build a more powerful strategy for reservoir prediction than would be possible by using either purely deterministic models or purely statistical methods.

We may have data about an uncertain reservoir property X_1 that follows a pdf $P_1(X_1)$. Another uncertain reservoir property of interest is X_2. But we have no data for X_2. We believe that X_1 and X_2 are related by a deterministic rock physics model: $X_2 = g(X_1)$. We can estimate the pdf of X_2 by simulating multiple realizations of X_1 drawn from $P_1(X_1)$, propagating them through the function $g(X_1)$ to compute multiple values of X_2, and then estimate the distribution of X_2, $P_2(X_2)$, by binning and smoothing the computed values. This is a derived distribution. Derived distributions extend the training data.

Ignoring the variability of rock properties in quantitative computations can drastically distort decisions. As an example of the pitfall of ignoring distribution, consider the following calculation. We would like to compute the normal-incidence reflectivity between an overlying shale layer and a packet of thinly bedded, sub-seismic-resolution sand/shale layers. The effective properties (velocity, density, and impedance) of the thin sand/shale layers are computed using the Backus average. The contrast between the impedance of the shale and the effective impedance of the sand/shale packet then gives us the normal-incidence reflectivity. The Backus average elastic modulus depends on the volumetric sand fraction. For normal-incidence propagation perpendicular to the layers, the effective P-wave modulus is given by a volumetric harmonic average of the sand and shale moduli. The effective density is given by the usual volumetric arithmetic average of sand and shale densities. By doing the calculations for all sand/shale ratios (0 to 1) we get a relation between sand ratio and normal-incidence reflectivity. We could do the computation in two different ways. We could take an average value for sand V_P, density, and impedance, and another average value for the shale V_P, density, and impedance, and use them in the equations for Backus average and reflectivity. The average values could be estimated from blocked well logs, for example. This average

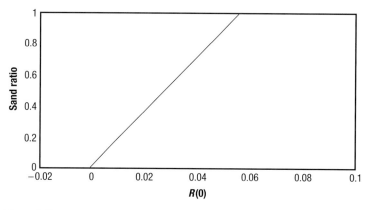

Figure 3.5 Relation between normal-incidence reflectivity, $R(0)$, and sand/shale ratio in very thin bedded sand/shale layers. The line is obtained using Backus average for thin layers, with blocked average values of sand and shale properties from well log. In this computation we ignored the variability of sand and shale velocity, density, and impedance.

computation, ignoring the variability, gives us the single line shown in Figure 3.5. One might use this line for interpreting observed reflectivities to estimate sand/shale ratio in the thinly layered packet. Instead of taking average sand and shale properties, another way to do the computation would be by Monte Carlo simulation. This would take into account the natural variability in the properties of the sand and shales. We draw a sample sand (V_P, density) and a sample shale (V_P, density) from the distributions observed in the well log. The simulated values are then used in the equations. A number of realizations are drawn to get the full distribution of reflectivity for every sand fraction. We bombard the equations with the whole range of possible inputs in accordance with their probability of occurrence. Now we see something very different (Figure 3.6). The average of the Monte Carlo simulations (solid curve) is not the same as the result from the average computation (solid straight line). The straight line would overpredict considerably the sand ratio for large values of reflectivity, and slightly underpredict the sand ratio at small values of reflectivity. The average computation also completely misses the negative branch of the relation between sand ratio and reflectivity. As we see from the distributions of the shale and sand impedances in Figure 3.7A, there is considerable overlap, leading to a small finite chance of getting negative reflectivity. The histogram of reflectivity from the Monte Carlo Backus calculation is shown in Figure 3.7B. Monte Carlo simulations also give us confidence intervals, such as the 10 percentile and 90 percentile curves shown in Figure 3.6. Computations using averages alone do not give any indication of the uncertainty due to the variability in the properties. The example described above was not a contrived example with cooked-up distributions, but an actual example, using log derived properties from a region where identification of thin sand/shale packets is a practical problem. The surprising outcome is an example of the flaw of averages (Savage, 2002, 2003). One should not expect to

3.2 Why quantify uncertainty?

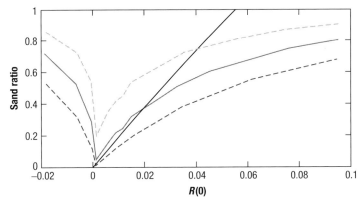

Figure 3.6 Relation between normal-incidence reflectivity and sand/shale ratio in very thin bedded sand/shale layers. The solid straight line is obtained using the Backus average for thin layers, with blocked average values of sand and shale properties from well log. Curves show the result of computations using Monte Carlo simulation to incorporate the variability of sand and shale velocity, density, and impedance. The solid curve is the mean of the distributions obtained from the simulations, and the upper and lower dashed curves are the 10 percentile and 90 percentile curves, respectively. We see clearly the pitfalls of ignoring variability. The calculations using average values can seriously overestimate the sand/shale ratio, leading to wrong decisions. It does not capture the negative reflectivity branch, and gives no indication of the uncertainty in the relation. The percentile curves obtained from Monte Carlo simulations are a measure of uncertainty in the relation, subject to the assumed model being correct.

get even correct average results using average values of inputs. Plugging in a single "best guess" input does not result in the "best guess" output. Mathematically, for a function $g(x)$, in general, $\langle g(x) \rangle \neq g(\langle x \rangle)$, unless $g(x)$ is linear. The symbol $\langle \ \rangle$ denotes the expectation operator. When there is a lot of variability, calculations using single point values are almost worthless. Simple simulations to take account of the variability can be easily performed using software such as Excel, MATLAB, or S. There is hardly any excuse for falling prey to the flaw of averages.

> Monte Carlo simulations, by taking into account whole distributions of values instead of single average values, help to avoid the flaw of averages. Calculations based on a single "best guess" input may not give the "best guess" output.

3.2.2 Describing the value of additional data with probability distributions

As already mentioned, additional data can sometimes bring in information that can help to reduce the uncertainty. For example, studies have shown that knowing V_S, in addition to V_P, can help to resolve ambiguities in reservoir facies versus fluid identification that arise from geologic variability.

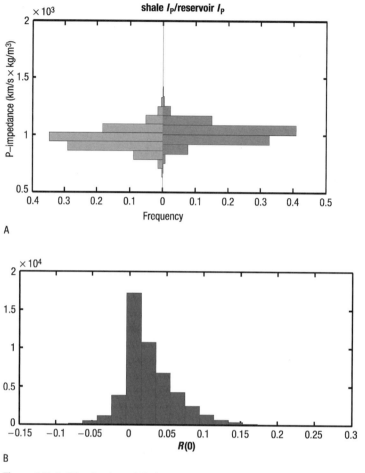

Figure 3.7 A, Distribution of shale and sand impedances used in the computation for Figure 3.6. On average the sand has slightly higher impedance than the shale, but there is considerable overlap. B, Histogram of reflectivity from Monte Carlo simulations. Because of the overlap in shale and sand impedances, there is a small but finite chance of negative reflectivity values.

Figure 3.8 illustrates an uncertainty in pore fluid identification that can occur from natural variability of rock properties such as porosity, sorting, and texture. On both the left and right figures, velocity data from gas- and water-saturated sandstones are plotted. In both, the porosities are large and the rocks are elastically soft, so there is a large separation between the two clouds in the two-dimensional (V_P, V_S) domain, as would be predicted by the Gassmann relations (Chapter 1). The biggest difference between the two cases is the *range* of porosities. On the left, the standard deviation in porosity is only 2%. On the right, it is twice as large. Because velocity is well correlated with porosity, we find a very narrow range of velocities on the left (standard deviation ~100 m/s), and nearly twice the range on the right.

3.2 Why quantify uncertainty?

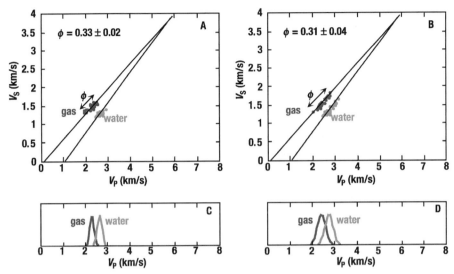

Figure 3.8 Changes in natural variability of porosity and velocity can become comparable to fluid effects.

The *variability* in velocity and porosity increases the uncertainty if we try to detect pore fluids using V_P only. Figure 3.8C shows the pdfs of V_P for the narrower porosity/velocity range. There is relatively little overlap between the gas- and water-saturated curves, so V_P would be sufficient for identifying the pore fluid, and V_S would add little value. Figure 3.8D shows the V_P pdfs for the broader porosity range. Even though the average shift in velocity between gas and water saturation is the same on the left (Figure 3.8C) and on the right (3.8D), and there is no measurement error in either, the broadening of the distributions would make fluid detection with V_P unreliable. *In this case adding shear information would be extremely valuable.*

A simple statement of the problem is that the change in P-velocity due to fluids is small compared with the naturally occurring variability in P-velocity due to porosity. Adding shear data essentially makes the ambiguity vanish.

In Figure 3.9 we explore how geologic variability might affect exploration or monitoring problems. For these plots, we have randomly sampled Han's sandstone data (Han, 1986) in the porosity range 15–31%. In 3.9A, we compare velocity data under both water- and gas-saturated conditions, at effective pressure of 10 MPa. The combination of high porosity and low effective stress makes the two data clouds fairly well separated, so that we would expect reasonable success in distinguishing gas vs. water from V_P and V_S data. It is fairly clear, however, that V_P alone would not be enough. The overlapping pdfs in the plot below Figure 3.9A show the distributions of V_P for the gas- and water-saturated conditions, and there is virtually no separation. In this case, *introduction of even noisy shear data, allowing us to work in two dimensions, would greatly improve our interpretation.*

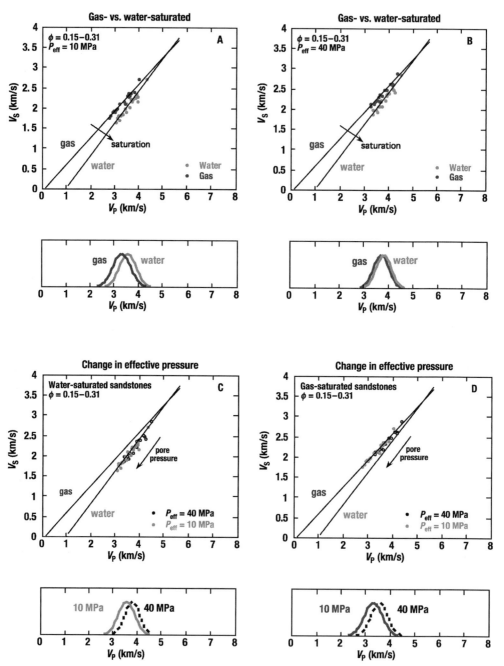

Figure 3.9 Subsets of Han's sandstone data. A, Data at 10 MPa, showing the separate gas- and water-saturated clouds. B, Data at 40 MPa, showing the gas- and water-saturated clouds, now with less separation. In both A and B, V_P alone would not be very valuable for separating the clouds, as seen by the overlap of the smoothed histograms. C, Water-saturated data, showing the overlapping clouds for high and low effective pressures. D, Gas-saturated data, showing the overlapping clouds for high and low effective pressures.

In Figure 3.9B, we compare data for the same rocks at 40 MPa effective pressure, under both water- and gas-saturated conditions. The larger effective pressure makes the rock elastically stiffer, and a stiffer rock will have a smaller seismic sensitivity to pore fluid changes, as predicted by Gassmann theory. Detection of the saturation difference using V_P alone would be hopeless, as shown by nearly perfect overlap of the V_P pdfs. Working in two dimensions with both V_P and V_S will be better, although we can expect more interpretation errors than in 3.9A, because the two clouds are closer together. We will show in Section 3.6 how to quantify these differences statistically, using discriminant analysis and Bayesian classification methods.

Figure 3.9C shows a comparison of water-saturated data at two different effective pressures. Figure 3.9D shows a similar comparison of two different effective pressures with gas saturation. The pressure change shows up as a shift in V_P which might be detectable in time-lapse seismic data, but would be difficult to detect in an exploration mode. Adding shear velocity does not make the change any more detectable, although it allows us to see that the saturation stayed constant and only the pressure changed in each case.

There are seismic attributes that implicitly include shear information and that can be derived from conventional P-wave 3D seismic data. The AVO gradient (proposed by Shuey, 1985) and various other forms of AVO attributes, as well as the far-offset elastic impedance (Connolly, 1998; Mukerji et al., 1998b; see Chapter 4) are examples of such "physical attributes" that indirectly contain shear-wave information. The elastic impedance (EI) is not an intrinsic property of the rocks, because it depends on the incidence angle of the waves. Nevertheless, it can be a very useful S-velocity-dependent attribute, which can be obtained from inversion of far-offset partial stacks. Shear-wave information may also be obtained more explicitly from multi-component surveys, which, however, are costlier than conventional single-component surveys. How do we assess the relative value of cheaper, indirect shear-wave information from far-offset data versus costlier, direct shear-wave information from multi-component data? We will describe how pdfs allow us to compute Shannon's information entropy. This information entropy gives us a quantitative measure of the *value* of different types of additional data in reducing the uncertainty of rock property estimation.

3.3 Statistical rock physics workflow

The statistical rock physics methodology described in this chapter may be broadly divided into four steps:
(1) First, well-log data are analyzed to obtain *facies definition*. This is done after appropriate corrections, including fluid substitution and shear velocity estimation when required. Basic rock physics relations such as velocity–porosity and V_P–V_S are defined for each facies (see Chapters 1 and 2).

(2) This is followed by Monte Carlo simulation of seismic rock properties (V_P, V_S, and density) and computations of the *facies-dependent statistical* pdfs for seismic attributes of interest. A key feature is the use of rock physics modeling to *extend* the pdfs to situations that are of interest but were not encountered in the wells (e.g. different fluid saturations, presence of fractures, different levels of diagenesis or cementation, different depths). Gassmann's equations can be used for fluid substitution. Lithology substitution can be done using the various rock models described in Chapter 2. The extended pdfs are the derived distributions. Using the derived pdfs of seismic attributes, feasibility evaluations are made about which set of seismic attributes contains the most information for the problem. Discriminating lithologies may require a different set of attributes than, say, discriminating fractured from nonfractured reservoir zones. Evaluation of the seismic attribute pdfs based on well-log data can be used to guide the choice of attributes to be extracted from the seismic data.

(3) The seismic attributes from *seismic inversion* or analyses (AVO analyses, impedance inversion, etc.) are used in a *statistical classification* technique to classify the voxels within the seismic attribute cube. Calibrating the attributes with the probability distributions defined at well locations allows us to obtain a measure of the probability of occurrence of each facies. Various standard statistical validation tests can be performed to obtain a measure of the classification success.

(4) Geostatistics is used to include the *spatial correlation*, represented by variograms or multiple-point spatial statistics, and the *small-scale variability*, which is not captured in seismic data because of their limited resolution. The probabilities obtained in the previous step from classification of seismic attributes alone are dependent on the local voxel values of the seismic attributes, and are not conditioned to the neighboring spatially correlated values. Hence, this final geostatistical step may be used to update the seismically derived probabilities by taking into account geologically reasonable spatial correlation and by conditioning to the facies and fluids observed at the well locations.

Four steps of the statistical rock physics methodology:
(1) Well-log analyses and facies definition
(2) Rock physics modeling, Monte Carlo simulation and pdf estimation
(3) Seismic inversion, calibration to well pdfs, and statistical classification
(4) Geostatistical simulations incorporating spatial correlation and fine-scale heterogeneity

Depending on the stage in the reservoir exploration, development, and production cycle, the steps outlined above may be modified.

Not all of the steps may be carried out in the initial exploration stages where there are few or no well data. In the exploration stage, the pdfs of just a few basic facies categories (say shale, oil sand, brine sand) may be estimated from wells and a quick classification may be done using seismic attributes derived at a few locations, e.g. a few AVO intercepts and gradients derived from a handful of CDP (common depth point) gathers. In some cases, at the early exploration stage, there may be no wells, and the pdfs of rock properties may be on the basis of analogous data from regions of similar geologic history. In the development stage, on the basis of more extensive well data, more facies categories may be defined (e.g. shale, unconsolidated sand, cemented sand, etc.). Seismic attributes extracted after careful inversions over a full 3D volume may be used in the classification. Geostatistical methods would be applied both in the characterization and production stages. Although we describe geostatistics as the fourth step, in practice geostatistical methods can be incorporated at all stages: in defining horizons and estimating reservoir structures, as well as in the seismic inversion itself. Our workflow described geostatistical simulations after the seismic inversion, but constrained by the probabilities derived from statistical classification of the seismic inversion results. As will be described briefly in Chapter 4, as an alternative, geostatistical seismic inversion can incorporate spatial correlation within the steps of the inversion itself. In the production monitoring stages, the lithologies stay constant and the changes in seismic signatures are due to production-related changes in fluid saturation and pressure. In this case, obviously, the pdfs would be conditioned to changes in pressure and saturation. Use would also be made of multi-phase flow simulation results to constrain the saturation-related changes in seismic attributes. In the production monitoring stage, much more care has to be given to derive the distributions not only for different fluid saturations, but also for different spatial scales of heterogeneous saturations.

3.3.1 Step 1: Identifying facies from well logs and geology

Usually the information from the wells is the most directly available observation of the reservoir. In many reservoir characterization projects the first step is to define, based on the well information, the facies that, a priori, we would like to be delineated in the reservoir. We will use the term *facies* for categorical groups, not necessarily only by lithology type, but also by some property or a collection of properties, for example a combination of lithology and pore fluids. Brine sands and oil sands would be considered as two different "facies" or categories. An example of three different facies might be: carbonates with gas-filled fractures, carbonates with brine-filled fractures, and unfractured carbonates.

Using the available information at the well – cores, thin sections, geology, logs, production data – a facies *indicator* is assigned to each depth. It is convenient to do this with one or a few key wells where the data and interpretation are most complete and reliable. The criteria to define the facies depend on the targeted objective. This could be to map different lithologies (sands/shale facies), to delineate fractures (unfractured/fractured facies), to identify fluids (brine-sand/oil-sand facies) or to monitor changes in pressures and/or temperatures in a reservoir. It is a common practice to initiate the facies definition with exploratory cross-plots between the different logs, looking for cluster separation. The gamma-ray, resistivity, density, and sonic logs are often very useful for defining different lithologies. Knowledge of background geology and core and thin-section information also play an important part in this step. Facies or category identification may be done purely statistically using what are known as "unsupervised learning" algorithms. However, this usually gives poorer results than supervised learning, where facies clusters are defined on the basis of expert knowledge such as petrophysical and geologic expertise. Plate 3.10A shows the result of facies definition in a set of well logs from the North Sea (see also Sections 2.4–2.5).

A critical, and obvious, point is that each facies is not a single rock, but a collection of geologically similar rocks that span a range of petrophysical and seismic properties. For example, in Plate 3.10 the oil sands span a range of velocity, density, and impedance, as do the water sands and the shales. This *intrinsic variability* of rock properties presents one of the biggest challenges of quantitative seismic interpretation: when does an observed attribute change indicate a significant change across facies rather than a minor fluctuation within a facies?

3.3.2 Step 2: Rock physics, Monte Carlo simulation and pdf estimation

Basic rock physics relations such as velocity–porosity behavior and V_P–V_S relations are defined for the facies. In wells with missing V_S, V_S prediction has to be carried out using V_P–V_S relations appropriate for the facies (see Chapters 1 and 2 for basic rock physics modeling). The V_P–V_S relations should ideally be locally calibrated.

The different physical conditions or facies of interest that we would like to identify may not always be adequately sampled in the initial well training data. It is often necessary to *extend the training data*, using rock physics to simulate different physical conditions theoretically. As we saw in Chapters 1 and 2, there are theoretical and empirical models to predict rock physics "*What ifs*" – the effects of changes in fluid saturation, temperature, pressure, and variations in sedimentological properties such as cementation, sorting, and lithology. With rock physics it is possible to translate production or geologic information into elastic properties that condition the seismic response.

A critical assumption of the well-calibration process is that the well-log data *extended by rock physics modeling* will be statistically representative of all the possible values of V_P, V_S, and density that might be encountered in the study area. This includes facies-to-facies variations as well as the intrinsic variability within each facies, and is not limited to the facies variability encountered in the wells alone. The subjective decision that the training data are a statistically representative set underlies almost all statistical classification techniques. With this assumption it is possible to explore the intrinsic variability of each facies in the V_P–V_S–density space using *correlated Monte Carlo simulation* (see Section 3.5). A large number of points are generated spanning the intrinsic variability while preserving the distributions and the V_P–V_S and V_P–density correlations of the original data. Using well data to estimate the intrinsic geologic variability implicitly relies on Walther's law in geology (Middleton, 1973) that relates vertical variability to lateral variability within conformable stratigraphic sequences. Thus we take the vertical variability in rock properties observed in the well and use this to infer the possible lateral variability that we might encounter between the wells for the same sequence. Deviated or near-horizontal wells may give a more direct inference of the lateral variability within the formation intersected by the well.

Next, to establish the link with the seismic information, seismic observables and attributes are theoretically calculated using the extended log-based training data. Monte Carlo realizations are drawn from the distributions of each facies defined above, and are used in deterministic models to compute seismic attributes such as AVO intercept and gradient, P- and S-wave impedance and others of interest. An attribute is any characteristic that can be extracted from the seismic data. Although the statistical methodology is completely general and can be applied to any collection of mathematically defined attributes, in this book only seismic attributes with some "physical meaning" are considered. This type of seismic attribute has a well-defined physical relation with the reservoir properties, and can be either calculated using well logs (V_P, V_S, density) or extracted from seismic data (e.g. with inversion, or AVO techniques).

Not all seismic attributes respond equally to different reservoir properties. For example, seismic P-wave velocity and S-wave velocity are not equally affected by changes in fluid saturations. Therefore the optimum seismic attribute or combination of seismic attributes to be used depends on *both* the particular reservoir and the targeted problem: lithology classification, fluid detection, fracture-zone identification, etc. Possibly the easiest (but not the most rigorously objective) way to select attributes when there are only a few of them is by visual inspection of color-coded comparative histogram plots or cross-plots of attributes. A more quantitative approach is described in Section 3.4 on information theory. Plate 3.10B shows an example of a cross-plot of acoustic impedance (AI) vs. elastic impedance at 30° (EI) calculated with well logs. As can be seen, there are three color-coded groups: oil sandstones, brine sandstones, and shales, clearly separated in this AI–EI plane. On the other hand, if a single attribute is used (equivalent to projecting the points over one of the axes), it is not possible to discriminate the three groups completely. The computation of seismic attributes and estimation of their pdfs from log data serve as a feasibility check to decide which attributes should be extracted from the field seismic data. In the initial exploration stages, this kind of feasibility study may also be used as a guide for designing surveys suitable for extracting the most promising attributes.

During this process of computing attributes it may be determined that not all of the a-priori defined facies, based on geologic, petrophysical and log data, can actually be separated in the seismic attributes space. In that case it is necessary to consider the union or division of some of the facies. Looking carefully at Plate 3.10B, we can identify different symbol shapes (triangles, circles, etc.) within each color-coded group. A priori, eight groups were defined, but it is clear that not all groups were separable with the proposed seismic attributes. In practice, splitting or combining categories can be done quantitatively using cluster analysis techniques. However, as mentioned earlier, completely unsupervised cluster analysis usually gives poorer results than supervised learning, where clusters are defined on the basis of expert subjective knowledge of petrophysics and geology. Subjective expert decisions play an important role in quantitative interpretation. When splitting or combining facies, it is not enough to analyze the attribute plots; it is also necessary to justify the decisions with geologic or production observations. This helps to prevent the analysis from being driven to wrong conclusions if there are problems with the data.

From the points computed in the seismic attribute space, the probability density functions (pdfs), either univariate (one attribute) or multivariate (combinations of attributes), are estimated for each defined facies. In the simplest sense, an empirical pdf can be thought of as a normalized and smoothed histogram. In practice, to obtain the pdfs it is necessary to discretize the space where they will be calculated, and use a kernel (window) function for smoothing, as described in Section 3.2.1. The monograph by Silverman (1986) gives a good description of density estimation. More recent techniques include the use of filtering in the wavelet-transform domain to estimate the smooth density function from the empirical histogram. Plate 3.10C shows a bivariate example of the results of density estimation using a smoothing window. In the pdf estimation, there has to be a compromise between the discretization and the smoothing. With too many cells, the pdfs would be too specific to the particularities of the input sample, and would not generalize to other data. With too much smoothing, the data variability would not be captured, and the discrimination between groups would be washed away. To choose these two parameters, a set of classification tests has to be done with a validation data subgroup. In spaces with few dimensions (few attributes), the pdf calculation is not very difficult, although there are some details (smoothness, grid definition, limit extrapolation, etc.) that have to be carefully handled. On the other hand, in spaces with high dimensionality, nonparametric pdf estimation is computationally demanding, and may not be very reliable because of sparse data. Instead of pdf-based classification, other classification methods, such as K-nearest neighbors, neural networks, or classification trees, have to be used in such situations.

3.3.3 Step 3: Seismic information and statistical classification of seismic attributes

Seismic attributes, which include reflectivities, velocities, impedances, and others, are derived from seismic data using different processing, analysis, or inversion techniques. How to obtain attributes from the seismic data is a topic of ongoing research and discussion. There are different algorithms for seismic inversion, each with its pros and cons, not discussed here (for more on this see Chapter 4). Velocity determination and migration, and careful seismic processing, are decisive in the usefulness of the estimated values of seismic attributes. In some cases the seismic acquisition and processing may affect the *absolute* values of seismic amplitudes but maintain their *relative* variations. With well calibration, these relative variations could still be of practical interest for discrimination and classification of reservoir properties. In general terms, having "good data" increases the probability of deriving reliable interpretations.

Some seismic attributes respond to the reservoir *interval* properties (e.g. acoustic impedance, elastic impedance). Others respond to *interface* properties caused by contrasts across layers (e.g. reflectivity, AVO attributes). Plate 3.11A presents an example of interface attributes extracted from a 3D seismic data set. The attributes shown are the popular AVO attributes defined by Shuey (1985): normal incidence P-to-P reflectivity $R(0)$ (AVO intercept) and G (AVO gradient). The topography follows the traveltime topography of the interpreted seismic horizon. The attributes were estimated along this horizon from seismic AVO analyses. Plate 3.11B, a different example, this time of interval attributes, shows acoustic and elastic impedance volumes resulting from impedance inversion of near-offset and far-offset seismic partial stacks.

The seismic attribute values and their pdfs derived from the seismic data inversion are usually not equal to the same attribute values computed theoretically from the well-log V_P, V_S, and density. The reasons for those differences include the simplifications of the models used to derive the analytical expressions (linearization of the reflectivity with the incidence angle, plane layers, single interface, small contrasts, etc.), imperfections in the data processing (residuals in attenuation corrections, diffraction, etc.), arbitrary scaling of the field amplitudes, and noisy data. Additionally, an important issue is that the measurement scales of the seismic and well logs are very different. The seismic wave responds to reservoir property averages that are not always well approximated by upscaling from the well logs. Often the Backus averaging of layer properties, prior to drawing realizations, might work well for upscaling; sometimes it may not, as the well log only samples heterogeneities along the well path, and lateral heterogeneities are missed. Some other attempts to account for the scaling issues may include the following. Instead of using individual Monte Carlo draws of rock properties from log-scale pdfs, averages over multiple draws can be used. The number of points to average would represent a scale equal to about 1/10 of the seismic wavelength. A more rigorous attempt would draw realizations not only of the point properties, but also their spatial distribution along the well. In essence this would create pseudo-logs with the appropriate variability in point properties as well as the appropriate spatial correlation. In addition to estimating the pdfs of rock properties, their spatial correlation has to be estimated from well logs. Two common ways of describing the spatial correlation are by variograms and by Markov transition matrices. Geostatistical simulations based on variograms or Markov chain models can be used to generate the pseudo-logs. Once multiple pseudo-logs are generated, synthetic seismograms are calculated and the desired attributes extracted from the synthetic seismogram. This process is carried out with a large number of realizations of the pseudo-logs to capture the variability of the seismic attributes. Upscaling is more rigorous in this procedure because of the wave propagation calculation. Takahashi (2000) shows an example of deriving distributions for normal-incidence reflectivity using variogram-based pseudo-log simulation and forward synthetic seismic modeling. Modeling seismic signatures of lithologic sequences using Markov chains has been presented by Godfrey *et al.* (1980), Velzeboer (1981), and Sinvhal *et al.* (1984), among others. Eidsvik *et al.* (2003) use Hidden Markov Chain models to estimate from well logs the parameters of the Markov transition matrix for stratigraphic sequences.

Because of discrepancies between logs and seismic data, it may not be possible in general to use directly the pdfs calculated with the well logs for classifying the attributes extracted from the seismic data. In order to avoid the differences between the computed and extracted attributes, the pdf-derived classification system should be calibrated with the attribute traces around the wells. An option, when there are few available well data, is to calibrate or scale the global pdfs derived from the seismic data with the corresponding global pdfs calculated from the well logs. A simple scaling might just consist of equalizing the histograms of well-derived and seismically derived attributes. For multiple attributes this can also include equalizing the covariance of the attributes. This recalibration of the seismic pdfs is based on the idea that the facies of interest (e.g. oil sands) are outliers, and the global pdfs are predominantly the shales.

Once we have the calibrated pdfs and the seismically derived attributes, we can go on to classify the volume or horizon of seismic attributes into different classes. These classes are the ones defined in the first step, and depend on the targeted problem: the classes could represent different facies, fluid types, fractured vs. unfractured rock, etc. There are many statistical methods for pattern recognition or attribute classification. Some examples include: discriminant analysis, K-nearest-neighbor classification, neural networks, classification trees, or Bayes classification with the estimated pdfs. There exists a vast literature on statistical pattern recognition and classification with applications in various fields. Two general texts that cover many of the algorithms are Duda *et al.* (2001) and Hastie *et al.* (2001). The book by Bishop (1995) describes classification using neural networks. We describe briefly the mathematical basis for some of the common classification techniques later in Section 3.6, after discussing the statistical rock physics workflow. Plate 3.12 shows examples of applying Bayesian classification for two different cases. With the Bayes classification method (Fukunaga, 1990; Houck, 1999), the conditional probability of each group given the combination of attributes is calculated, and the sample is classified as belonging to the group that has the highest probability. Bayesian classification, by working with the conditional probabilities, provides not only the most likely classification but also an estimate of the different kinds of errors in the classification process. Plate 3.12A is the result of classifying the AVO attributes $R(0)$ and G shown in Plate 3.11A. Plate 3.12B shows iso-probability surfaces obtained after Bayesian classification of near- and far-offset impedances shown in

Plate 3.11B. It is important to keep in mind that such probability surface visualizations do not show the actual sand (or shale) bodies but show the probability that the bodies have that spatial location and distribution.

3.3.4 Step 4: Geostatistics

By including geostatistical techniques of stochastic simulation in the analysis, we can further take into account geologically realistic spatial correlations and spatial uncertainty of reservoir properties. Geostatistical simulations can also attempt to reproduce the expected small-scale variability that cannot be detected with seismic data, but is seen in the well-log data. Geostatistical analysis requires estimation of spatial variograms (or spatial auto- and cross-correlations) that measure how different reservoir properties are correlated in space. Modern geostatistical techniques not only use the traditional two-point spatial correlation, but can also incorporate multi-point spatial statistics.

As an example, we show results from a particular geostatistical technique known as indicator simulation. In this technique the facies are represented by binary-valued indicator random variables. Indicator random variables take the value 1 when the facies is present, and 0 otherwise. Indicator simulation generates multiple equiprobable realizations of facies in the reservoir after incorporating the results from seismic attribute classification as soft indicators. Plate 3.13 (top) presents a particular vertical section of the multiple equiprobable volumes (realizations) generated using indicator simulation. The figure clearly shows the characteristic spatial variability of the stochastic process. For this example, the probabilities derived from the seismic acoustic and elastic impedance attribute volumes of Plate 3.12 were used as soft indicators. Soft indicators take values between 0 and 1. The Markov–Bayes indicator formalism (Deutsch and Journel, 1996) was used to obtain the *posterior conditional pdfs*, incorporating the spatial correlation through the indicator variograms. Plate 3.13 (bottom) shows the result of this updating of the *prior pdfs*, the probability of a facies given the seismic attributes, to the *posterior pdfs*, the probability of a facies given seismic attributes, the spatial correlation, and the facies indicator data from the wells. The colors in the section shown in Plate 3.13 correspond to the probability that each point belongs to a particular facies, oil sands in this case. The probabilities are calculated from the statistics of a large number of geostatistical realizations. As was mentioned, this type of result is an extension of the facies classification process described before, where the spatial correlation and small-scale variability were included. In exploration situations with sparse well data, often the updating may not change the pdfs very much. Some applications (e.g. risk assessment for well placement) will require the pdfs, whereas in some other applications (e.g. reservoir flow simulations) it will be necessary to have stochastic realizations of reservoir properties drawn from the pdfs. Geostatistics provides powerful tools for spatial data integration, and can play an important role at various stages of reservoir exploration and development. In the early stages, seismic traveltime data and sparse well horizon markers can be combined geostatistically to delineate reservoir architecture. With more well data, core measurements, and multiple seismic attributes, it is possible to obtain geostatistical simulations of reservoir properties such as lithofacies, porosity, and permeability. Geostatistics can be used in seismic impedance inversions to impart the appropriate spatial correlation structure to the inversions. In the later stages, geostatistics provides tools to incorporate production data into the analysis. One of the pitfalls of using geostatistics is that users may apply it in a black-box mode without understanding the underlying spatial models. Users may naively throw in disparate data without accounting for the physics that relates the data to the reservoir properties of interest. This leads to poor results. One of the main benefits of geostatistics when properly applied is that it provides ways to estimate *joint spatial uncertainty*. For mathematical background and details the reader is referred to the literature on geostatistics. The books by Isaaks and Srivastava (1989) and Kitanidis (1997) provide good introductory treatments. Software is available in the book by Deutsch and Journel (1996). More comprehensive texts on geostatistics include Journel and Huijbregts (1978), Cressie (1993), Goovaerts (1997) and Chiles and Delfiner (1999). Rubin (2003) gives an excellent treatment of spatial stochastic methods for quantifying uncertainty in groundwater and solute transport.

After classification, it is imperative to cross-check the results with geologic, production, and other reservoir data, as it is not always possible to include all kinds of available information (especially the subjective information of experienced veterans, encoded in their "natural neural networks") in the analysis.

3.4 Information entropy: some simple examples

Statistical information theory gives us simple yet powerful tools to quantify the information that each attribute can bring to discriminate the different facies (Mavko and Mukerji, 1998b).

The information entropy, $H(X)$, is a statistical parameter that quantifies the intrinsic variability of a random variable X. The concept of information entropy, which originated in statistics and communication theory, has found applications in diverse fields such as computational chemistry, linguistics, bioinformatics, and genetics. It can be computed from the pdf, $P(X)$, as follows (Cover and Thomas, 1991):

$$H(X) = -\sum_{i} p(x_i) \ln[p(x_i)]$$

when X takes discrete values. For continuous X the corresponding equation is:

$$H(X) = -\int_{-\infty}^{\infty} \ln[p(X)] p(X) \, dX$$

3.4 Information entropy: some simple examples

In this equation all X values where the pdf $P(X)$ is zero are excluded from the integral. For a given variance, the pdf with maximum entropy is the Gaussian distribution. For a bounded pdf, the one with maximum entropy is the uniform distribution. For a known mean, the pdf with maximum entropy is a truncated exponential. In practice, for continuous variable X, the values may be discretized into N bins or intervals and the entropy computed using the discrete summation formula given above. One extreme case is when each interval is equally likely. Then $P(X_i) = 1/N$ for all i and $H(X) = \ln N$. Note that this depends on the discretization level N. So a standardized measure of entropy would be $H_N(X) = H(X)/\ln N$ with values ranging between 0 and 1.

Consider a categorical variable C that takes two values, say, shale or sand: $C = \{$shale, sand$\}$. Let $P(C) = \{1/2, 1/2\}$ be the probabilities for C being shale or sand, respectively. In other words, both events, shale or sand, are equally likely, and we are not sure which one occurs. The entropy, H, is given by the summation of $-[P \log P]$, summed over all the different categories (in this case 2). So, $H = -[(1/2)\log(1/2) + (1/2)\log(1/2)] = 1$ (using log to base 2) or $H = 0.693$ (using natural log). When the log is computed in base 2, entropy is measured in units of "bit." One can also compute H using natural logarithms, in which case the unit of entropy is termed "nat," while logarithms to base 10 give units of "dit." Now let us compare this entropy of 1 bit to another situation when $P(C) = \{9/10, 1/10\}$, i.e., shale is much more likely. This might be, for example, the updated posterior probability after measuring some seismic attribute, such as impedance. Now $H = -[(9/10)\log(9/10) + (1/10)\log(1/10)] = 0.469$ bit or 0.325 nat. In this situation, the entropy or "disorder" is reduced, compared with the first case, because now both events are not equally likely; one of the events (shale) is much more likely to occur. There is now more predictability, less disorder, and less uncertainty.

While this was a very simple univariate example of calculating and comparing entropies from probabilities, the same principle carries forward to multivariate data types and multivariate probability distribution functions. Consider another simple univariate example to compare variance and entropy as measures of uncertainty. Now let the random variable X take the discrete values

$$X = \{-2, -1, 0, 1, 2\}$$

with two different probability functions:

$P_1(X) = \{0.0909, 0.1818, 0.4545, 0.1818, 0.0909\}$ and

$P_2(X) = \{0.03, 0.44, 0.06, 0.44, 0.03\}$

The variance (square of standard deviation) is about the same (~ 1.1) for both cases:

Var1 $= 1.09$, Var2 $= 1.12$

But the uncertainty is actually greater for case 1 as seen by computing the entropy:

Entropy1 $= 1.41$, Entropy2 $= 1.10$

The variance is a measure of deviation from central tendency and may not always be a complete measure of the uncertainty.

Using the concept of Shannon's information entropy, it is possible to select the "best" attributes as those that most reduce the uncertainty in the identification of reservoir properties.

The quantity of information about a reservoir property X (e.g. porosity) contained in an attribute A (e.g. seismic impedance) can be defined as:

$$I(X|A) = H(X) - H(X|A)$$

where $H(X)$ is the *information entropy*. Information entropy quantifies the intrinsic variability of X, before observing the attribute A. It can be computed from the pdf, $P(X)$. The quantity $H(X|A)$ is the conditional mean entropy of X given A, that is, the average uncertainty of X after observing A. The conditional mean entropy is calculated from the *conditional pdf* of X given A, $P(X|A)$. The information $I(X|A)$ can be interpreted as the reduction in the uncertainty of the reservoir property X, due to observing the attribute A. Therefore, a quantitative criterion to select the best attribute (or combination of attributes) is the one (or ones) that maximize $I(X|A)$ (Takahashi et al., 1999).

The reduction in information entropy and uncertainty by additional data can be shown by the example in Figure 3.14. The relationships among porosity, V_P, and V_S of a particular reservoir are described by the trivariate pdf shown in Figure 3.14 (top). Conditioning of porosity information by velocities is summarized in Figure 3.14 (bottom). The unconditional marginal prior pdf of porosity describing the distribution of all porosities in the reservoir changes to narrower and taller conditional pdfs, $P(\text{porosity} | V_P)$, and $P(\text{porosity} | V_P, V_S)$ upon inclusion of velocity information. The velocity observations decrease the spread and variability (and hence uncertainty) about porosity. The information entropy quantifies this decrease in uncertainty. The prior information entropy of porosity computed from its unconditional pdf is 3.44. This decreases to 3.06 with V_P alone, and to 2.89 with both V_P and V_S. Calculations such as these can be used to select the best set of attributes that contain the most information about the targeted rock property of interest. While linear measures of uncertainty, such as variance and covariance, can also be used, their nonlinear counterparts, such as entropy and relative entropy, go beyond, and generalize the linear uncertainty measures in several ways. Uncertainties for categorical variables (e.g. shale/sand categories) can be estimated using entropy. The variance and covariance are not properly defined for categorical variables. The entropy is a better measure of uncertainty when pdfs have multiple modes, since variance is just a measure of deviation from the mean value, while the entropy takes the full distribution into account. Entropy captures nonlinear co-dependence whereas the covariance captures only linear dependence. Entropy measures can be estimated from nonparametric pdfs, are invariant to linear and nonlinear coordinate transforms,

3.4 Information entropy: some simple examples

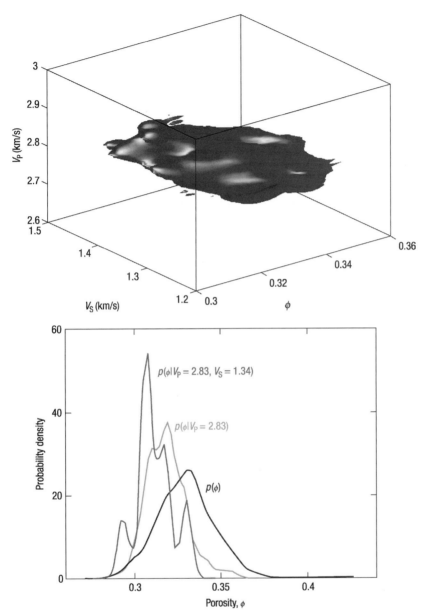

Figure 3.14 Trivariate pdf of porosity, V_P, and V_S (top). Conditioning of porosity pdf (bottom) by V_P and V_S information, corresponding to the trivariate pdf.

and hence offer a more flexible representation of the state of information about the rock property. Nonetheless, linear measures of variance are useful as quick checks in the preliminary steps of analyzing data variability, before going on to the nonlinear measures. A pitfall of using entropy as a measure of uncertainty is that entropy does not depend on the actual values of the variables but only on the pdfs.

3.5 Monte Carlo simulation

Statistical simulation has become a popular numerical method with which to tackle many probabilistic problems. One of the steps in many statistical simulation procedures is to draw samples X_i such that they follow a desired probability distribution function $F(x)$. Once a large number of samples have been drawn, the desired function is computed from these. This procedure is often termed *Monte Carlo simulation*, a term made popular by physicists working on the bomb during the Second World War. Discussions of Monte Carlo procedures applied to modeling and financial evaluation of oil prospects may be found in Newendorp and Schuyler (2000) and Harbaugh *et al.* (1995). In general, Monte Carlo simulation can be a very difficult problem, especially when X is multivariate with correlated components, and $F(x)$ is a complicated function. Sometimes $F(x)$ may be known only up to a constant. Markov chain Monte Carlo is a sequential procedure that allows one to draw samples x_i using the properties of Markov chains. This is a powerful, numerically intensive procedure that has been applied in various fields of science and engineering to solve complicated probabilistic problems. The book by Liu (2001) gives a readable modern account of Markov chain Monte Carlo methods and applications. In this book we talk about simpler Monte Carlo methods for drawing random samples with desired distributions.

For the simple case of a univariate X, and a completely known $F(x)$ (either analytically or numerically), drawing x_i amounts to first drawing uniform random variates u_i between 0 and 1, and then evaluating the inverse of the desired distribution function at these u_i: $x_i = F^{-1}(u_i)$. A simple proof follows. Let $F_U(u)$ and $F_X(x)$ be the cumulative probability distribution functions (cdfs) of U and X respectively, defined as:

$$F_U(u) = P(U \leq u)$$
$$F_X(x) = P(X \leq x)$$

These are monotonic increasing functions and take values between 0 and 1. In the above equations P denotes probability. Since U is uniform between 0 and 1, $F_U(u) = u$. Now

$$P(X \leq x) = P(F_X^{-1}(U) \leq x) = P(U \leq F_X(x))$$

since F_X is a monotonic increasing function. Since $F_U(u) = u$, we can rewrite the right-hand side of the above equation as:

$$P(U \leq F_X(x)) = F_U(F_X(x)) = F_X(x)$$

Hence the simulated X indeed follow the desired distribution $F_X(x)$.

Often $F^{-1}(X)$ may not be known analytically. In this case the inversion can be easily done by table-lookup and interpolation from the numerically evaluated $F(X)$.

3.5 Monte Carlo simulation

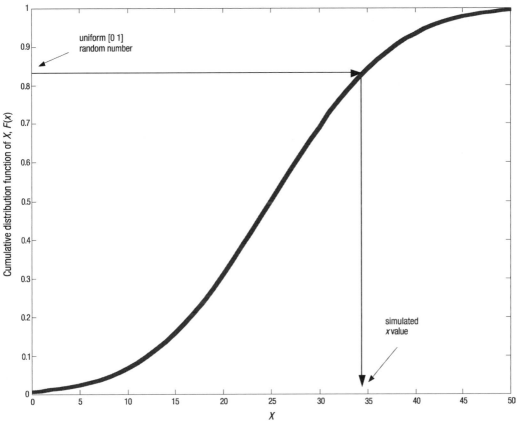

Figure 3.15 Schematic diagram of a univariate Monte Carlo simulation using the cdf of the variable to be simulated.

Graphically univariate Monte Carlo simulation may be described as shown in Figure 3.15. Many computer packages these days have random number generators not only for uniform and normal (Gaussian) distributions, but also for a large number of well-known analytically defined statistical distributions. The simple technique described above also works well for data-derived, *nonparametric* distribution functions where the distribution function is only known numerically.

How do we draw two random variables that are correlated? For example, we have drawn V_P using the technique described above. Now we need a V_S to go along with that V_P. We cannot draw V_S independently from the marginal unconditional distribution of V_S, because that would ignore the correlation between V_P and V_S. One simple approach is to use a V_P–V_S regression that may have been derived from the log data or borrowed from the literature. Chapter 1 gave some examples of V_P–V_S regressions.

> A simple procedure for correlated Monte Carlo draws is as follows:
> - draw a V_P sample from the V_P distribution
> - compute a V_S from the drawn V_P and the V_P–V_S regression
> - add to the computed V_S a random Gaussian error with zero mean and variance equal to the variance of the residuals from the V_P–V_S regression
>
> This gives a random, correlated (V_P, V_S) sample. A better approach is to draw V_S from the conditional distributions of V_S for each given V_P value, instead of using a V_P–V_S regression.

Given sufficient V_P–V_S training data, the conditional distributions of V_S for different V_P bins can be pre-computed. For each randomly drawn V_P, the V_S is drawn from the corresponding conditional distribution associated with the V_P value.

When using log data to derive V_P–V_S correlations, care must be taken to see that there are no depth mismatches between the V_P and V_S logs. Slight depth mismatches can give rise to poor correlations. Different regressions should be derived for different lithologies (shales versus sands, for example), and for different pore fluids. A usual practice is to derive a V_P–V_S relation for a reference fluid (say brine), then do Monte Carlo simulations and fluid substitution (via Gassmann's equations; see Chapter 1) to simulate V_P–V_S data for other fluids not present in the training data.

After extending the data by Monte Carlo simulations, it is very important to check that the simulated and original data (for the same facies and fluids) do indeed have similar statistical distributions. This can be done by plotting comparative histograms, quantile–quantile plots and scatter plots of the original log data and the simulated data.

> When doing correlated Monte Carlo simulations:
> - data correlations should be specific to a particular facies and pore fluids
> - be careful of depth mismatch in log data as this may mask data correlations
> - check that the simulated data have a statistical distribution similar to the original data for the same facies and fluid conditions.

3.6 Statistical classification and pattern recognition

The typical scenario of the statistical classification problem is as follows. There is a set of input variables or predictors (sometimes also called "attributes" or "features") that influence one or more "outcomes" or "responses." The inputs and outcomes can be either quantitative or categorical or a combination of both types. For example, the inputs could be seismic P and S impedances and the outputs could be lithofacies classes: sand or shale. In a well-log classification problem the inputs might be measured log curves such as gamma ray, density, V_P and resistivity, and the outputs could again be different lithofacies categories. Often when the inputs and outputs are quantitative variables, the problem is termed a regression problem; when we have categorical outputs it is termed a classification problem. In a broader sense, however, the problems have much in common, and it is possible to express a regression problem as a classification problem or vice versa.

The goal in classification is to predict the outcome (e.g. shale or sand) based on the observed inputs (e.g. P and S impedance). We have a training data set where we have observed both the inputs and the outcomes. For example, this could be from well-log measurements. Using the training data we have to devise a classification rule or prediction model that will allow us to predict the outcomes for new data where the outcomes are unknown. This is called supervised learning. We have a training data set with known outcomes that can help us to come up with the classification rule. In unsupervised learning, we have only observed input features, with no measurements of the outputs. Unsupervised learning then tries to cluster the data into groups that are statistically different from each other. We will discuss mostly supervised classification. Usually some calibration or supervision gives better results than completely unsupervised classification. Unsupervised classification may give rise to classes that are optimally distinct according to some chosen statistical criteria but may not have any significance from a geologic or reservoir production viewpoint. Nevertheless unsupervised classification may sometimes be useful for pointing out outliers of data clusters. These outliers may be related to artifacts such as data processing noise, or they may be pointers to classes not yet considered in the classification scheme.

For any given reservoir, we believe it is useful to quantify the "best-case" classification uncertainty that we would have if we could measure V_P, V_S, and density *error-free*. In this case, the interpretation accuracy will be limited by geologic parameters, such as mineralogy, pore stiffness, fluid contrasts, and shaliness. This is the "intrinsic resolvability" of the reservoir parameters. The value of quantifying the best-case uncertainty is that we will be able to identify and avoid hopeless classification problems right from the start. These will be the field problems where no amount of geophysical investment will allow accurate rock physics interpretation (Figure 3.16).

For most other situations, we can estimate how the uncertainty worsens compared to the best case (1) when measurement errors are introduced (2), when we drop from three parameters to two (e.g. V_P, V_S), or (3) when using alternative pairs of attributes (R_0, G, or ρV_P, ρV_S, etc.). *We believe that this kind of analysis can be helpful in the decision-making process to find the most cost-effective use of seismic data.*

There is a wide variety of algorithms for statistical classification. Here we will discuss a few well-known, simple, yet often very effective methods. For excellent coverage of many modern classification methods the reader is referred to Hastie *et al.* (2001) and Duda *et al.* (2000).

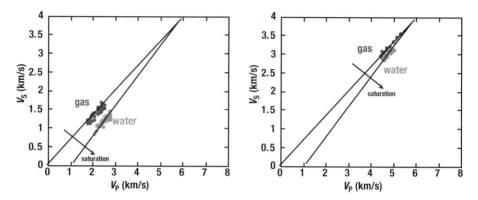

Figure 3.16 Estimating the best-case, error-free rock properties can be useful. On the left, soft, high-porosity rocks with good potential for detecting pressure and saturation changes. There is a lot of overlap in V_P only, but not in the 2D V_P–V_S plane. On the right, low-porosity stiff rocks offering little chance of detecting fluid and pressure effects. The problem is close to hopeless, suggesting that special shear acquisition would be a waste.

3.6.1 Discriminant analysis

The first method we discuss is the traditional classification method based on simple linear (or quadratic) discriminant analysis (Davis, 2002; Doveton, 1994). This method uses only the means and covariances of the training data. The underlying assumption is that the input features follow a Gaussian distribution. New samples are classified according to the minimum Mahalanobis distance to each cluster in the training data (Duda and Hart, 1973; Fukunaga, 1990). The Mahalanobis distance is defined as:

$$M^2 = (\mathbf{x} - \boldsymbol{\mu}_i)^T \Sigma^{-1} (\mathbf{x} - \boldsymbol{\mu}_i)$$

where $\mathbf{x}$ is the sample feature vector (measured attribute), $\boldsymbol{\mu}_i$ are the vectors of the attribute means for the different categories or facies classes, and Σ is the covariance matrix of the training data. The Mahalanobis distance can be interpreted as the usual Euclidean distance scaled by the covariance, in order to decorrelate and normalize the components of the feature vector. So, a given voxel in a seismic cube, with observed seismic attribute $\mathbf{x}$, would be classified as the facies to which it is "nearest" in terms of the Mahalanobis distance. When the covariance matrices for all the classes are taken to be identical, the minimum Mahalanobis distance classifications give rise to linear discriminant surfaces in the feature space. More generally, with different covariance matrices for each category we have quadratic discriminant surfaces. If the classes have unequal prior probabilities, then the decision has to be biased in favor of the class that is more likely. This is done by adding the term $\ln[P(\text{class})]$ to the right-hand side of the equation for the Mahalanobis distance, where $P(\text{class})$ is the prior probability for each class. Linear and quadratic discriminant classifiers are simple classifiers and often produce very good results, performing amongst the top few classifier algorithms. Even better performance may be achieved by generalizations of linear discriminant analysis, such as flexible discriminant analysis (FDA) and mixture discriminant analysis (MDA) described in Hastie et al. (2001).

Now let us look at an example of discriminant classification of facies using P-wave acoustic impedance and elastic impedance (AI–EI) as the input features. Discriminant analysis can be done with any number of attributes, not just two. Here we show an example with two attributes for ease of plotting. Plate 3.17A shows the scatter plot of AI versus EI(30°) for three facies: shales, brine sands, and oil sands. This constitutes the training data and was generated by extension of the well-log data by Monte Carlo simulation. The minimum Mahalanobis distance discriminant is plotted in Plate 3.17B. For more than two attributes, the discriminants will be surfaces or hyper-surfaces in the high-dimension attribute space. Because of the overlap among the groups, classification will not be perfect. To compute the classification success rate we exclude one sample from the training data, and then classify that sample based on the remaining training data. This is referred to as the "leave-one-out" jackknife technique and is done successively for all samples in the training data. Validation can also be done by having two subsets of the data: a training subset and a validation subset. Using validation methods one can also estimate the elements of the classification confusion matrix P_{ij} (also called Bayesian confusion matrix in Bayesian classification). The ijth element of P_{ij} is the conditional probability that the true facies is "i" when the predicted facies is "j": $P_{ij} = P(\text{true} = i | \text{predicted} = j)$. In general this is not a symmetric matrix. By keeping track of the results in the leave-one-out jackknife we can estimate P_{ij}. The diagonal elements of the matrix are the success rates for each facies, while the off-diagonal elements are probabilities of misclassification. Figure 3.18 shows the confusion matrix and the success rate for classification of the three facies based on the discriminant function shown in Plate 3.17B. Overall, the classification is quite good, with a success rate of 80–90%. The confusion matrix is very useful in understanding the different kinds of errors. For example, in this case, we see that when the classifier predicts oil sand, there is a 7% chance that actually it is brine sand, and about 2% chance that it is truly shale. When the prediction is shale, there is very little chance (<1%) that it is an oil sand, although it could be (17%) a brine sand. In the figure, the elements of P_{ij} have been rounded off. The columns actually add up to 1. How certain are we about the success rates themselves? What is the variability in the success rate given the training data we have? Could the minimum success rate go down to as low as 60%? We can attempt to answer these questions by bootstrap techniques. Bootstrap is a very powerful computational statistical method for assigning measures of accuracy to statistical estimates (Efron, 1979; Efron and Tibshirani, 1993). The general idea is to make multiple replicates of the data by drawing from the original data with replacement. Each of the bootstrap data replicates has the same number of elements as the original data set, but since they are drawn with replacement, some of the data may be represented more than once in the replicate data sets, while others might be missing. The statistic of interest is computed

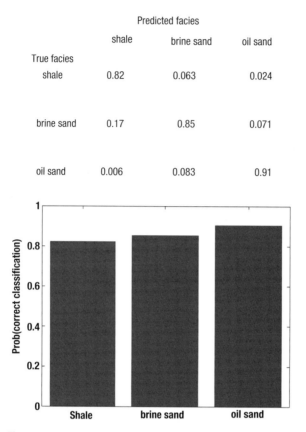

Figure 3.18 Classification confusion matrix for minimum Mahalanobis distance (top) and success rate.

on all of the replicate bootstrap data sets. The distribution of the bootstrap replicates of the statistic is a measure of the accuracy of the statistic. By using a simple bootstrap analysis, we can estimate the variability of the probabilities in P_{ij}. For the AI–EI classification of the three facies, bootstrapping gives histograms for each element of P_{ij} as shown in Figure 3.19. From these histograms we can estimate distributions for the probability of correct classification (Figure 3.20). Figure 3.20A shows that the probability of correctly classifying the brine sands can vary from 70% to about 90% with a peak around 85%. The variability in the probability of correctly classifying oil sands is less, as indicated by the more peaked distribution. We can also examine the distribution of probabilities for different kinds of errors: P(oil sand | brine sand) and P(brine sand | oil sand). These are shown in Figure 3.20B. The probability of a dry hole is estimated by the sum of the error probabilities, P(shale | oil sand) + P(brine sand | oil sand). Since the bootstrap analysis gives us multiple replicates of these probabilities, we can plot the cumulative distribution function of the sum of the two probabilities (Figure 3.20C). The cdf plot indicates that there is very little chance (<5%) that the probability

3.6 Statistical classification and pattern recognition

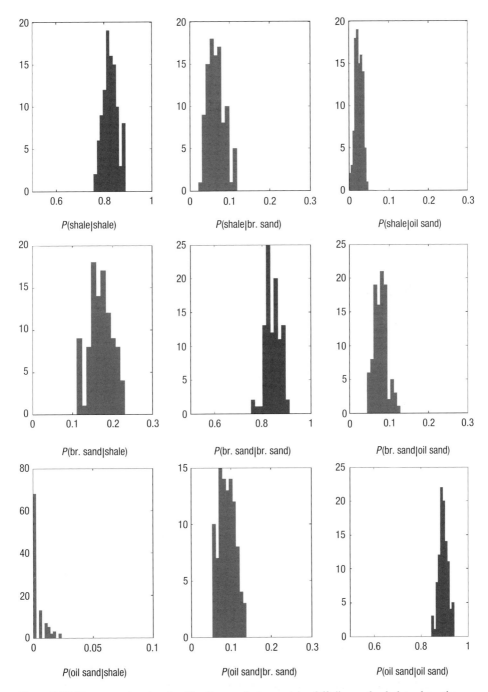

Figure 3.19 Bootstrapping the classification confusion matrix. Off-diagonal subplots show the distribution of probabilities for different types of errors. Diagonal subplots show distribution of probability of correct classification.

144 Statistical rock physics

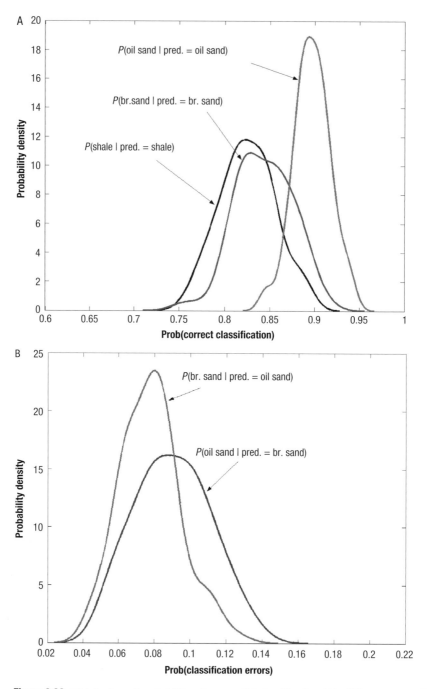

Figure 3.20 Distribution of probability of successful classification (A); different types of misclassification (B); and the risk of dry hole (C).

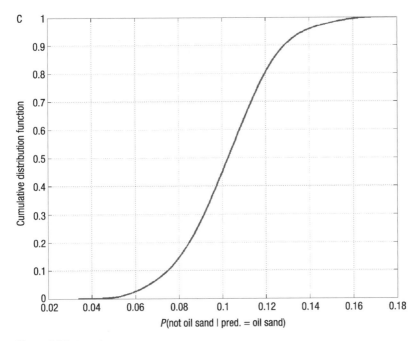

Figure 3.20 (*cont.*)

of a dry hole will be less than 6%. It is very likely (>90%) that the probability of a dry hole is 15% or less, and there is some small chance (5%) that it could be as high as 18%. All of these specific uncertainty estimates are valid only for this particular example and do not apply in general. The techniques, however, are very general and widely applicable. Estimation of the classification confusion matrix can be done for any classification method, not just a minimum-distance discriminator. The confusion matrix can be very useful in testing the feasibility of different attributes, and in selecting the appropriate sets of attributes to use in the classification, based on the classification success rate, and error rates.

Probably one of the oldest and most well-known techniques for multivariate statistical analysis is principal component analysis (PCA) introduced by Pearson around 1901. What is the use of PCA in the context of discriminant analysis and classification? By itself PCA is not a classification technique, but a data description technique. Unlike the classification problem where there is a unique response variable or dependent variable (the class) and a set of predictor variables, in PCA all the variables are treated on an equal footing as components of a multivariate random vector. The basic concept behind PCA is to reduce the dimensionality of data with a large number of interrelated variables, but at the same time retain most of the variation present in the data. The data are transformed to a new set of uncorrelated variables, the principal components (PCs), which are linear combinations of the old variables. The principal components are ordered so that the first few retain most of the variation present in all of the original variables. Thus the most obvious way of using PCA in classification problems is to reduce the dimensions by choosing to work with just the first few principal components rather than the full set of variables in the discriminant analysis. For example, one might use the first two PCs to make 2D graphical plots that capture most of the behavior of the multivariate data. Use of PCA helps to orthogonalize a classification or regression problem by replacing the set of correlated original variables by the uncorrelated PCs. Computing the principal components consists of solving an eigenvalue–eigenvector problem for the covariance matrix or the correlation matrix. In most realistic cases, the sample covariance matrix estimated from the data is used, in place of the unknown true covariance. The kth principal component of a multivariate random vector $\mathbf{x}$ is given by $p^{(k)} = \mathbf{v}^{(k)'}\mathbf{x}$, where $\mathbf{v}^{(k)}$ is the eigenvector of the covariance matrix, $\Sigma = \mathbf{x}'\mathbf{x}$, corresponding to the kth largest eigenvalue λ_k. Instead of the covariance matrix, it is more common to use the correlation matrix $C = \tilde{\mathbf{x}}'\tilde{\mathbf{x}}$ to compute the eigenvectors. Here $\tilde{x}_i = x_i/\sigma_i$ are the standardized variables with each component normalized by the standard deviation of that component. The benefits of using the correlation matrix to compute PCs are that sensitivity to choice of measurement units is reduced, and PCs from different sets of random variables are more directly comparable. Covariance matrix PCs depend sensitively on the measurement units of each element of $\mathbf{x}$. If the variables have large disparities in variance, those variables with the largest variance will dominate the PCs. Standardizing the variables solves this problem. Most modern numerical software packages for multivariate data analysis provide tools for principal component analysis. Since the principal components are just a linear transformation of the original variables, they do not contain any new information. Furthermore, the principal components may not have a clearly interpretable physical meaning. Jolliffe (2002) gives an excellent authoritative and accessible account of principal component analysis and its various applications.

> There are some pitfalls in using principal components for discriminant analysis and classification of groups. The covariance or correlation matrix may not be the same for all groups, and the PCs from different groups may not be directly comparable. PCs computed from the overall covariance matrix are more useful than those based on within-group covariances. However, the overall-covariance PCs will work well only when the within-group variation is much less than the between-group variation. There is no guarantee that the first few PCs with the largest variance also have the best discriminatory power to separate the groups. The first few PCs will only be useful for classification if the within-group and between-group variation is along the same direction. Omitting the low-variance PCs may throw away information about variations between the groups.

3.6.2 Bayesian classification

A fundamental approach to the classification problem is provided by Bayesian decision theory (Duda *et al.*, 2000). Our use of the Bayes method for rock physics classification problems is closely linked to the data clouds corresponding to different reservoir *states*. We begin by describing each of the data clouds as having been drawn from a probability density function (pdf). For a cloud corresponding to gas-saturated rocks we describe the distribution of P- and S-wave velocity data points by the function $P(V_P, V_S \mid \text{gas})$; for a cloud corresponding to water-saturated rocks, $P(V_P, V_S \mid \text{water})$; and so on, for each cloud representing a different reservoir state of interest (lithofaces, saturation, pore pressure, etc.). The pdfs are sometimes called the *state-conditional probability density functions* (Duda *et al.*, 2000), since they describe the expected distribution of P- and S-wave velocities, once we specify the saturation state of the reservoir. This approach uses the complete probability distribution functions of the input features, and hence assumes that all the pdfs are known. In practice they must be estimated from the training data. The state-conditioned pdfs are highly site-dependent, varying with rock type, depth, age, etc., and are most often determined empirically from well logs. The approach we use is to make cross-plots of sonic V_P vs. V_S, searching for clusters when the data are sorted by saturation, pore pressure, lithofacies, etc. We have to make a judgement of how many clusters to separate. In some North Sea reservoirs we have good separation in the (V_P, V_S) domain between oil-saturated unconsolidated sands, water-saturated unconsolidated sands, oil-saturated cemented sands, water-saturated cemented sands, silty shales, and shales – six different states! In other reservoirs there is too much overlap, so we only try to separate sands from shales – two different states. We will refer to this number of clusters that covers all possibilities in our data as N. As stressed earlier, rock physics transformations can be used to extend the log-based training data – for example, using Gassmann's relations to map the log data to other saturation states not observed in the wells. (Additional transformations to different ranges of porosity, pore pressure, or shaliness can be applied, although these should be chosen using carefully considered rock physics principles!) Once the data are cross-plotted, we define bins in V_P and V_S and create a two-dimensional histogram of the (V_P, V_S) data in each cloud. The histograms can be smoothed to approximate a continuous pdf function, and finally normalized to ensure that the volume under the pdf is unity. (More formal approaches exist for pdf estimation.)

An important consideration is to upscale the log-derived cross-plots and pdfs to the 3D seismic scale. A simple, though crude, procedure is to smooth the well-log slownesses before plotting, at $\sim\lambda/4$, to approximate the spatial averaging of the seismic wave.

We can similarly plot the clouds corresponding to the water- and gas-saturated states (or any other reservoir state of interest such as lithofacies or pore pressure) in the AVO domain $(R(0), G)$ with pdfs $P(R(0), G \mid \text{water})$, $P(R(0), G \mid \text{gas})$, impedance domain, λ–μ domain, and so on. *This operation of representing the reservoir states as distributions in different seismic attribute planes will be very important for our discussion, since our goal is to quantify the relative values (and cost) of different seismic acquisition and interpretation strategies.* Making cross-plots and estimating pdfs for other attributes might require a bit of calculation. For example, if we wish to plot in the AVO ($R(0)$, G) domain, then we can randomly draw (density, V_P, V_S) data from the logs and compute the corresponding ($R(0)$, G) values using the Zoeppritz equations (see Chapter 4).

Let x denote the univariate or multivariate input. This could be V_P and gamma ray for log classification, or AVO gradient and intercept for classifying seismic attributes, and so on. Let c_j, with $j = 1, \ldots, N$, denote the N different states or classes. It might be helpful to think of the specific example when $N = 2$ and the two classes are $c_1 =$ shale and $c_2 =$ sand. The different classes may have different prior probabilities of occurrence $P(c_j)$. For example, shales are more likely than sands. Statisticians sometimes argue about the problem of unknown priors. A practical approach, the empirical Bayes approach is to use the existing data to estimate the prior. Any other relevant information can also be used to start with any reasonable prior pdf. The different classes will have different distributions of the input features, although the distributions may have some overlap. The distribution may be expressed as $P(x \mid c_j)$, the class-conditional (or state-conditional) pdfs. Plate 3.21 shows the estimated class-conditional bivariate pdfs for $R(0)$ and G, for two different classes, brine sands and oil sands. The pdfs were estimated by Monte Carlo simulations from V_P, V_S, and density well logs.

Bayes' formula allows us to express the probability of a particular class given an observed x as:

$$P(c_j \mid x) = \frac{P(x, c_j)}{P(x)} = \frac{P(x \mid c_j)P(c_j)}{P(x)}$$

where we have used the usual notation for probabilities: $P(x, c_j)$ denotes the joint probability of x and c_j; $P(x \mid c_j)$ denotes the conditional probability of x given c_j. The Bayes relation converts the prior probability, $P(c_j)$, of a particular class (before having observed any x) to the posterior probability given an observed x. The class-conditional pdf, $P(x \mid c_j)$, is estimated from the training data or from a combination of training data and forward models. In the above equation, $P(c_j)$ describes the a-priori probability that the reservoir is in state "c_j." For example, examination of logs might show that shale is encountered 60% of the time and sand 40% of the time in the interval of interest. Then, if no additional data are available, we might estimate $P(\text{shale}) \approx 0.60$ and $P(\text{sand}) \approx 0.40$. Other estimates of the a-priori probabilities might come from the sedimentological model. The point is that the a-priori probabilities quantify our expectation of the reservoir state, *before* we look at any seismic data. Estimates of the a-priori probabilities during production, such as $P(\text{water})$, $P(\text{gas})$, $P(\text{high pressure})$, $P(\text{low pressure})$ might be taken from reservoir flow simulations through stochastically simulated reservoir models. Again, the idea is to quantify our expectation of the reservoir

3.6 Statistical classification and pattern recognition

state as best we can, *before* we look at any seismic data. When we do not have much of an expectation (we really don't know what to expect) then a reasonable guess is "fifty-fifty," each state is equally likely, or $P(\text{state}_i) = 1/N$, where N is the number of different states that we are trying to distinguish.

Finally, $P(x)$ is the marginal or unconditional pdf of the seismic observable values across all N reservoir states. It can be written as

$$P(X) = \sum_{i=1}^{N} P(X \mid \text{state}_i) P(\text{state}_i)$$

and serves as a normalization to ensure that $\sum_{i=1}^{N} P(\text{state}_i \mid X) = 1$. Specifically, if we assume that there are only two possible reservoir states, "sand" and "shale," and two seismic observables, V_P and V_S, then

$$P(V_P, V_S) = P(V_P, V_S \mid \text{sand}) P(\text{sand}) + P(V_P, V_S \mid \text{shale}) P(\text{shale})$$

This simply describes the distribution of seismic data points that we observe. The denominator, $P(x) = \sum_j P(x \mid c_j) P(c_j)$, is a scale factor and does not play any important role in the Bayes decision rule for classification.

> **Bayes' decision rule says:**
>
> classify as class c_k if $P(c_k \mid x) > P(c_j \mid x)$ for all $j \neq k$.

This is equivalent to choosing c_k when $P(x \mid c_k) P(c_k) > P(x \mid c_j) P(c_j)$ for all $j \neq k$.

For example, if there are only $N = 2$ states, "sand" and "shale," and two seismic observables V_P and V_S, then we classify each new data point as:

sand if $P(\text{sand} \mid V_P, V_S) > P(\text{shale} \mid V_P, V_S)$
shale if $P(\text{shale} \mid V_P, V_S) > P(\text{sand} \mid V_P, V_S)$

The procedure is illustrated in Figure 3.22, for a simple one-dimensional case with a single seismic observable V_P. In Figure 3.22A, two state-conditional pdfs are shown, corresponding to the distributions of V_P in gas-saturated rocks and water-saturated rocks. The area under each of the pdfs is unity. The a-priori probabilities were assumed to be $P(\text{gas}) = 0.4$ and $P(\text{water}) = 0.6$. That is, if we do not look at any seismic data, we generally expect to find gas 40% of the time and water 60% of the time (only for this example!). In Figure 3.22B, we show the product of the a-priori probabilities with the state-conditional probabilities, $P(V_P \mid \text{gas}) P(\text{gas})$ and $P(V_P \mid \text{water}) P(\text{water})$. Figure 3.22C shows the Bayes probabilities of gas and water, given some observation V_P. For the specific observation value highlighted with the vertical line, $V_P = 3.15$ km/s, the probabilities are $P(\text{gas} \mid V_P = 3.15) = 0.85$ and $P(\text{water} \mid V_P = 3.15) = 0.15$. Hence, we would interpret this point in the seismic survey to be gas-saturated. Gas is much more likely than water given this velocity.

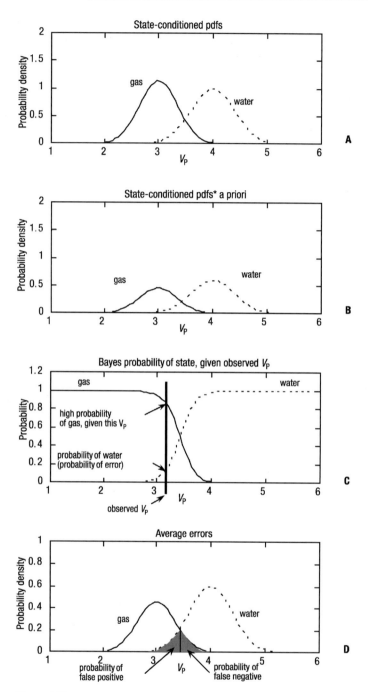

Figure 3.22 A, State-conditional probabilities of P-wave velocity assuming that the reservoir rocks are either gas-saturated or water-saturated. B, The same curves as in A, multiplied respectively by the a-priori probabilities $P(\text{gas}) = 0.4$ and $P(\text{water}) = 0.6$. C, The Bayes probability of gas and of water, as a function of measured V_P. D, Areas under the curves used to represent interpretation errors.

3.6 Statistical classification and pattern recognition

The Bayes decision rule is the optimal one that minimizes the misclassification error and maximizes the posterior probability. This is the classification criterion when the loss associated with each type of misclassification is the same. It does not matter whether a shale is misclassified as oil sand (dry well) or an oil sand is classified as shale (bypassed oil). Both have equal losses associated with them. In real-world decision analyses the losses of dry well and bypassed oil are different. The final decision therefore takes into account the loss function associated with each type of misclassification as well as the risk-aversion policy of the company. Some companies may be more risk-averse than others. The decision rule is then modified from the Bayes rule, which minimizes misclassification error, to one that minimizes the expected loss or maximizes the expected profit (see e.g. Berger (1985), Harbaugh *et al.* (1995)).

The misclassification error is given by the area of overlap of the class-conditional pdfs. Bayes' theory provides a straightforward approach to estimating seismic/rock physics interpretation errors – *the risk of being wrong*. Classification errors occur when the sample is classified as class c_j when it is actually c_k. Consider again the two-class water/gas example shown in Figure 3.22. For any observed value of velocity, we would interpret the reservoir state using the Bayes decision criterion as

gas   if $P(\text{gas} \mid V_P) > P(\text{water} \mid V_P)$

water   if $P(\text{water} \mid V_P) > P(\text{water} \mid V_P)$

For the particular velocity example, $V_P = 3.15$ km/s, in Figure 3.22C, we interpreted the reservoir as having gas, because the Bayes probability of gas was higher.

Nevertheless, we might be wrong. Bayes' theory tells us that there is a small but finite probability, $P(\text{water} \mid V_P = 3.15) = 0.15$ of finding water at $V_P = 3.15$ km/s. Hence, we write the probability of wrongly interpreting that measured velocity as indicating gas as:

$$P(\text{error} \mid V_P) = P(\text{water} \mid V_P = 3.15) = 0.15$$

More generally,

$P(\text{error} \mid V_P) = P(\text{water} \mid V_P),$ if $P(\text{gas} \mid V_P) > P(\text{water} \mid V_P)$

$P(\text{error} \mid V_P) = P(\text{gas} \mid V_P),$ if $P(\text{water} \mid V_P) > P(\text{gas} \mid V_P)$

The average probability of making *any* interpretation error can be written as

$$P(\text{any error}) = \int\limits_{\substack{P(\text{gas} \mid V_P) \\ > P(\text{water} \mid V_P)}} P(V_P \mid \text{water}) P(\text{water}) \, dV_P \\ + \int\limits_{\substack{P(\text{water} \mid V_P) \\ > P(\text{gas} \mid V_P)}} P(V_P \mid \text{gas}) P(\text{gas}) \, dV_P$$

For geophysical work it is valuable to make the distinction between *false positives* and *false negatives*. In the example that we have been talking about, any V_P that falls in the region where $P(\text{gas} \mid V_P) > P(\text{water} \mid V_P)$ would be interpreted as a *positive* hydrocarbon indicator. Hence, any of these interpretations that are in error are *false positives*. These have particular cost/risk implications, such as drilling a dry well. On the other hand, any V_P that falls in the region where $P(\text{water} \mid V_P) > P(\text{gas} \mid V_P)$ would be interpreted as a *negative* hydrocarbon indicator. Any of these interpretations that are in error are *false negatives*. These have different cost/risk implications, such as missing potential reserves.

We will define the probability of a false positive as the fraction of all V_P observations falling in the region $P(\text{gas} \mid V_P) > P(\text{water} \mid V_P)$ that actually correspond to water:

$$P(\text{false positive}) = \frac{\int\limits_{\substack{P(\text{gas} \mid V_P) \\ > P(\text{water} \mid V_P)}} P(V_P \mid \text{water}) P(\text{water}) \, dV_P}{\int\limits_{\substack{P(\text{gas} \mid V_P) \\ > P(\text{water} \mid V_P)}} P(V_P) \, dV_P}$$

The denominator indicates the fraction of all observations that fall in the region interpreted as gas. Similarly, we define the probability of a false negative as the fraction of all V_P observations falling in the region $P(\text{water} \mid V_P) > P(\text{gas} \mid V_P)$ that actually correspond to gas:

$$P(\text{false\_negative}) = \frac{\int\limits_{\substack{P(\text{water} \mid V_P) \\ > P(\text{gas} \mid V_P)}} P(V_P \mid \text{gas}) P(\text{gas}) \, dV_P}{\int\limits_{\substack{P(\text{water} \mid V_P) \\ > P(\text{gas} \mid V_P)}} P(V_P) \, dV_P}$$

For multi-class problems, it is easier to compute the probability of correct classification as

$$P(\text{correct}) = \sum_{j=1}^{N} \int_{\Omega_j} p(x \mid c_j) P(c_j) \, dx$$

where Ω_j is the region in feature space assigned to class c_j. The Bayes error is then $1 - P(\text{correct})$. Theoretically the Bayes error is the minimum possible error, given that the pdfs are known. In practice, the pdfs are not known but estimated. Hence we can only approach the theoretical minimum Bayes error rate. For a simple nearest-neighbor rule classification, the upper bound on the error rate is twice the minimum Bayes error rate. Because the error rate is always less than or equal to twice the minimum Bayes error, any classification system involving complicated rules and lots of data can at most cut the error rate by half (Duda *et al.*, 2000). Roughly speaking, the nearest neighbor contains at least half of the classification information.

The Bayes approach allows us to recursively update facies probabilities estimated from some earlier classification using old data, x_n, to posterior probabilities as new

3.6 Statistical classification and pattern recognition

data, x_{n+1}, become available. For example, we may have the probabilities for each class from observed post-stack P-impedance data. Denote these probabilities by $P(c_j \mid x_n)$. Later, suppose new information in the form of S-impedance (denoted by x_{n+1}) becomes available after pre-stack inversion. Bayes' formula allows us to update $P(c_j \mid x_n)$ to obtain $P(c_j \mid x_{n+1}, x_n)$ as follows:

$$P(c_j \mid x_{n+1}, x_n) = \frac{P(x_{n+1}, x_n, c_j)}{P(x_{n+1}, x_n)} = \frac{P(x_{n+1} \mid c_j, x_n)}{P(x_{n+1} \mid x_n)} P(c_j \mid x_n)$$

This expression incorporates all dependencies between new and old data. Often the two pieces of information may not be independent. For example, P- and S-wave impedances in a fixed lithology are correlated. If x_{n+1} and x_n are independent the expression simplifies to:

$$P(c_j \mid x_{n+1}, x_n) = \frac{P(x_{n+1} \mid c_j)}{P(x_{n+1})} P(c_j \mid x_n)$$

The class-conditional probabilities, $P(x_{n+1} \mid c_j, x_n)$ are estimated from data and forward modeling.

What about measurement uncertainties? Measurement errors arise from many sources: navigation errors, cable feathering, hardware calibration, noise, processing difficulties, velocity estimation, and so on. We do not observe x directly but can get only a noise-contaminated estimate $\hat{x}$ of x. The distribution of c_j given the estimated $\hat{x}$ may be written as:

$$P(c_j \mid \hat{x}) = \int P(c_j \mid x) P(x \mid \hat{x}) \, dx = \int \frac{P(x \mid c_j) P(c_j)}{P(x)} P(x \mid \hat{x}) \, dx$$

The *geologic uncertainty* arising from natural variability is captured by the conditional distribution $P(x \mid c_j)$. Within the same class c_j there is not a fixed value of the attribute x but a distribution of possible values, and different classes can have different, possibly overlapping distributions of x values. The *measurement uncertainty* is captured by the conditional distribution $P(x \mid \hat{x})$, the probability that the true value is x given the estimate $\hat{x}$. If enough training data (e.g. well data) exist, then it may be possible to estimate $P(x \mid \hat{x})$ from data alone. For example the P-impedance obtained from a seismic inversion may be compared to the measured P-impedance at the wells. In other cases, without enough training data, the error distribution has to be modeled. The inversion process used to get the estimate can give us error distributions, under some approximations (e.g. Gaussian errors). Stochastic inversion methods and statistical bootstrap techniques may also be used to estimate the distribution of $\hat{x}$ and thus quantify the measurement uncertainty. These are usually computationally intensive and involve running the forward model many times.

From the point of view of statistical classification, we can think of measurement errors as broadening the clouds that we wish to separate. Figure 3.23A shows a graph

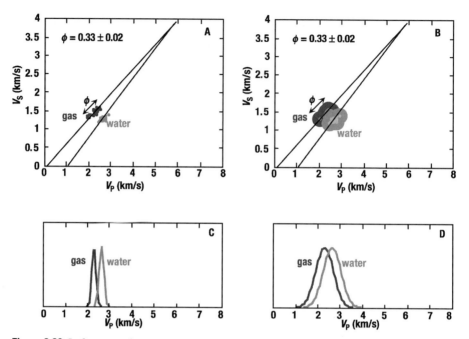

Figure 3.23 Left: gas- and water-saturated sandstones, error-free. Right: effect of measurement error can be simulated by convolving with the error pdf.

of gas- and water-saturated sandstones in the (V_P, V_S) domain (same as Figure 3.8A). The data are fairly free of errors, and the clouds are well separated.

If we assume random velocity errors with standard deviations $\pm \Delta V_P$, $\pm \Delta V_S$, then effectively each data point in Figure 3.23A should be replaced by a fuzzy cloud of possible measured values. Hence, the effect of random measurement errors can be approximated by convolving the error-free log-derived pdfs with the measurement error pdf, which we can approximate as a two-dimensional Gaussian distribution, having standard deviations $\pm \Delta V_P$, $\pm \Delta V_S$. Figures 3.23B and 3.23D show the result. The gas- and water-saturated clouds are broadened in both V_P and V_S. The clouds significantly overlap in the (V_P, V_S) plane, as well as in V_P only.

> The rock physics of error analysis is fairly simple. The most critical part is estimation of errors expected for each type of acquisition being considered.

When using multiple attributes in the classification, it should be kept in mind that the errors may be correlated. A simple model of uncorrelated identically distributed Gaussian error may not hold. Different seismic attributes have different measurement errors associated with them. For example P–P AVO attributes (R_0, G) have errors associated with noise, amplitude-picking, phase changes with offset, velocity estimation, non-hyperbolic moveout, anisotropy, fitting a $\sin^2 \theta$ function to the amplitudes, etc. Houck

3.6 Statistical classification and pattern recognition

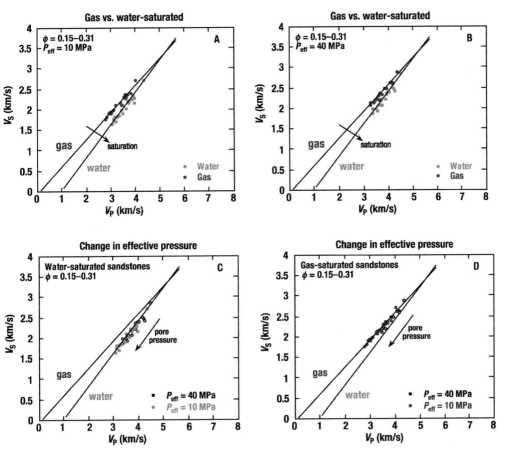

Figure 3.24 Subsets of Han's sandstone data. A, Data at 10 MPa, showing the separate gas- and water-saturated clouds. B, Data at 40 MPa, showing the separate gas- and water-saturated clouds. C, Water-saturated data, showing the overlapping clouds for high and low effective pressures. D, Gas-saturated data, showing the overlapping clouds for high and low effective pressures.

(2002) gives an analysis of correlated errors in P–P AVO attributes. The V_P/V_S ratios determined from comparing interval times on P-wave and converted S-wave stacks have errors associated with moveout, migration difficulties, anisotropy, time-picking, correlating the P and S events, etc. A critical part of acquisition decision-making is forward modeling to estimate these errors.

Let us now see how to apply the Bayes formalism to choose among different seismic attributes and acquisition schemes. Figure 3.24 illustrates gas- and water-saturated sandstone data in the (V_P, V_S) domain. In Figure 3.24A, the high porosity and low effective stress make separation of the fluid states fairly good. In Figure 3.24B, the higher effective stress stiffens the rock, leading to less fluid sensitivity and more overlap of the clouds. In Figure 3.24C, high and low effective pressures are shown, both with water saturation. The separation is poor. In Figure 3.24D, high and low effective pressures are shown, both with gas saturation. Again the separation is poor.

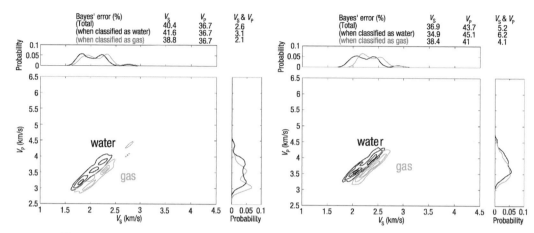

Figure 3.25 State-conditional pdfs and classification error rates for distinguishing gas and water using velocities. Left: rocks at 10 MPa. Right: rocks at 40 MPa. Table shows expected interpretation error rates using V_P only, V_S only, and $V_P + V_S$. In this case shear data add tremendous value.

We can illustrate a procedure for quantifying these differences using estimates of Bayes misclassification errors (Takahashi, 2000). Figure 3.25 shows the gas- and water-saturated data clouds at effective pressures of 10 MPa (left) and 40 MPa (right). The contours represent the state-conditional pdfs, estimated from the data that we showed in Figures 3.24A and 3.24B. The single attribute pdfs for V_P and V_S are shown above and to the right of each cross-plot.

A table of expected interpretation errors is also shown, assuming that these calibration pdfs would be used to interpret observed velocities, according to the Bayes decision criterion. In the table, the "total error" is what we defined as the average probability of making any error. Beneath that is the probability of being wrong when we classify data as indicating water (false negative); last is the probability of being wrong when classifying data as gas (false positive). The columns show each of the errors when using V_S only, V_P only, and both V_P and V_S. This gives us objective measures of the value of each seismic attribute for separating water sands from gas sands for this particular case.

Comparing the examples in Figure 3.25, we see that for both high- and low-pressure situations, the probability of false positives or false negatives is very high if only a single velocity is used, but the errors drop dramatically when both V_P and V_S are used. We also can quantify that most of the error rates go up a few percent at higher pressure, when the rocks are stiffer. The exception is that there is a slight reduction in errors at high pressure when only V_S is used. For fluid detection in these rocks, V_S adds tremendous value.

Figure 3.26 shows the problem of distinguishing gas and water for the same rocks as in Figure 3.25, except that here the AVO attributes, intercept (R_0) and gradient (G), are considered (see Chapter 4 for AVO attributes). We see again that classification is easier

3.6 Statistical classification and pattern recognition

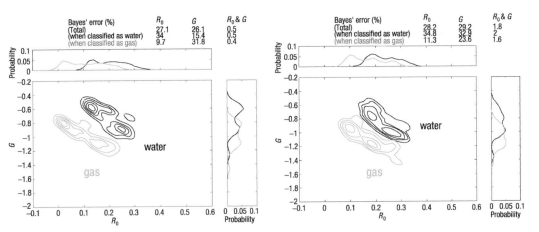

Figure 3.26 State-conditional pdfs and classification error rates for distinguishing gas and water, using AVO intercept and gradient. Left: rocks at 10 MPa. Right: rocks at 40 MPa. Table shows expected interpretation error rates using R_0 only, G only, and $R_0 + G$. In this case shear data add tremendous value.

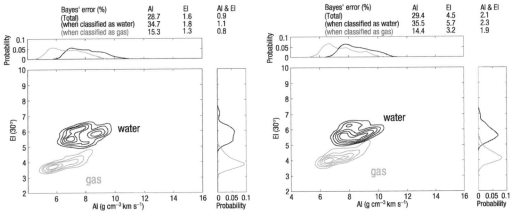

Figure 3.27 State-conditional pdfs and classification error rates for distinguishing gas and water, using acoustic impedance and elastic impedance at 30°. Left: rocks at 10 MPa. Right: rocks at 40 MPa. Table shows expected interpretation error rates using AI only, EI only, and AI + EI. In this case shear data add tremendous value. In fact EI alone is almost as good as AI + EI.

for the low-effective-pressure rocks. Again two attributes lead to substantially greater error reduction (tremendous value in shear-related data). Comparing with Figure 3.25, we can see that for these rocks and fluids AVO will lead to more accurate interpretation than V_P, V_S. However, in these examples we have not included the measurement errors for velocity or AVO attributes.

Figure 3.27 shows the same gas vs. water problem, using acoustic impedance (AI) and elastic impedance (EI) at 30°. Elastic impedance is a pseudo-impedance that will yield

158 Statistical rock physics

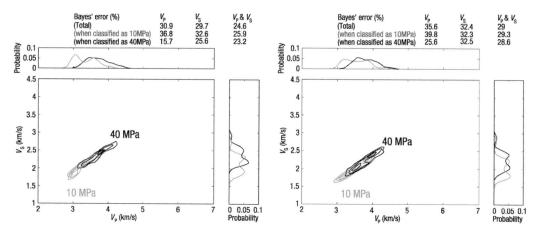

Figure 3.28 Detecting pressure differences for rocks at 10 MPa effective pressure (gray) and 40 MPa effective pressure (black). Gas-saturated on the left, brine-saturated on the right. Error rates are fairly high. Adding V_S only slightly reduces the errors relative to using V_P alone.

the far-offset amplitudes if we simply take its difference across layer boundaries, just as we would for normal-incidence reflectivity. Although EI depends on V_P, V_S, and density, it is not entirely a material property, since it also depends on the angle of incidence. We see again that using two attributes (AI + EI) gives the best discrimination of the fluid, very comparable to using AVO and slightly better than using V_P, V_S. However, we get the surprising result that using the single attribute EI does almost as well! Again, we caution that acquisition errors are not included.

Figure 3.28 shows the problem of seismically detecting rocks at low effective pressure (10 MPa) vs. high effective pressure (40 MPa). Rocks on the left are gas-saturated; on the right, water-saturated. Detection error rates are fairly high (~25–40%). Adding V_S reduces the errors only slightly.

Figure 3.29 illustrates the effect of measurement error. Both plots show the problem of separating gas vs. water using AI and EI, as in Figure 3.27. The left is error-free. The right has a large increase in measurement errors for both AI and EI, simulated by convolving with an error pdf. This leads to substantial overlap of the data clouds, and increases the classification error rates.

Figure 3.30 is the same as Figure 3.25, except that here we include the error rates when all three seismic attributes are measured, V_P, V_S, and density. This is the most information we could hope for. Dropping from three attributes to two has some effect.

One of the important steps in Bayes classification is to estimate the class-conditioned pdfs. If the pdfs are Gaussian, then the Bayes decision rule is equivalent to discriminant analysis based on Mahalanobis distances. However, Bayes classification is not limited to Gaussian pdfs, but in principle can be applied to any parametric or nonparametric pdfs as long as they can be estimated. Estimation of pdfs can become a problem in high dimensions. For low dimensions a simple approach is based on smoothing of histograms and scatter plots. The raw histograms obtained from the training data are smoothed to get

3.6 Statistical classification and pattern recognition

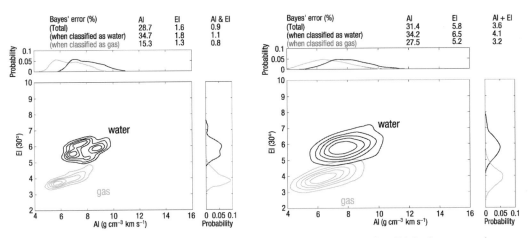

Figure 3.29 Separating gas- from water-saturated rocks. Left: error-free. Right: large errors in estimating both AI and EI.

pdfs that capture the overall general trend without fitting the specific idiosyncrasies of the data. Over-smoothing can give poor classification results by increasing the overlap between classes. The appropriate smoothing may be determined by dividing the whole training sample into two sets, one for the pdf estimation, and another, for validation. The smoothing that gives the best overall success rate with the validation data is selected.

Figure 3.31 shows a comparison of the state-conditioned pdfs for brine sands, gas sands, and shales from an Australian field and a North Sea field. We simply wish to illustrate that:

> Different geologic settings must be separately calibrated, and can lead to different interpretational problems. The classification success depends on
> - the categories we wish to separate (water/gas; sand/shale, etc.)
> - the reservoir rock and fluid properties and their variability, and
> - the seismic attributes used for the classification.

3.6.3 Neural network classification

Neural networks represent yet another way to classify observations. This amounts essentially to a nonlinear regression. The features or inputs are the independent variables, and the class categories are the desired dependent output variables or targets. This approach can be useful when the discriminant surfaces are highly nonlinear and are not well approximated by the simple linear discriminant analysis. Amongst others, Baldwin *et al.* (1989, 1990), and Rogers *et al.* (1992) used neural networks to classify porosity and density logs and lithologies. Harris *et al.* (1993) trained networks to classify lithology from borehole imagery data. Avseth and Mukerji (2002) compared neural networks, Bayesian classification and discriminant analysis for classifying lithologies

Statistical rock physics

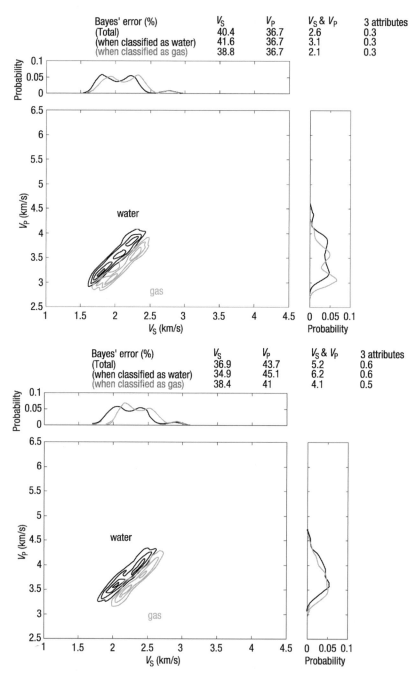

Figure 3.30 Distinguishing gas from water using velocities, the same as Figure 3.25. Top: 10 MPa. Bottom: 40 MPa. The table shows the complete best-case error if V_P, V_S, and density are all measured without errors.

3.6 Statistical classification and pattern recognition

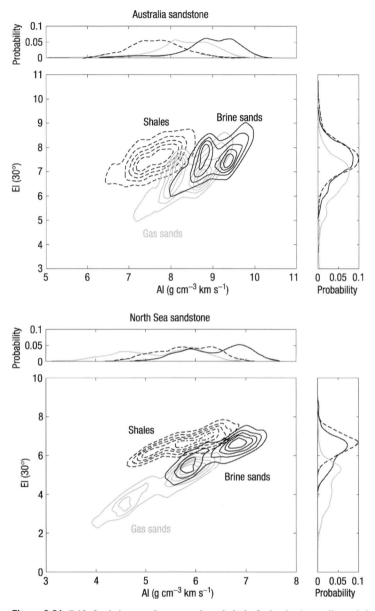

Figure 3.31 Pdfs for brine sand, gas sand, and shale facies in Australia and the North Sea.

from well-log data. Other examples of various applications in geosciences, and a practical introduction to neural network theory, are given in Dowla and Rogers (1995).

While there are various kinds of neural networks, we will describe the popular single-hidden-layer, feed-forward architecture with weight adaptation by back-propagation. The network consists of an input layer, and output layer, and a hidden layer with multiple nodes. So for example if we wanted to classify six different facies using V_P and gamma ray as the inputs, the input layer would have two nodes, the output layer would have six nodes, and the hidden layer would have any number of multiple nodes as selected by the user. Having too many nodes in the hidden layer would lead to overfitting the data. There are no hard and fast rules to pick the number of nodes to use, although there are some limiting analytical guidelines (e.g., Lin and Lee, 1996). In practice, the choice is made on the basis of trial and experiment, balancing among computation and training time, convergence, and network performance. In the neural network, derived "hidden" features are created by taking weighted linear combinations of the inputs and passing them through a nonlinear transfer function. Often a sigmoid function, $1/(1+ e^{-z})$, is used as the nonlinear transfer function. The weighted linear combination with the nonlinear transfer function forms the hidden layer. In general there can be multiple hidden layers. The desired outputs are then modeled as a function of linear combinations of the derived features. The unknown parameters of the neural network, the weights, are obtained by a process of training the neural network using the training data set. This amounts to minimizing some measure (e.g. sum-of-squared error) of misfit between the desired output and the output of the neural network. The minimization is done by gradient descent. This procedure is known as back-propagation in neural network literature. Ordinary gradient descent can be slow if the learning constant parameter is small, and can oscillate too much if the parameter is set too large. Gradient descent with momentum helps to solve this problem by adding a fractional (<1) contribution from the previous time step to each weight change during the training session. The weight update scheme is implemented by:

$$\Delta w(t) = -\eta \nabla E(t) + \alpha \Delta w(t-1)$$

where E is the error between the desired and actual network output, η is the learning parameter, α is the momentum parameter (<1), and Δw is the weight update. Figure 3.32 shows an example of the decrease in the error as the network goes towards convergence in one training session. Error plots help to monitor the network convergence during training. The weights are initialized to some small random values near zero. Large initial weights can lead to poor solutions as the network gets into the nonlinear region too quickly. With small weights the network starts out nearly linear, and then nonlinearity is gradually introduced as the weights are adapted during training. It is best to scale and standardize the inputs so that they have similar ranges of values. This ensures that all inputs have an equal effect on the weights. The final solution is dependent on the starting weights. Hence a number of initial random starting weights should be tried, and the one with the least error should be chosen. Another approach is to use a set of neural nets, and take the final classification as the average over the outputs of the set.

One of the complaints about neural networks is that these models are not easily interpretable. The inputs and their linear combinations enter nonlinearly into the model. The role of individual input features is unclear and neural networks are not very effective for interpretation of the processes that gave rise to the data.

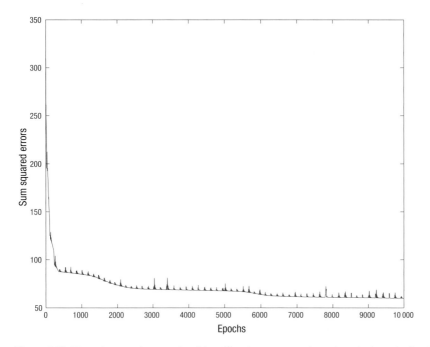

Figure 3.32 Neural network error plot. Plots like these are monitored to obtain an indication of the convergence of the network.

3.6.4 Comparison of three different classification methods

Now let us look at a simple example (Avseth and Mukerji, 2002) using three different methods to classify six different facies based on well-log measurements of V_P and gamma ray. Plate 3.33 shows the classification results in a type-well (in the target zone: 2100–2300 m), using different methods. The figure compares discriminant classification (MLDA), Bayes' rule classification (PDF) and neural network classification (NN). For the neural network classification, a simple feed-forward back-propagation network was used. There were six nodes in the hidden layer with a sigmoid transfer function. For our lithofacies classification problem the input vector consisted of the V_P and gamma-ray value from the log at each depth point. The desired output was a six-element binary vector (corresponding to six lithofacies classes) with a "1" at the position corresponding to the facies numeric code, and zero elsewhere. The weight update was done using conjugate gradient descent with momentum (Lin and Lee, 1996). The classification results from the three different methods are very similar, but some important differences occur. Note the thin dark stripe close to depth sample 600 in the PDF classification. This interval is actually classified as "zero," meaning no facies is recognized by the Bayes method. Taking a look at the core section from the well, we find that this interval corresponds to a 1-m-thick debris-flow unit, which is not represented in the training data.

164 Statistical rock physics

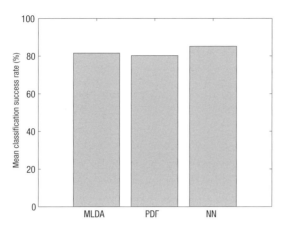

Figure 3.34 Mean classification success rate.

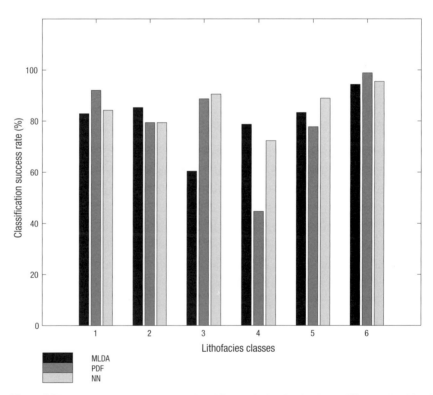

Figure 3.35 Classification success rate for different facies for the three different classification methods.

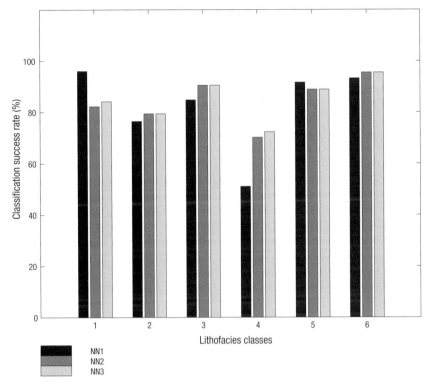

Figure 3.36 Classification success rate for neural network classifications with different weights.

Figures 3.34 and 3.35 show the overall classification success rates of the different methods. In general, all the methods give about 80% success rate (Figure 3.34). Note that here we have assumed that the training data set itself is error-free with perfect classification.

The neural network method can be tuned to give slightly different weights to the different facies. Figure 3.36 shows the classification success rate for three different networks. NN1 was trained with a training set biased towards Facies 1. This causes poor performance for Facies 4, but gives almost 100% success rate in classifying Facies 1. The other two networks (NN2 and NN3) had more evenly biased training data, but had different initial weights. The similar results for both networks show that consistency can be achieved with proper selection of training data and network architecture.

3.7 Discussion and summary

We have presented in this chapter concepts and methodologies of applied statistical rock physics used to quantify and reduce uncertainties in the reservoir characterization. Pictorially, the steps are described in Figure 3.37: facies definition, and rock physics

166 Statistical rock physics

Logs + rock physics + geology

Monte Carlo – probability distributions

Seismic inversions – near and far-offset attributes

Integrated statistical classification – facies probability maps

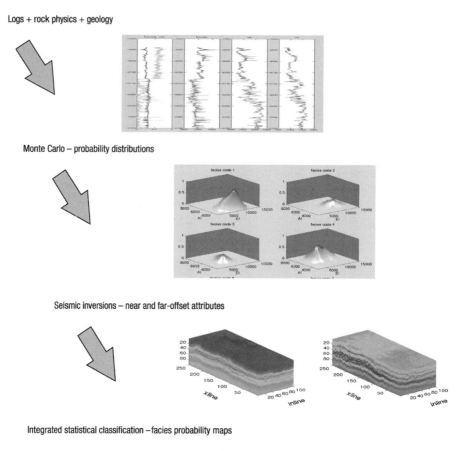

Figure 3.37 A schematic cartoon of the steps in a statistical rock physics workflow for quantitative seismic interpretation.

modeling from geology and well logs; Monte Carlo extension of data, and estimation of facies-conditioned pdfs of seismic attributes; selection of attribute or attribute combinations based on information content and classification success rates for the targeted problem; estimation of attributes from seismic data using various inversion methods; statistical classification of the volumes of seismic attributes into different facies categories based on the facies-conditioned, calibrated pdfs, and integrating small-scale spatial variability using geostatistics. The final products of this integrated technique are the spatial probability maps of reservoir fluids and facies, and stochastic realizations of the reservoir properties. In this way, not only do we obtain the most probable facies, but also we can quantify the uncertainty of the interpretation. In Chapter 5 we will present case studies that apply the methods described in this chapter.

Some of the emerging and future trends in applied statistical rock physics will include strategies for better understanding and integrating qualitative, "fuzzy" geologic information in terms of probabilities for quantitative seismic interpretation. On the deterministic side, we need a better understanding of the physics behind various attributes, especially attributes based on wave attenuation (Q) and mode conversions (P-to-S). Integrated 3D visualizations will have to combine not only reservoir architecture and lithologies but also their probabilities of occurrence. There will be an emphasis on quantifying uncertainties and risks associated with the geophysical interpretation, in order to incorporate them into the decision-making process. Interpretation based on Monte Carlo simulations, rather than single-point estimates, will become routine. Finally, with the advent of continuously monitored intelligent oil fields, on-line rock physics will be needed to quantify the uncertainties in real time, updating the interpretation risks as new data come in.

> Most of these endeavors will require analyzing and understanding large amounts of disparate data types: geological, geophysical, and production data. When using statistical pattern-recognition techniques it is wise to keep in mind some of the myths and pitfalls of these methods. It is a myth that the more attributes we throw in, the more effective will be the statistical effort. More attributes are useful only if they can contribute more information about the goal of the data-mining exercise. Otherwise they can do more harm than good. No statistical data-manipulation technique is so powerful that it can substitute for expertise in reservoir analysis and physical understanding of reservoir processes. The best strategies, we feel, combine the strengths of computational statistics with deterministic physics-based models and subjective human knowledge. Uncertainty estimates are always subjective.

4 Common techniques for quantitative seismic interpretation

There are no facts, only interpretations. *Friedrich Nietzsche*

4.1 Introduction

Conventional seismic interpretation implies picking and tracking laterally consistent seismic reflectors for the purpose of mapping geologic structures, stratigraphy and reservoir architecture. The ultimate goal is to detect hydrocarbon accumulations, delineate their extent, and calculate their volumes. Conventional seismic interpretation is an art that requires skill and thorough experience in geology and geophysics.

Traditionally, seismic interpretation has been essentially qualitative. The geometrical expression of seismic reflectors is thoroughly mapped in space and traveltime, but little emphasis is put on the physical understanding of seismic amplitude variations. In the last few decades, however, seismic interpreters have put increasing emphasis on more quantitative techniques for seismic interpretation, as these can validate hydrocarbon anomalies and give additional information during prospect evaluation and reservoir characterization. The most important of these techniques include post-stack amplitude analysis (bright-spot and dim-spot analysis), offset-dependent amplitude analysis (AVO analysis), acoustic and elastic impedance inversion, and forward seismic modeling.

These techniques, if used properly, open up new doors for the seismic interpreter. The seismic amplitudes, representing primarily contrasts in elastic properties between individual layers, contain information about lithology, porosity, pore-fluid type and saturation, as well as pore pressure – information that cannot be gained from conventional seismic interpretation.

4.2 Qualitative seismic amplitude interpretation

Until a few decades ago, it would be common for seismic interpreters to roll out their several-meters-long paper sections with seismic data down the hallway, go down on their knees, and use their colored pencils to interpret the horizons of interest in order to map geologic bodies. Little attention was paid to amplitude variations and their interpretations. In the early 1970s the so-called "bright-spot" technique proved successful in areas of the Gulf of Mexico, where bright amplitudes would coincide with gas-filled sands. However, experience would show that this technique did not always work. Some of the bright spots that were interpreted as gas sands, and subsequently drilled, were found to be volcanic intrusions or other lithologies with high impedance contrast compared with embedding shales. These failures were also related to lack of wavelet phase analysis, as hard volcanic intrusions would cause opposite polarity to low-impedance gas sands. Moreover, experience showed that gas-filled sands sometimes could cause "dim spots," not "bright spots," if the sands had high impedance compared with surrounding shales.

With the introduction of 3D seismic data, the utilization of amplitudes in seismic interpretation became much more important. Brown (see Brown *et al.*, 1981) was one of the pioneers in 3D seismic interpretation of lithofacies from amplitudes. The generation of time slices and horizon slices revealed 3D geologic patterns that had been impossible to discover from geometric interpretation of the wiggle traces in 2D stack sections. Today, the further advance in seismic technology has provided us with 3D visualization tools where the interpreter can step into a virtual-reality world of seismic wiggles and amplitudes, and trace these spatially (3D) and temporally (4D) in a way that one could only dream of a few decades ago. Certainly, the leap from the rolled-out paper sections down the hallways to the virtual-reality images in visualization "caves" is a giant leap with great business implications for the oil industry. In this section we review the qualitative aspects of seismic amplitude interpretation, before we dig into the more quantitative and rock-physics-based techniques such as AVO analysis, impedance inversion, and seismic modeling, in following sections.

4.2.1 Wavelet phase and polarity

The very first issue to resolve when interpreting seismic amplitudes is what kind of wavelet we have. Essential questions to ask are the following. What is the defined polarity in our case? Are we dealing with a zero-phase or a minimum-phase wavelet? Is there a phase shift in the data? These are not straightforward questions to answer, because the phase of the wavelet can change both laterally and vertically. However, there are a few pitfalls to be avoided.

First, we want to make sure what the defined standard is when processing the data. There exist two standards. The American standard defines a black peak as a "hard" or "positive" event, and a white trough as a "soft" or a "negative" event. On a near-offset stack section a "hard" event will correspond to an increase in acoustic impedance with depth, whereas a "soft" event will correspond to a decrease in acoustic impedance with depth. According to the European standard, a black peak is a "soft" event, whereas a white trough is a "hard" event. One way to check the polarity of marine data is to look at the sea-floor reflector. This reflector should be a strong positive reflector representing the boundary between water and sediment.

> **Data polarity**
>
> - American polarity: An increase in impedance gives positive amplitude, normally displayed as black peak (wiggle trace) or red intensity (color display).
> - European (or Australian) polarity: An increase in impedance gives negative amplitude, normally displayed as white trough (wiggle trace) or blue intensity (color display).
>
> (Adapted from Brown, 2001a, 2001b)

For optimal quantitative seismic interpretations, we should ensure that our data are zero-phase. Then, the seismic pick should be on the crest of the waveform corresponding with the peak amplitudes that we desire for quantitative use (Brown, 1998). With today's advanced seismic interpretation tools involving the use of interactive workstations, there exist various techniques for horizon picking that allow efficient interpretation of large amounts of seismic data. These techniques include manual picking, interpolation, autotracking, voxel tracking, and surface slicing (see Dorn (1998) for detailed descriptions).

For extraction of seismic horizon slices, autopicked or voxel-tracked horizons are very common. The obvious advantage of autotracking is the speed and efficiency. Furthermore, autopicking ensures that the peak amplitude is picked along a horizon. However, one pitfall is the assumption that seismic horizons are locally continuous and consistent. A lateral change in polarity within an event will not be recognized during autotracking. Also, in areas of poor signal-to-noise ratio or where a single event splits into a doublet, the autopicking may fail to track the correct horizon. Not only will important reservoir parameters be neglected, but the geometries and volumes may also be significantly off if we do not regard lateral phase shifts. It is important that the interpreter realizes this and reviews the seismic picks for quality control.

4.2.2 Sand/shale cross-overs with depth

Simple rock physics modeling can assist the initial phase of qualitative seismic interpretation, when we are uncertain about what polarity to expect for different lithology boundaries. In a siliciclastic environment, most seismic reflectors will be associated with sand–shale boundaries. Hence, the polarity will be related to the contrast in impedance between sand and shale. This contrast will vary with depth (Chapter 2). Usually, relatively soft sands are found at relatively shallow depths where the sands are unconsolidated. At greater depths, the sands become consolidated and cemented, whereas the

4.2 Qualitative seismic amplitude interpretation

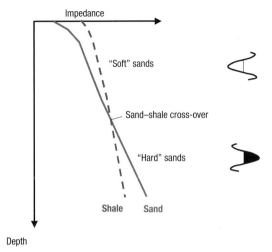

Figure 4.1 Schematic depth trends of sand and shale impedances. The depth trends can vary from basin to basin, and there can be more than one cross-over. Local depth trends should be established for different basins.

shales are mainly affected by mechanical compaction. Hence, cemented sandstones are normally found to be relatively hard events on the seismic. There will be a corresponding cross-over in acoustic impedance of sands and shales as we go from shallow and soft sands to the deep and hard sandstones (see Figure 4.1). However, the depth trends can be much more complex than shown in Figure 4.1 (Chapter 2, see Figures 2.34 and 2.35). Shallow sands can be relatively hard compared with surrounding shales, whereas deep cemented sandstones can be relatively soft compared with surrounding shales. There is no rule of thumb for what polarity to expect for sands and shales. However, using rock physics modeling constrained by local geologic knowledge, one can improve the understanding of expected polarity of seismic reflectors.

"Hard" versus "soft" events

During seismic interpretation of a prospect or a proven reservoir sand, the following question should be one of the first to be asked: what type of event do we expect, a "hard" or a "soft"? In other words, should we pick a positive peak, or a negative trough? If we have good well control, this issue can be solved by generating synthetic seismograms and correlating these with real seismic data. If we have no well control, we may have to guess. However, a reasonable guess can be made based on rock physics modeling. Below we have listed some "rules of thumb" on what type of reflector we expect for different geologic scenarios.

> **Typical "hard" events**
>
> - Very shallow sands at normal pressure embedded in pelagic shales
> - Cemented sandstone with brine saturation
> - Carbonate rocks embedded in siliciclastics
> - Mixed lithologies (heterolithics) like shaly sands, marls, volcanic ash deposits
>
> **Typical "soft" events**
>
> - Pelagic shale
> - Shallow, unconsolidated sands (any pore fluid) embedded in normally compacted shales
> - Hydrocarbon accumulations in clean, unconsolidated or poorly consolidated sands
> - Overpressured zones
>
> **Some pitfalls in conventional interpretation**
>
> - Make sure you know the polarity of the data. Remember there are two different standards, the US standard and the European standard, which are opposite.
> - A hard event can change to a soft laterally (i.e., lateral phase shift) if there are lithologic, petrographic or pore-fluid changes. Seismic autotracking will not detect these.
> - A dim seismic reflector or interval may be significant, especially in the zone of sand/shale impedance cross-over. AVO analysis should be undertaken to reveal potential hydrocarbon accumulations.

4.2.3 Frequency and scale effects

Seismic resolution

Vertical seismic resolution is defined as the minimum separation between two interfaces such that we can identify two interfaces rather than one (Sheriff and Geldhart, 1995). A stratigraphic layer can be resolved in seismic data if the layer thickness is larger than a quarter of a wavelength. The wavelength is given by:

$$\lambda = V/f \tag{4.1}$$

where V is the interval velocity of the layer, and f is the frequency of the seismic wave. If the wavelet has a peak frequency of 30 Hz, and the layer velocity is 3000 m/s, then the dominant wavelength is 100 m. In this case, a layer of 25 m can be resolved. Below this thickness, we can still gain important information via quantitative analysis of the interference amplitude. A bed only $\lambda/30$ in thickness may be detectable, although its thickness cannot be determined from the wave shape (Sheriff and Geldhart, 1995).

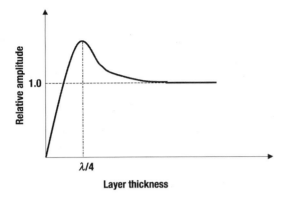

Figure 4.2 Seismic amplitude as a function of layer thickness for a given wavelength.

The horizontal resolution of unmigrated seismic data can be defined by the Fresnel zone. Approximately, the Fresnel zone is defined by a circle of radius, R, around a reflection point:

$$R \approx \sqrt{\lambda z/2} \tag{4.2}$$

where z is the reflector depth. Roughly, the Fresnel zone is the zone from which all reflected contributions have a phase difference of less than π radians. For a depth of 3 km and velocity of 3 km/s, the Fresnel zone radius will be 300–470 m for frequencies ranging from 50 to 20 Hz. When the size of the reflector is somewhat smaller than the Fresnel zone, the response is essentially that of a diffraction point. Using prestack migration we can collapse the diffractions to be smaller than the Fresnel zone, thus increasing the lateral seismic resolution (Sheriff and Geldhart, 1995). Depending on the migration aperture, the lateral resolution after migration is of the order of a wavelength. However, the migration only collapses the Fresnel zone in the direction of the migration, so if it is only performed along inlines of a 3D survey, the lateral resolution will still be limited by the Fresnel zone in the cross-line direction. The lateral resolution is also restricted by the lateral sampling which is governed by the spacing between individual CDP gathers, usually 12.5 or 18 meters in 3D seismic data. For typical surface seismic wavelengths (~50–100 m), lateral sampling is not the limiting factor.

Interference and tuning effects

A thin-layered reservoir can cause what is called event tuning, which is interference between the seismic pulse representing the top of the reservoir and the seismic pulse representing the base of the reservoir. This happens if the layer thickness is less than a quarter of a wavelength (Widess, 1973). Figure 4.2 shows the effective seismic amplitude as a function of layer thickness for a given wavelength, where a given layer has higher impedance than the surrounding sediments. We observe that the amplitude increases and becomes larger than the real reflectivity when the layer thickness is between a half and a quarter of a wavelength. This is when we have constructive interference between the top and the base of the layer. The maximum constructive interference occurs when the bed thickness is equal to $\lambda/4$, and this is often referred to as the *tuning thickness*. Furthermore, we observe that the amplitude decreases and approaches zero for layer thicknesses between one-quarter of a wavelength and zero thickness. We refer to this as destructive interference between the top and the base. Trough-to-peak time measurements give approximately the correct gross thicknesses for thicknesses larger than a quarter of a wavelength, but no information for thicknesses less than a quarter of a wavelength. The thickness of an individual thin-bed unit can be extracted from amplitude measurements if the unit is thinner than about $\lambda/4$ (Sheriff and Geldhart, 1995). When the layer thickness equals $\lambda/8$, Widess (1973) found that the composite response approximated the derivative of the original signal. He referred to this thickness as the *theoretical threshold of resolution*. The amplitude–thickness curve is almost linear below $\lambda/8$ with decreasing amplitude as the layer gets thinner, but the composite response stays the same.

4.2.4 Amplitude and reflectivity strength

"Bright spots" and "dim spots"

The first use of amplitude information as hydrocarbon indicators was in the early 1970s when it was found that bright-spot amplitude anomalies could be associated with hydrocarbon traps (Hammond, 1974). This discovery increased interest in the physical properties of rocks and how amplitudes changed with different types of rocks and pore fluids (Gardner *et al.*, 1974). In a relatively soft sand, the presence of gas and/or light oil will increase the compressibility of the rock dramatically, the velocity will drop accordingly, and the amplitude will decrease to a negative "bright spot." However, if the sand is relatively hard (compared with cap-rock), the sand saturated with brine may induce a "bright-spot" anomaly, while a gas-filled sand may be transparent, causing a so-called dim spot, that is, a very weak reflector. It is very important before starting to interpret seismic data to find out what change in amplitude we expect for different pore fluids, and whether hydrocarbons will cause a relative dimming or brightening compared with brine saturation. Brown (1999) states that "*the most important seismic property of a reservoir is whether it is bright spot regime or dim spot regime.*"

One obvious problem in the identification of dim spots is that they are dim – they are hard to see. This issue can be dealt with by investigating limited-range stack sections. A very weak near-offset reflector may have a corresponding strong far-offset reflector. However, some sands, although they are significant, produce a weak positive near-offset reflection as well as a weak negative far-offset reflection. Only a quantitative analysis of the change in near- to far-offset amplitude, a gradient analysis, will be able to reveal the sand with any considerable degree of confidence. This is explained in Section 4.3.

> **Pitfalls: False "bright spots"**
>
> During seismic exploration of hydrocarbons, "bright spots" are usually the first type of DHI (direct hydrocarbon indicators) one looks for. However, there have been several cases where bright-spot anomalies have been drilled, and turned out not to be hydrocarbons.
>
> Some common "false bright spots" include:
> - Volcanic intrusions and volcanic ash layers
> - Highly cemented sands, often calcite cement in thin pinch-out zones
> - Low-porosity heterolithic sands
> - Overpressured sands or shales
> - Coal beds
> - Top of salt diapirs
>
> Only the last three on the list above will cause the same polarity as a gas sand. The first three will cause so-called "hard-kick" amplitudes. Therefore, if one knows the polarity of the data one should be able to discriminate hydrocarbon-associated bright spots from the "hard-kick" anomalies. AVO analysis should permit discrimination of hydrocarbons from coal, salt or overpressured sands/shales.
>
> A very common seismic amplitude attribute used among seismic interpreters is reflection intensity, which is root-mean-square amplitudes calculated over a given time window. This attribute does not distinguish between negative and positive amplitudes; therefore geologic interpretation of this attribute should be made with great caution.

"Flat spots"

Flat spots occur at the reflective boundary between different fluids, either gas–oil, gas–water, or water–oil contacts. These can be easy to detect in areas where the background stratigraphy is tilted, so the flat spot will stick out. However, if the stratigraphy is more or less flat, the fluid-related flat spot can be difficult to discover. Then, quantitative methods like AVO analysis can help to discriminate the fluid-related flat spot from the flat-lying lithostratigraphy.

One should be aware of several pitfalls when using flat spots as hydrocarbon indicators. Flat spots can be related to diagenetic events that are depth-dependent. The boundary between opal-A and opal-CT represents an impedance increase in the same way as for a fluid contact, and dry wells have been drilled on diagenetic flat spots. Clinoforms can appear as flat features even if the larger-scale stratigraphy is tilted. Other "false" flat spots include volcanic sills, paleo-contacts, sheet-flood deposits and flat bases of lobes and channels.

> **Pitfalls: False "flat spots"**
>
> One of the best DHIs to look for is a flat spot, the contact between gas and water, gas and oil, or oil and water. However, there are other causes that can give rise to flat spots:
> - Ocean bottom multiples
> - Flat stratigraphy. The bases of sand lobes especially tend to be flat.
> - Opal-A to opal-CT diagenetic boundary
> - Paleo-contacts, either related to diagenesis or residual hydrocarbon saturation
> - Volcanic sills
>
> Rigorous flat-spot analysis should include detailed rock physics analysis, and forward seismic modeling, as well as AVO analysis of real data (see Section 4.3.8).

Lithology, porosity and fluid ambiguities

The ultimate goal in seismic exploration is to discover and delineate hydrocarbon reservoirs. Seismic amplitude maps from 3D seismic data are often *qualitatively* interpreted in terms of lithology and fluids. However, rigorous rock physics modeling and analysis of available well-log data is required to discriminate fluid effects *quantitatively* from lithology effects (Chapters 1 and 2).

The "bright-spot" analysis method has often been unsuccessful because lithology effects rather than fluid effects set up the bright spot. The consequence is the drilling of dry holes. In order to reveal "pitfall" amplitude anomalies it is essential to investigate the rock physics properties from well-log data. However, in new frontier areas well-log data are sparse or lacking. This requires rock physics modeling constrained by reasonable geologic assumptions and/or knowledge about local compactional and depositional trends.

A common way to extract porosity from seismic data is to do acoustic impedance inversion. Increasing porosity can imply reduced acoustic impedance, and by extracting empirical porosity–impedance trends from well-log data, one can estimate porosity from the inverted impedance. However, this methodology suffers from several ambiguities. Firstly, a clay-rich shale can have very high porosities, even if the permeability is close to zero. Hence, a high-porosity zone identified by this technique may be shale. Moreover, the porosity may be constant while fluid saturation varies, and one simple impedance–porosity model may not be adequate for seismic porosity mapping.

In addition to lithology–fluid ambiguities, lithology–porosity ambiguities, and porosity–fluid ambiguities, we may have lithology–lithology ambiguities and fluid–fluid ambiguities. Sand and shale can have the same acoustic impedance, causing no reflectivity on a near-offset seismic section. This has been reported in several areas of the world (e.g. Zeng *et al.*, 1996; Avseth *et al.*, 2001b). It is often reported that fluvial channels or turbidite channels are dim on seismic amplitude maps, and the

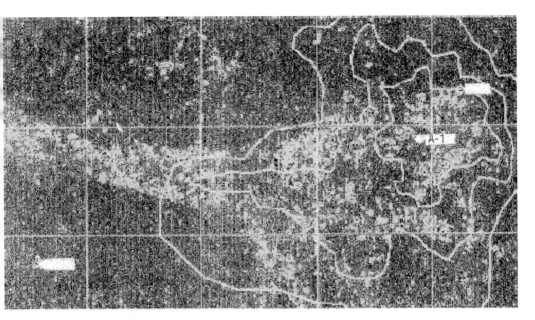

Plate 1.1 Seismic P–P amplitude map over a submarine fan. The amplitudes are sensitive to lithofacies and pore fluids, but the relation varies across the image because of the interplay of sedimentologic and diagenetic influences. Blue indicates low amplitudes, yellow and red high amplitudes.

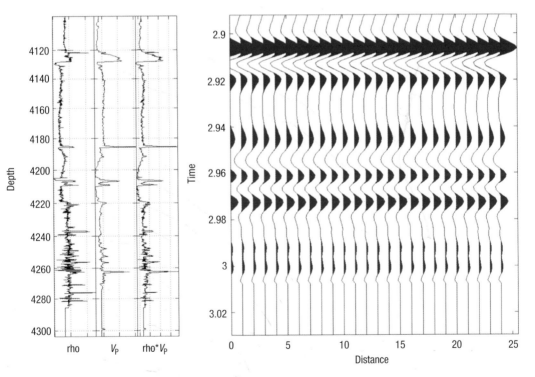

Plate 1.30 Top left, logs penetrating a sandy turbidite sequence; top right, normal-incidence synthetics with a 50 Hz Ricker wavelet. Bottom: increasing water saturation S_w from 10% to 90% (oil API 35, GOR 200) increases density and V_P (left), giving both amplitude and traveltime changes (right).

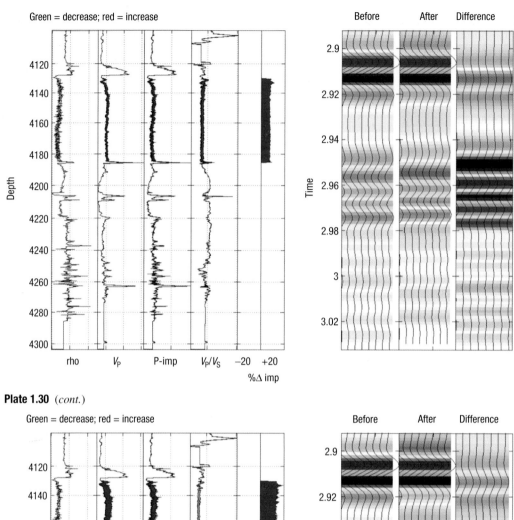

Plate 1.31 Top: reducing porosity by 3% by increasing cement. Bottom: reducing porosity by 3% by going to poor sorting.

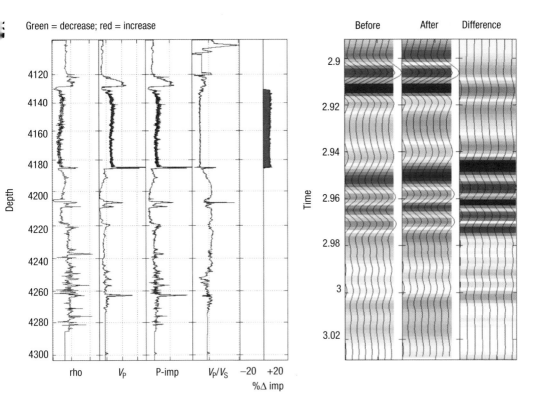

Plate 1.31 (cont.)

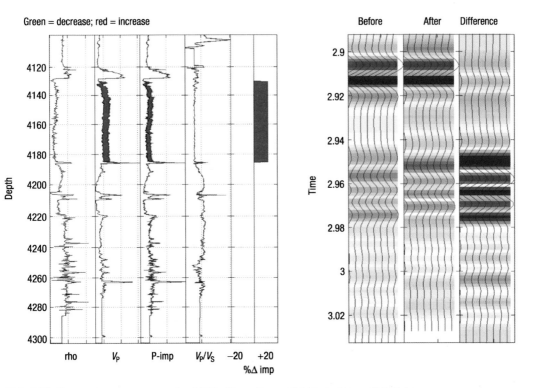

Plate 1.32 Decreasing pore pressure by 5 MPa (from $P_{\text{eff}} = 10$ MPa to $P_{\text{eff}} = 15$ MPa).

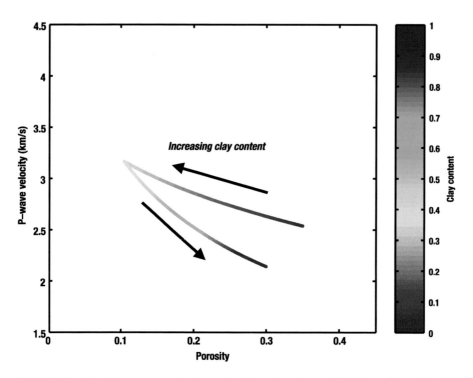

Plate 2.9 Trend in V_P versus porosity for sand–shale mix, using modified lower bound Hashin–Shtrikman model. We observe the same V-shape as observed in data by Marion (1992); see Figure 2.8.

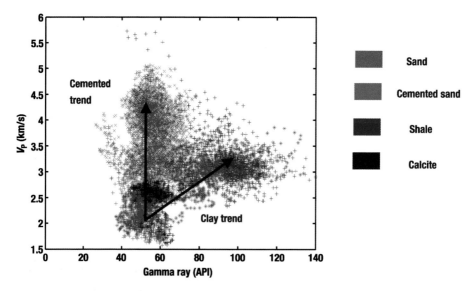

Plate 2.12 Example of velocity versus gamma-ray data for sands and silty shales, offshore Brazil. All the data are from a relatively thin depth interval, yet the sands span a great range in velocities. Some of the sands are completely unconsolidated whereas some are almost completely cemented. The local source of a carbonate reef (calcite) has caused some of the sands to become well cemented.

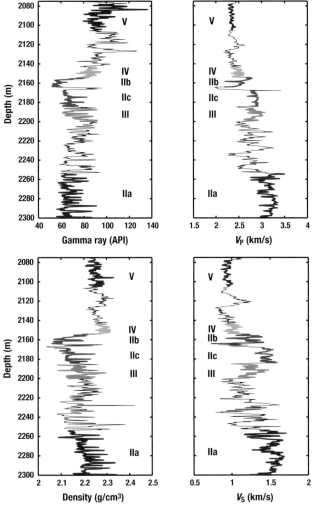

Plate 2.31 Lithofacies interpretation in type-well, representing training data for further classification.

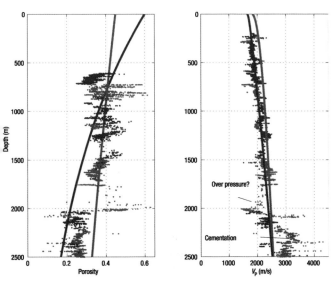

Plate 2.42 Porosity versus depth (left) and velocity versus depth (right). Empirical porosity–depth models for sands and shale have been calibrated to North Sea data (the calibration was done in a different well than the data here), and then Hertz–Mindlin theory has been applied to calculate expected velocity–depth trends. Deviations from these unconsolidated trends can be related to cementation (interval 2200–2500 m), overpressure (interval 1900–2000 m), or other factors like gas, borehole wash-outs, tectonic uplifts, calcareous or volcanic lithologies, etc.

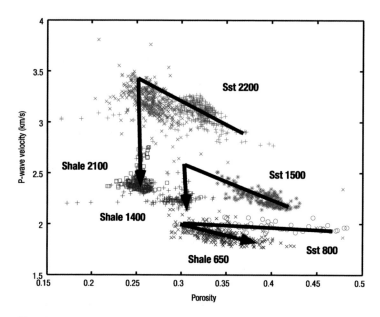

Plate 2.43 Velocity–porosity cross-plot for sands (red) and shales (blue) at different depth levels in the North Sea, with superimposed paths assumed to correspond to gradually increasing clay content, from clean sands (0%) to pure shales (100%).

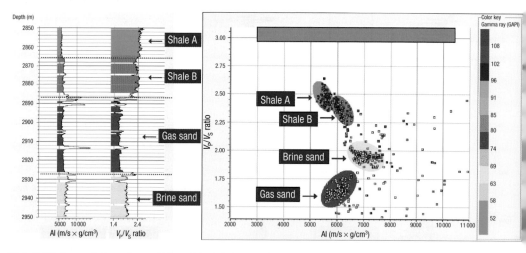

Plate 2.47 Acoustic impedance (AI) and V_P/V_S logs (left) and V_P/V_S vs. AI cross-plot (right). The logs are color-coded based on the populations defined in the cross-plot domain, and the cross-plot points are color-coded using the gamma-ray log (not shown). The interpretation is based on all available log data.

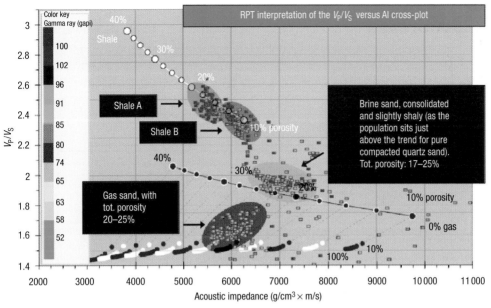

Plate 2.48 Cross-plot of V_P/V_S vs. AI, with theoretical rock physics trends for pure shale and clean compacted brine-filled quartz sand superimposed. The trends are plotted as functions of the total porosities. The effects of different gas saturations are added below the brine-sand trend. The color-coding is the same as in Plate 2.47.

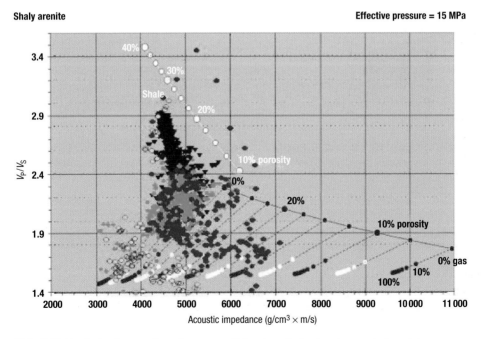

Plate 2.49 Rock physics templates for unconsolidated sands (top) and cemented sandstone (bottom). Data from the same basin and same stratigraphic level, but with different burial depths are superimposed on the templates. The fluid effects are completely different in the two different situations, because the deeper sands have been cemented, unlike the shallower sands. (Color codes: blue = shale, green = shaly sand, cyan = brine sand, red = oil sand, yellow = gas sand.)

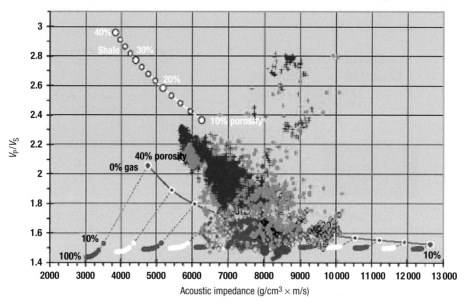

Plate 2.49 (*cont.*)

Plate 3.1 Iso-probability surface of 75% probability of oil-sand occurrence in a North Sea reservoir. The lateral extent is about 12 km along the long dimension. The total vertical extent is about 100 m. The probability estimates are obtained by combining well-log data, rock physics models, seismic impedance inversions, and statistical pattern recognition. This is a typical result from a statistical rock physics workflow.

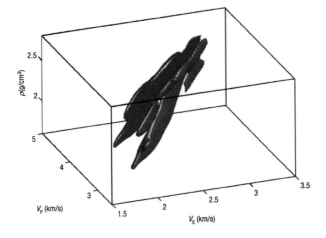

Plate 3.4 Iso-surfaces of trivariate nonparametric pdf estimate for V_P, V_S, and density. Blue indicates brine sands, and red indicates gas sands.

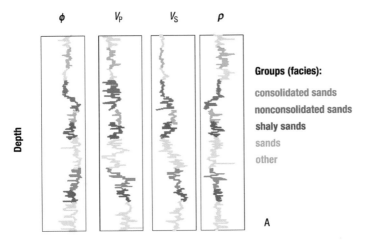

Groups (facies):
consolidated sands
nonconsolidated sands
shaly sands
sands
other

A

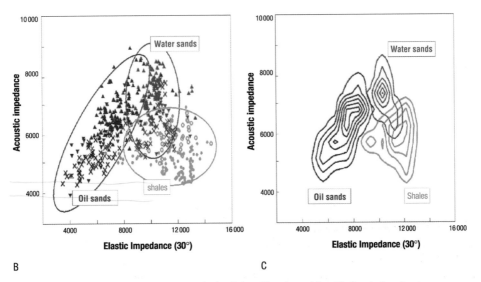

B

C

Plate 3.10 A, "Classified" well logs (each depth level has been identified as belonging to a particular facies). B, Acoustic and elastic (30°) impedance calculated theoretically from well logs. The color of each point corresponds to the facies to which it belongs. C, Probability density function (pdf) contours generated with the data of B extended by Monte Carlo simulation.

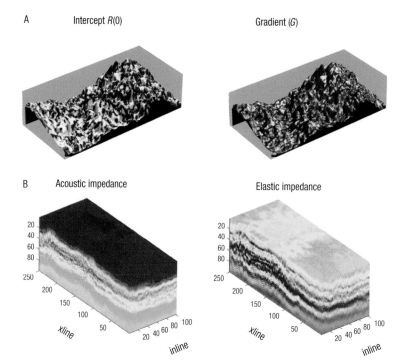

Plate 3.11 A, P-wave AVO attributes as defined by Shuey: $R(0)$ (intercept) and G (gradient) extracted from seismic data. The topography follows the traveltime interpretation of the seismic horizon along which the reflectivity and gradient were estimated from AVO analyses of pre-stack data. B, Acoustic and elastic impedance (at 30°) volumes. These two attributes respond to the reservoir interval properties. The far-offset elastic impedance implicitly contains shear-wave information. These were estimated by impedance inversion of partial stacks.

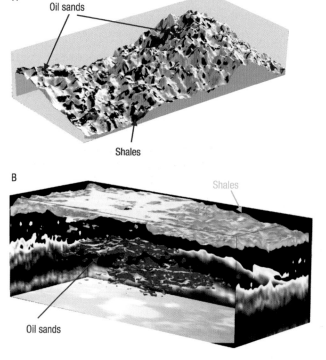

Plate 3.12 A, Areas with more probability of finding oil sands (red) and shales (blue), resulting from the Bayesian classification using the P-wave AVO attributes, $R(0)$ and G, shown in Plate 3.11A. The topography follows the interpretation (traveltime) of the seismic horizon (amplitudes) used to calculate the attributes. B, Iso-probability surfaces resulting from applying statistical classification (nonparametric Bayesian) using the seismic attributes acoustic and elastic impedance, shown in Plate 3.11B.

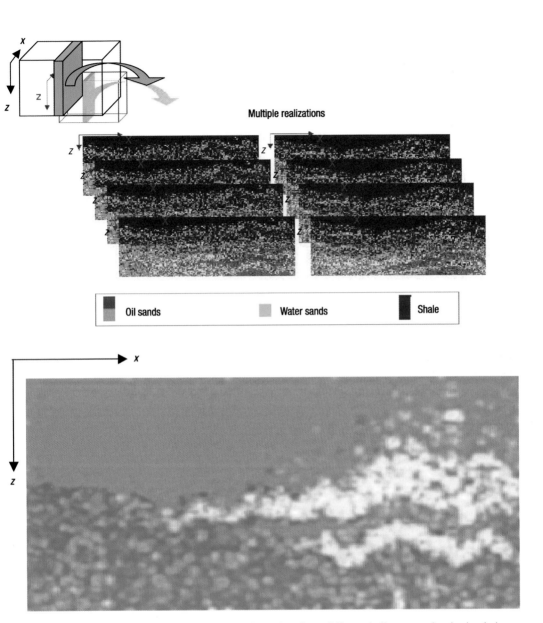

Plate 3.13 Top: A vertical section (at the same position) taken from different indicator stochastic simulation realizations (top). The red colors correspond to the oil-sand facies. Bottom: One vertical section from the 3D probability volume cube, showing probability of finding oil sands. These probabilities are obtained by averaging over multiple geostatistical simulations. The yellow color indicates areas with higher probabilities. The geostatistical simulation updates the seismically derived probability (e.g. Plate 3.12) by incorporating the spatial correlation and small-scale variability seen in well logs.

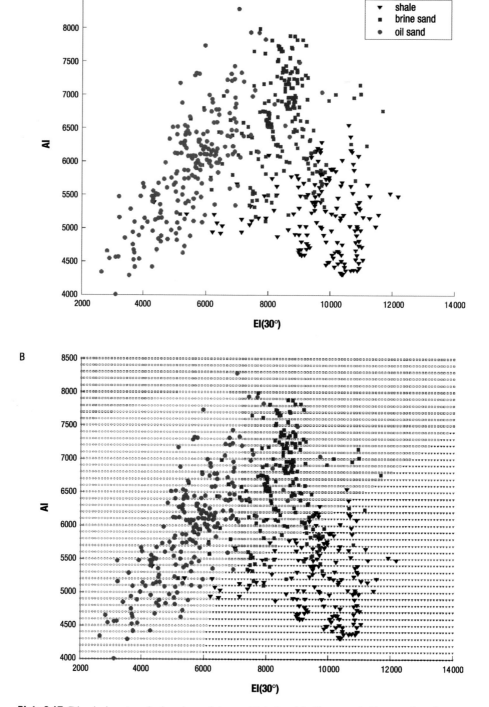

Plate 3.17 Discriminant analysis using minimum Mahalanobis distance. A, Scatter plot of far-offset elastic impedance [EI(30°)] and acoustic impedance (AI), showing the training data for three classes. B, Three different domains separated out by the discriminant function. A new unclassified observation of EI–AI will be classified according to the region in which it lies. Even in the training data all points will not be correctly classified because of overlap. We see that some blue points fall within the red domain and vice versa. The misclassification error rate for these data using the discriminant function plotted in B is about 13%.

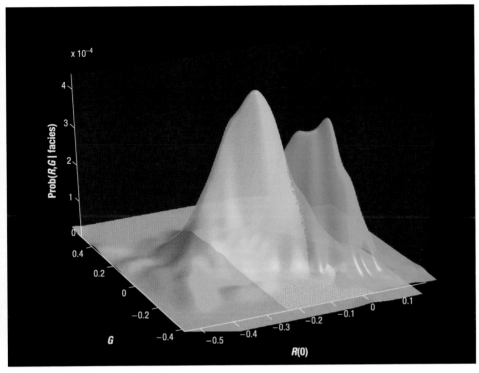

Plate 3.21 Class-conditioned pdf of $R(0)$ and G for different facies based on well-log data. The blue surface represents the pdf for brine sands, and the yellow surface represents the pdf for oil sands.

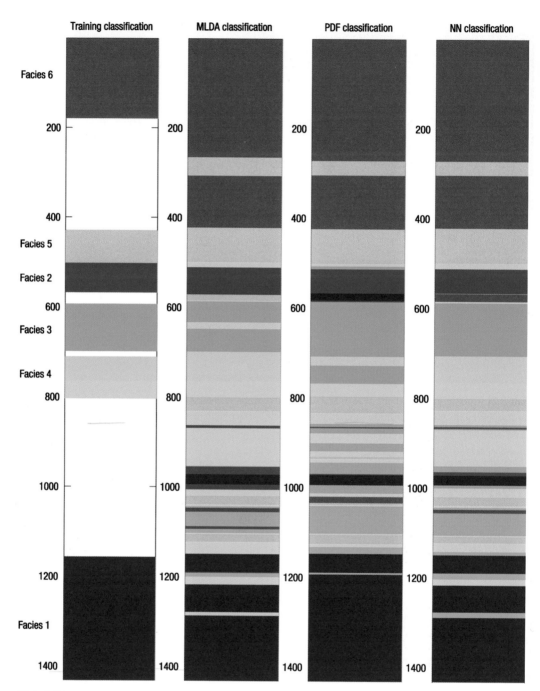

Plate 3.33 Comparing discriminant analysis, Bayes' rule and neural network classification results in a type-well. The depth axis is annotated with sample number. Sample number 1 is located at about 2075 m and sample number 1400 is located at about 2300 m.

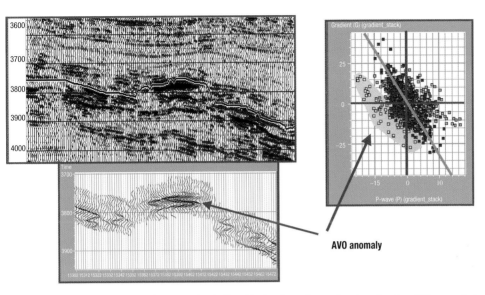

Plate 4.10 Inverted $R(0)$ and G from a seismic line. Projection of a predominantly class III anomaly is confined to the top of a local structural high.

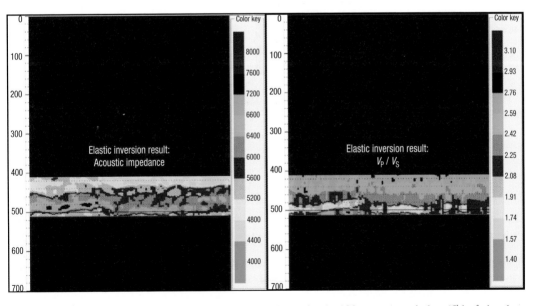

Plate 4.26 Elastic inversion results, AI (left) and V_P/V_S (right), for the 100-ms target window (Ødegård and Avseth, 2004).

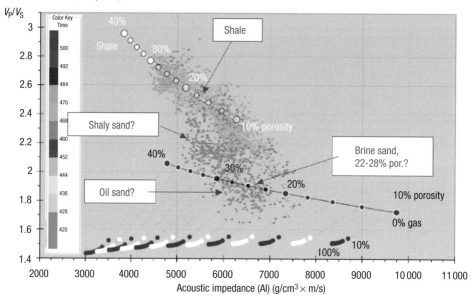

Plate 4.27 Cross-plot of V_P/V_S vs. AI for the elastic inversion results shown in Plate 4.26, with the selected rock physics template superimposed. A rough interpretation is indicated.

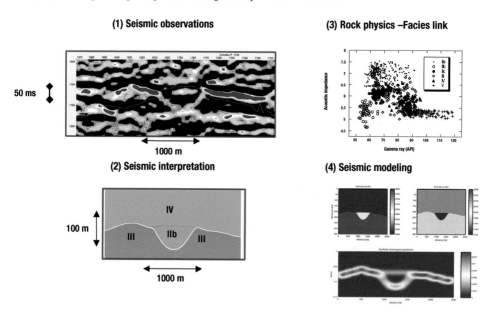

Plate 4.28 Simple seismic modeling exercise demonstrating the value of linking depositional facies to rock physics properties. (1) A turbidite channel-levee complex is observed in real seismic data (near-offset stack). (2) A plausible geologic interpretation in terms of seismic lithofacies (see Chapter 2 for definition of seismic lithofacies) is represented by clean channel sands that are laterally confined by heterolithic levee deposits. The channel axis is dim, whereas the levees are bright. Without good rock physics understanding, the seismic signatures observed in the real data might have been interpreted as a shale-filled channel axis with adjacent sand-rich levee deposits. However, rock physics analysis and seismic modeling show that the bright levees can be represented by hard heterolithics (shaly sands), while the dim channel can be filled with clean sands. This is an example from a Tertiary turbidite system in the North Sea.

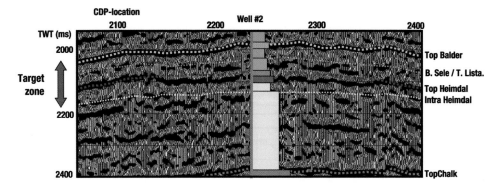

Plate 4.29 Seismic stack section intersecting a North Sea well (same well as in Plate 2.31), and superimposed facies observation at well location (from top to base: facies V = olive green, tuff = brown, facies IV = green, facies IIb-oil = orange, facies IIc-oil = red, facies III = light green, facies IIa-brine = yellow, chalk = blue). Seismic interpretation combined with well-log facies and rock physics analysis constrains the synthetic seismic modeling.

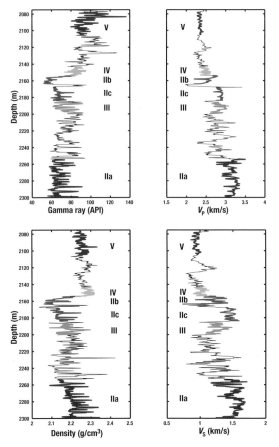

Plate 5.2 Lithofacies interpretation in type-well (Well 2). From upper left to lower right: gamma ray, V_P, density, and V_S logs. The zones that are not interpreted have been left out because of ambiguous and/or transitional characteristics.

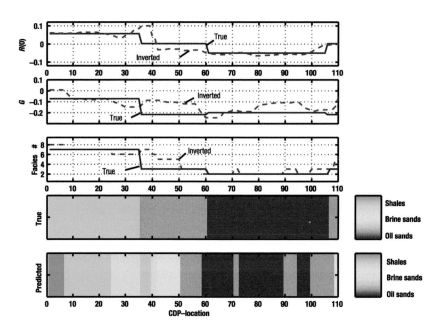

Plate 5.12 Seismic lithofacies prediction based on AVO inversion along the Top Heimdal horizon.

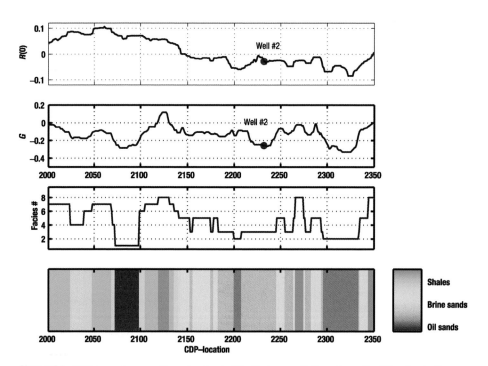

Plate 5.14 AVO inversion results and seismic lithofacies prediction along the 2D seismic line intersecting Well 2 (Figure 5.13).

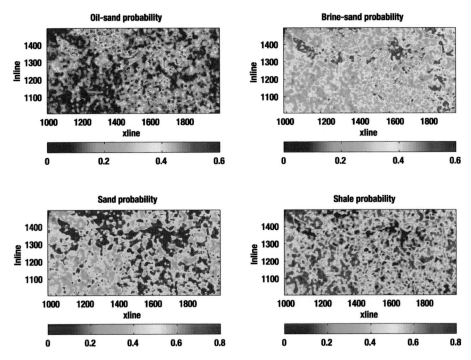

Plate 5.20 Probability maps of different grouped lithofacies.

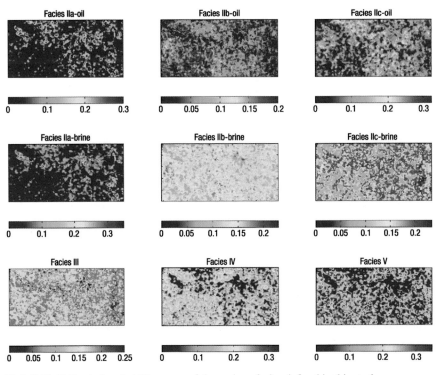

Plate 5.21 Estimated probability maps of the various facies defined in this study.

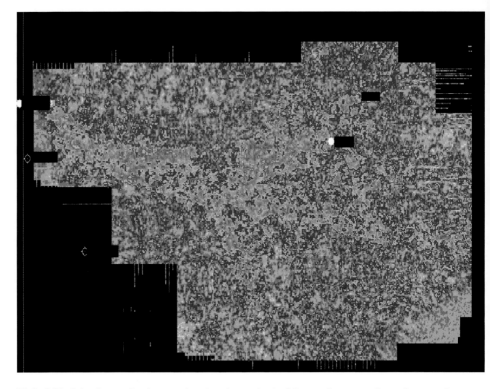

Plate 5.22 Seismic amplitude map showing the geological feature interpreted as a deep marine turbidite sediment system. What does it mean quantitatively in terms of lithology and pore fluid?

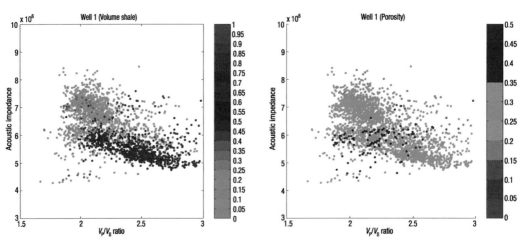

Plate 5.24 Rock properties as a function of seismic parameters: acoustic impedance vs. V_P/V_S color-coded to volume shale (left) and porosity (right). On the left the shalier points (purple and red) have both a lower acoustic impedance and higher V_P/V_S ratio than the sands (green). The porosity is relatively constant (right).

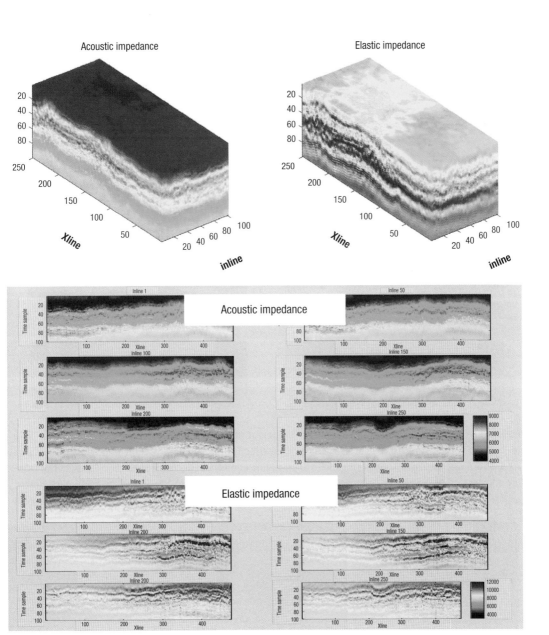

Plate 5.27 Acoustic (left) and elastic (right) impedance cubes (top) from inversion of near- and far-offset partial stacks, and examples of vertical sections from the inverted 3D cubes. (The 3D cubes and vertical sections have different color scales.)

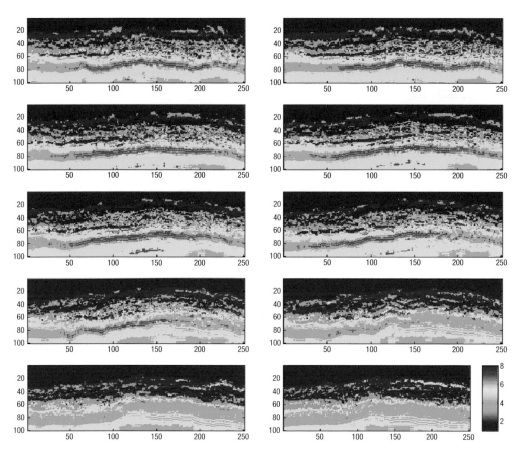

Plate 5.30 Facies classification from Mahalanobis linear distance method on a sub-cube of the seismic acoustic and elastic impedance data set. Vertical numbering is time samples from top of the sub-cube. Sample interval is 4 ms. Horizontal numbering refers to cross-line numbers in the original seismic data set. The facies are color-coded. Facies 1 and 2 are shaly facies, 3 to 5 are brine sands and 6 to 8 are the corresponding oil sands.

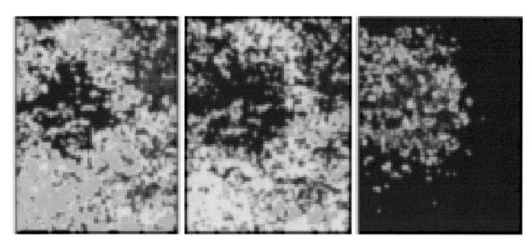

Plate 5.32 Horizontal time slices showing the conditional probabilities of occurrence of shales (left), brine sands (center), and oil sands (right) within the seismic grid. Blue indicates low probability and yellows and reds indicate high probabilities.

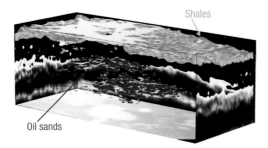

Plate 5.33 Iso-probability surfaces (80% probability) showing probable spatial distribution of oil-sand bodies (red) and the overlying shales (blue). Vertical dimension is 100 time samples at 4 ms per sample, and the longer horizontal dimension is about 250 cross-lines at 50-m spacing for a total distance of about 12.5 km.

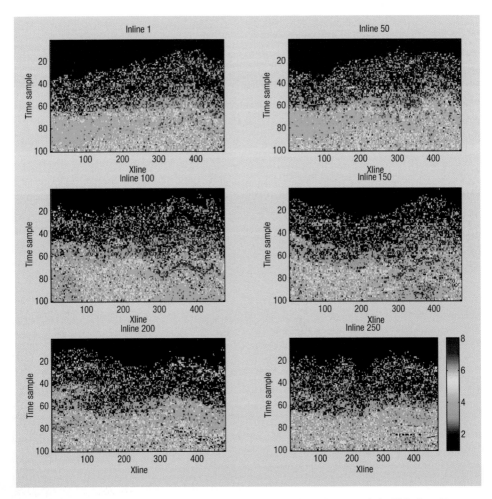

Plate 5.36 Simulated realizations of facies along different vertical sections. A geostatistical Markov–Bayes indicator simulation algorithm was used for generating the realizations. As is often the case with stochastic simulations, the realizations show more heterogeneity at finer spatial scales than present in the seismic impedances used as conditioning data for the simulations.

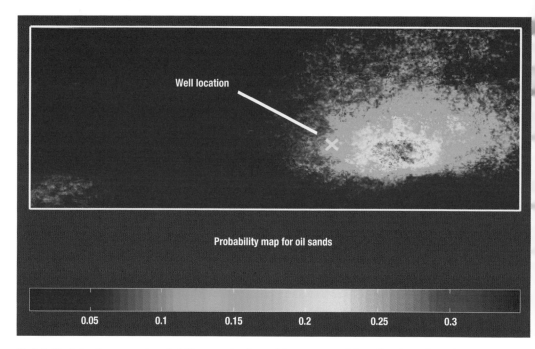

Plate 5.38 A depth-averaged probability map (estimated from near- and far-offset impedance interpretation) of oil sands, with the location of the new well. The choice of location of the well was independent of the probability maps from this analysis.

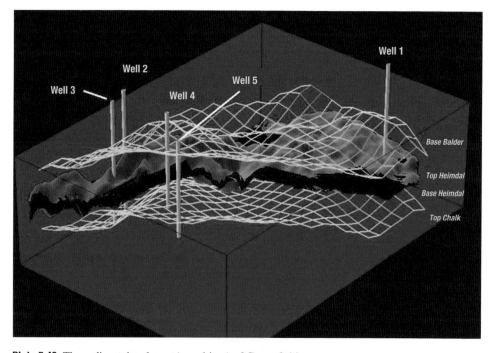

Plate 5.40 Three-dimensional map (traveltime) of Grane field.

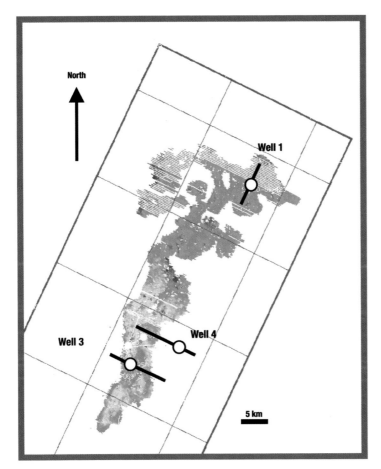

Plate 5.41 Map of Grane field. The reservoir extent is based on conventional seismic interpretation. Black lines indicate 2D seismic lines considered in this article. They intersect Well 1, Well 3, and Well 4.

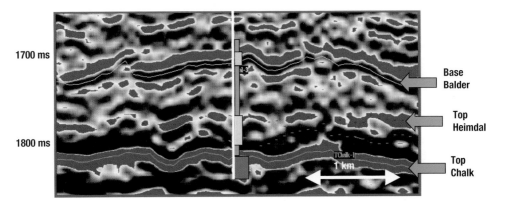

Plate 5.44 Seismic stack section intersecting Well 1. Important seismic reflectors include Base Balder, Top Heimdal, and Top Chalk. The lithofacies column in the type-well is superimposed.

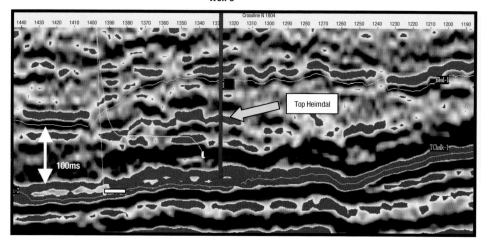

Plate 5.48 Seismic stack section intersecting Well 3. This well encountered thick reservoir sand with oil saturation (top reflector indicated by arrow). The internal positive reflector beneath the top is related to rock texture change. (CDP spacing is 25 m.)

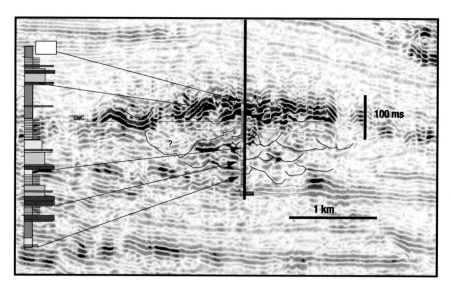

Plate 5.56 Seismic section intersecting a well penetrating a turbiditic gas and oil field, offshore West Africa. Gas was encountered in the upper sandy interval, whereas oil was found in the middle sand interval. Brine was encountered in the lower sandy interval. (Compare with well logs in Figure 5.57.)

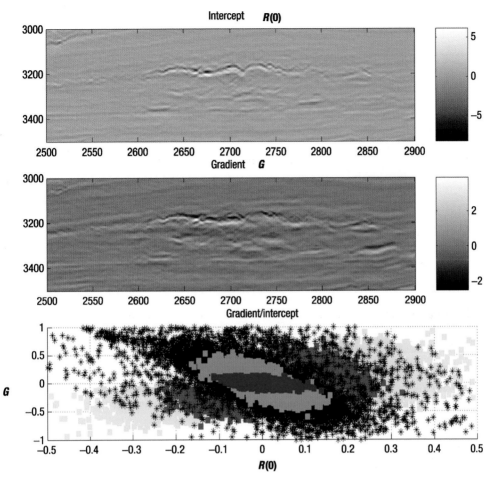

Plate 5.61 Calibration of AVO attributes extracted from pre-stack seismic data with modeled AVO scatter plots in Figure 5.60. The black stars represent the AVO attributes estimated from the real data (i.e., cross-plot of each sample in upper $R(0)$ versus lower G section), while the colored dots represent the modeled AVO responses for the various interface categories. The modeled heterolithics (i.e., shaly sands) and shales in green and blue, respectively, represent the modeled background trend, which has been calibrated with the background trend in the real data.

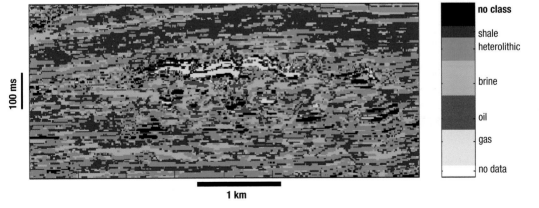

Plate 5.62 The most likely lithology and pore fluid along the seismic section in Plate 5.56.

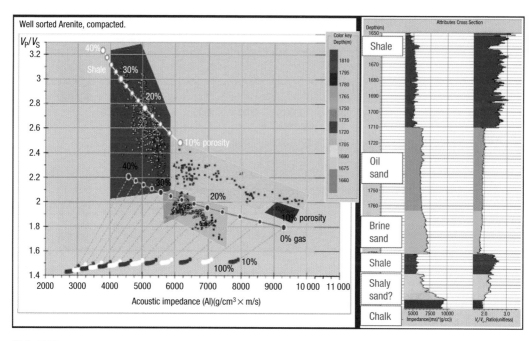

Plate 5.63 AI and V_P/V_S logs (right) and V_P/V_S vs. AI cross-plot (left). The logs are color-coded based on the populations defined in the cross-plot domain. The interpretation is based on all available log data.

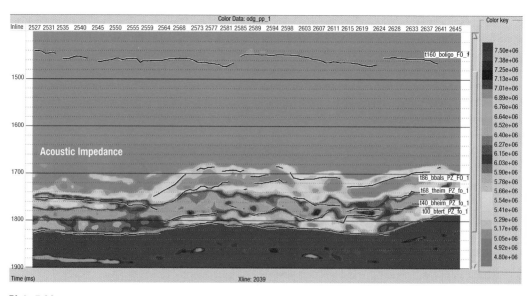

Plate 5.64 Cross-section of acoustic impedance estimated from pre-stack seismic inversion.

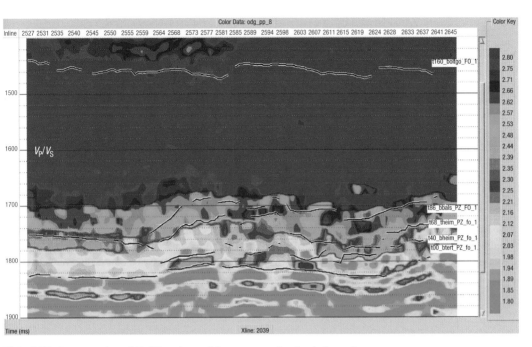

Plate 5.65 Cross-section of V_P/V_S estimated from pre-stack seismic inversion.

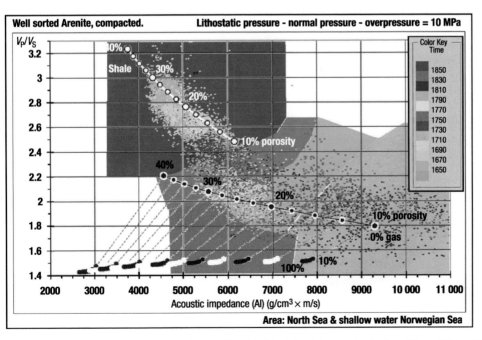

Plate 5.66 Cross-plot of acoustic impedance versus V_P/V_S derived from seismic data (Plates 5.64 and 5.65) superimposed onto the same RPT that was validated with well-log data (Plate 5.63). Note that we only include data within the target zone (1650–1850 ms).

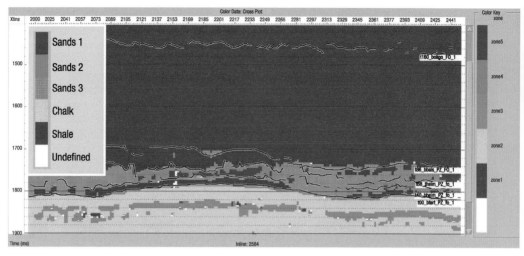

Plate 5.67 Vertical section of RPT classified lithofacies across the Grane field. We observe the good quality "sands" (Sands 1 and Sands 2) in the central part, with poorer quality sands and/or silty shales (Sands 3) to the sides. The sands are capped by massive shales. We also detect the presence of a thin shale layer between the base of the reservoir sands and underlying chalk deposits. Some marls within the chalks have been erroneously classified as Sands 2.

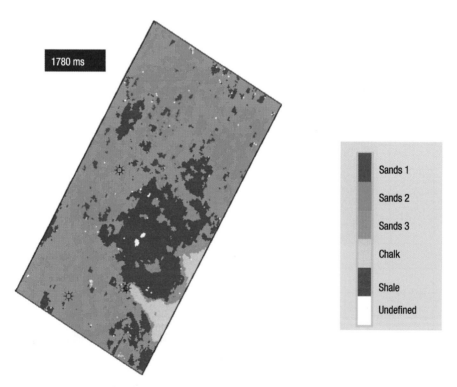

Plate 5.68 Time-slice map view of RPT classified lithofacies (at 1780 ms) showing the lateral distribution of various facies.

Seismic data analysis using RPT plots

- Well-log analysis / rock physics template verification

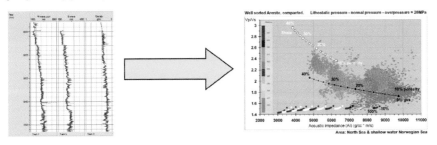

- Interpretation of elastic inversion results using the selected rock physics template(s)

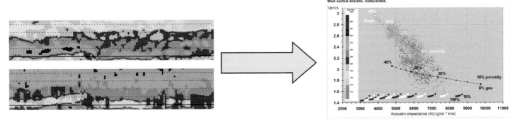

Plate 6.2 Workflow for RPT analysis.

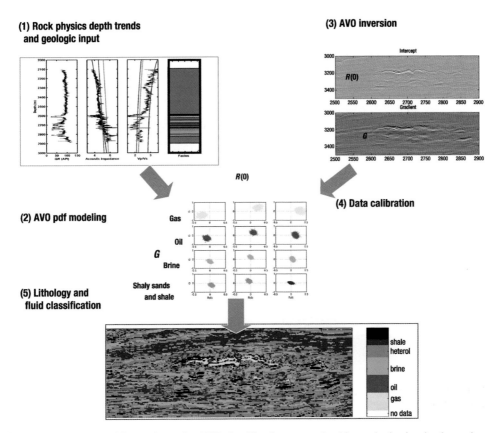

Plate 6.3 Overview of the workflow scheme for AVO classification constrained by rock physics depth trends.

Logs + rock physics + geology

Monte Carlo – probability distributions

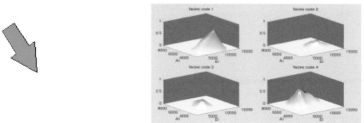

Seismic inversions – near- and far-offset attributes

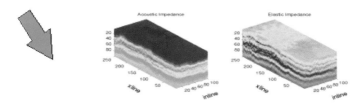

Integrated statistical classification – facies probability maps

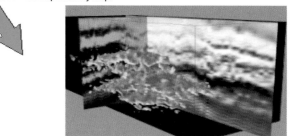

Plate 6.4 Schematic workflow for seismic reservoir characterization constrained by statistical rock physics and facies analysis of well-log data.

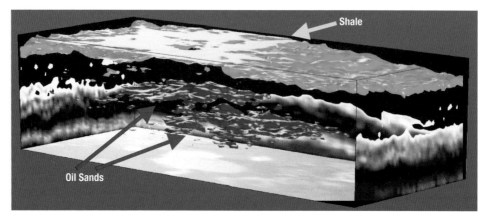

Plate 6.5 3D visualization of iso-probable surfaces (Mukerji *et al.*, 2001).

interpretation is usually that the channel is shale-filled. However, a clean sand filling in the channel can be transparent as well. A geological assessment of geometries indicating differential compaction above the channel may reveal the presence of sand. More advanced geophysical techniques such as offset-dependent reflectivity analysis may also reveal the sands. During conventional interpretation, one should interpret top reservoir horizons from limited-range stack sections, avoiding the pitfall of missing a dim sand on a near- or full-stack seismic section.

Facies interpretation

Lithology influence on amplitudes can often be recognized by the pattern of amplitudes as observed on horizon slices and by understanding how different lithologies occur within a depositional system. By relating lithologies to depositional systems we often refer to these as lithofacies or facies. The link between amplitude characteristics and depositional patterns makes it easier to distinguish lithofacies variations and fluid changes in amplitude maps.

Traditional seismic facies interpretation has been predominantly qualitative, based on seismic traveltimes. The traditional methodology consisted of purely visual inspection of geometric patterns in the seismic reflections (e.g., Mitchum *et al.*, 1977; Weimer and Link, 1991). Brown *et al.* (1981), by recognizing buried river channels from amplitude information, were amongst the first to interpret depositional facies from 3D seismic amplitudes. More recent and increasingly quantitative work includes that of Ryseth *et al.* (1998) who used acoustic impedance inversions to guide the interpretation of sand channels, and Zeng *et al.* (1996) who used forward modeling to improve the understanding of shallow marine facies from seismic amplitudes. Neri (1997) used neural networks to map facies from seismic pulse shape. Reliable quantitative lithofacies prediction from seismic amplitudes depends on establishing a link between rock physics properties and sedimentary facies. Sections 2.4 and 2.5 demonstrated how such links might be established. The case studies in Chapter 5 show how these links allow us to predict lithofacies from seismic amplitudes.

Stratigraphic interpretation

The subsurface is by nature a layered medium, where different lithologies or facies have been superimposed during geologic deposition. Seismic stratigraphic interpretation seeks to map geologic stratigraphy from geometric expression of seismic reflections in traveltime and space. Stratigraphic boundaries can be defined by different lithologies (facies boundaries) or by time (time boundaries). These often coincide, but not always. Examples where facies boundaries and time boundaries do not coincide are erosional surfaces cutting across lithostratigraphy, or the prograding front of a delta almost perpendicular to the lithologic surfaces within the delta.

There are several pitfalls when interpreting stratigraphy from traveltime information. First, the interpretation is based on layer boundaries or interfaces, that is, the contrasts between different strata or layers, and not the properties of the layers themselves. Two layers with different lithology can have the same seismic properties; hence, a lithostratigraphic boundary may not be observed. Second, a seismic reflection may occur without a lithology change (e.g., Hardage, 1985). For instance, a hiatus with no deposition within a shale interval can give a strong seismic signature because the shales above and below the hiatus have different characteristics. Similarily, amalgamated sands can yield internal stratigraphy within sandy intervals, reflecting different texture of sands from different depositional events. Third, seismic resolution can be a pitfall in seismic interpretation, especially when interpreting stratigraphic onlaps or downlaps. These are essential characteristics in seismic interpretation, as they can give information about the coastal development related to relative sea level changes (e.g., Vail *et al.*, 1977). However, pseudo-onlaps can occur if the thickness of individual layers reduces beneath the seismic resolution. The layer can still exist, even if the seismic expression yields an onlap.

Pitfalls

There are several pitfalls in conventional seismic stratigraphic interpretation that can be avoided if we use complementary quantitative techniques:
- Important lithostratigraphic boundaries between layers with very weak contrasts in seismic properties can easily be missed. However, if different lithologies are transparent in post-stack seismic data, they are normally visible in pre-stack seismic data. AVO analysis is a useful tool to reveal sands with impedances similar to capping shales (see Section 4.3).
- It is commonly believed that seismic events are time boundaries, and not necessarily lithostratigraphic boundaries. For instance, a hiatus within a shale may cause a strong seismic reflection if the shale above the hiatus is less compacted than the one below, even if the lithology is the same. Rock physics diagnostics of well-log data may reveal nonlithologic seismic events (see Chapter 2).
- Because of limited seismic resolution, false seismic onlaps can occur. The layer may still exist beneath resolution. Impedance inversion can improve the resolution, and reveal subtle stratigraphic features not observed in the original seismic data (see Section 4.4).

Quantitative interpretation of amplitudes can add information about stratigraphic patterns, and help us avoid some of the pitfalls mentioned above. First, relating lithology to seismic properties (Chapter 2) can help us understand the nature of reflections, and improve the geologic understanding of the seismic stratigraphy. Gutierrez (2001) showed how stratigraphy-guided rock physics analysis of well-log data improved the sequence stratigraphic interpretation of a fluvial system in Colombia using impedance inversion of 3D seismic data. Conducting impedance inversion of the seismic data will give us layer properties from interface properties, and an impedance cross-section can reveal stratigraphic features not observed on the original seismic section. Impedance inversion has the potential to guide the stratigraphic interpretation, because it is less oscillatory than the original seismic data, it is more readily correlated to well-log data, and it tends to average out random noise, thereby improving the detectability of laterally weak reflections (Gluck *et al.*, 1997). Moreover, frequency-band-limited impedance inversion can improve on the stratigraphic resolution, and the seismic interpretation can be significantly modified if the inversion results are included in the interpretation procedure. For brief explanations of different types of impedance inversions, see Section 4.4. Forward seismic modeling is also an excellent tool to study the seismic signatures of geologic stratigraphy (see Section 4.5).

Layer thickness and net-to-gross from seismic amplitude

As mentioned in the previous section, we can extract layer thickness from seismic amplitudes. As depicted in Figure 4.2, the relationship is only linear for thin layers in pinch-out zones or in sheet-like deposits, so one should avoid correlating layer thickness to seismic amplitudes in areas where the top and base of sands are resolved as separate reflectors in the seismic data.

Meckel and Nath (1977) found that, for sands embedded in shale, the amplitude would depend on the net sand present, given that the thickness of the entire sequence is less than $\lambda/4$. Brown (1996) extended this principle to include beds thicker than the tuning thickness, assuming that individual sand layers are below tuning and that the entire interval of interbedded sands has a uniform layering. Brown introduced the "composite amplitude" defined as the absolute value summation of the top reflection amplitude and the base reflection amplitude of a reservoir interval. The summation of the absolute values of the top and the base emphasizes the effect of the reservoir and reduces the effect of the embedding material.

Zeng *et al.* (1996) studied the influence of reservoir thickness on seismic signal and introduced what they referred to as *effective reflection strength*, applicable to layers thinner than the tuning thickness:

$$R_e = \frac{Z_{ss} - Z_{sh}}{Z_{sh}} \cdot h \tag{4.3}$$

where Z_{ss} is the sandstone impedance, Z_{sh} is the average shale impedance and h is the layer thickness. A more common way to extract layer thickness from seismic amplitudes is by linear regression of relative amplitude versus net sand thickness as observed at wells that are available. A recent case study showing the application to seismic reservoir mapping was provided by Hill and Halvatis (2001).

Vernik *et al.* (2002) demonstrated how to estimate net-to-gross from P- and S-impedances for a turbidite system. From acoustic impedance (AI) versus shear impedance (SI) cross-plots, the net-to-gross can be calculated with the following formulas:

$$N/G = \frac{\int_{Z_{top}}^{Z_{base}} V_{sand} \, dZ}{\Delta Z} \qquad (4.4)$$

where V_{sand} is the oil-sand fraction given by:

$$V_{sand} = \frac{SI - bAI - a_0}{a_1 - a_0} \qquad (4.5)$$

where b is the average slope of the shale slope (b_0) and oil-sand slope (b_1), whereas a_0 and a_1 are the respective intercepts in the AI–SI cross-plot.

> Calculation of reservoir thickness from seismic amplitude should be done only in areas where sands are expected to be thinner than the tuning thickness, that is a quarter of a wavelength, and where well-log data show evidence of good correlation between net sand thickness and relative amplitude.
>
> It can be difficult to discriminate layer thickness changes from lithology and fluid changes. In relatively soft sands, the impact of increasing porosity and hydrocarbon saturation tends to increase the seismic amplitude, and therefore works in the same "direction" to layer thickness. However, in relatively hard sands, increasing porosity and hydrocarbon saturation tend to decrease the relative amplitude and therefore work in the opposite "direction" to layer thickness.

4.3 AVO analysis

In 1984, 12 years after the bright-spot technology became a commercial tool for hydrocarbon prediction, Ostrander published a break-through paper in *Geophysics* (Ostrander, 1984). He showed that the presence of gas in a sand capped by a shale would cause an amplitude variation with offset in pre-stack seismic data. He also found that this change was related to the reduced Poisson's ratio caused by the presence of gas. Then, the year after, Shuey (1985) confirmed mathematically via approximations of the Zoeppritz equations that Poisson's ratio was the elastic constant most directly related to the offset-dependent reflectivity for incident angles up to 30°. AVO technology, a commercial tool for the oil industry, was born.

The AVO technique became very popular in the oil industry, as one could physically explain the seismic amplitudes in terms of rock properties. Now, bright-spot anomalies could be investigated before stack, to see if they also had AVO anomalies (Figure 4.3). The technique proved successful in certain areas of the world, but in many cases it was not successful. The technique suffered from ambiguities caused by lithology effects,

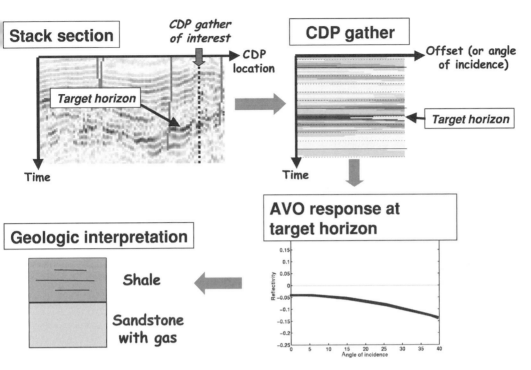

Figure 4.3 Schematic illustration of the principles in AVO analysis.

tuning effects, and overburden effects. Even processing and acquisition effects could cause false AVO anomalies. But in many of the failures, it was not the technique itself that failed, but the use of the technique that was incorrect. Lack of shear-wave velocity information and the use of too simple geologic models were common reasons for failure. Processing techniques that affected near-offset traces in CDP gathers in a different way from far-offset traces could also create false AVO anomalies. Nevertheless, in the last decade we have observed a revival of the AVO technique. This is due to the improvement of 3D seismic technology, better pre-processing routines, more frequent shear-wave logging and improved understanding of rock physics properties, larger data capacity, more focus on cross-disciplinary aspects of AVO, and last but not least, more awareness among the users of the potential pitfalls. The technique provides the seismic interpreter with more data, but also new physical dimensions that add information to the conventional interpretation of seismic facies, stratigraphy and geomorphology.

In this section we describe the practical aspects of AVO technology, the potential of this technique as a direct hydrocarbon indicator, and the pitfalls associated with this technique. Without going into the theoretical details of wave theory, we address issues related to acquisition, processing and interpretation of AVO data. For an excellent overview of the history of AVO and the theory behind this technology, we refer the reader to Castagna (1993). We expect the future application of AVO to expand on today's common AVO cross-plot analysis and hence we include overviews of important contributions from the literature, include tuning, attenuation and anisotropy effects on AVO. Finally, we elaborate on probabilistic AVO analysis constrained by rock physics models. These comprise the methodologies applied in case studies 1, 3 and 4 in Chapter 5.

4.3.1 The reflection coefficient

Analysis of AVO, or amplitude variation with offset, seeks to extract rock parameters by analyzing seismic amplitude as a function of offset, or more correctly as a function of reflection angle. The reflection coefficient for plane elastic waves as a function of reflection angle at a single interface is described by the complicated Zoeppritz equations (Zoeppritz, 1919). For analysis of P-wave reflections, a well-known approximation is given by Aki and Richards (1980), assuming weak layer contrasts:

$$R(\theta_1) \approx \frac{1}{2}\left(1 - 4p^2 V_S^2\right)\frac{\Delta \rho}{\rho} + \frac{1}{2\cos^2\theta}\frac{\Delta V_P}{V_P} - 4p^2 V_S^2 \frac{\Delta V_S}{V_S} \tag{4.6}$$

where:

$$p = \frac{\sin\theta_1}{V_{P1}} \qquad \theta = (\theta_1 + \theta_2)/2 \approx \theta_1$$
$$\Delta\rho = \rho_2 - \rho_1 \qquad \rho = (\rho_2 + \rho_1)/2$$
$$\Delta V_P = V_{P2} - V_{P1} \qquad V_P = (V_{P2} + V_{P1})/2$$
$$\Delta V_S = V_{S2} - V_{S1} \qquad V_S = (V_{S2} + V_{S1})/2$$

In the formulas above, p is the ray parameter, θ_1 is the angle of incidence, and θ_2 is the transmission angle; V_{P1} and V_{P2} are the P-wave velocities above and below a given interface, respectively. Similarly, V_{S1} and V_{S2} are the S-wave velocities, while ρ_1 and ρ_2 are densities above and below this interface (Figure 4.4).

The approximation given by Aki and Richards can be further approximated (Shuey, 1985):

$$R(\theta) \approx R(0) + G\sin^2\theta + F(\tan^2\theta - \sin^2\theta) \tag{4.7}$$

where

$$R(0) = \frac{1}{2}\left(\frac{\Delta V_P}{V_P} + \frac{\Delta \rho}{\rho}\right)$$

$$G = \frac{1}{2}\frac{\Delta V_P}{V_P} - 2\frac{V_S^2}{V_P^2}\left(\frac{\Delta \rho}{\rho} + 2\frac{\Delta V_S}{V_S}\right)$$

$$= R(0) - \frac{\Delta \rho}{\rho}\left(\frac{1}{2} + \frac{2V_S^2}{V_P^2}\right) - \frac{4V_S^2}{V_P^2}\frac{\Delta V_S}{V_S}$$

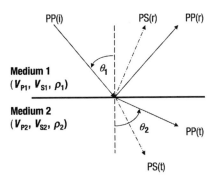

Figure 4.4 Reflections and transmissions at a single interface between two elastic half-space media for an incident plane P-wave, PP(i). There will be both a reflected P-wave, PP(r), and a transmitted P-wave, PP(t). Note that there are wave mode conversions at the reflection point occurring at nonzero incidence angles. In addition to the P-waves, a reflected S-wave, PS(r), and a transmitted S-wave, PS(t), will be produced.

and

$$F = \frac{1}{2}\frac{\Delta V_P}{V_P}$$

This form can be interpreted in terms of different angular ranges, where $R(0)$ is the normal-incidence reflection coefficient, G describes the variation at intermediate offsets and is often referred to as the AVO gradient, whereas F dominates the far offsets, near critical angle. Normally, the range of angles available for AVO analysis is less than 30–40°. Therefore, we only need to consider the two first terms, valid for angles less than 30° (Shuey, 1985):

$$R(\theta) \approx R(0) + G\sin^2\theta \tag{4.8}$$

The zero-offset reflectivity, $R(0)$, is controlled by the contrast in acoustic impedance across an interface. The gradient, G, is more complex in terms of rock properties, but from the expression given above we see that not only the contrasts in V_P and density affect the gradient, but also V_S. The importance of the V_P/V_S ratio (or equivalently the Poisson's ratio) on the offset-dependent reflectivity was first indicated by Koefoed (1955). Ostrander (1984) showed that a gas-filled formation would have a very low Poisson's ratio compared with the Poisson's ratios in the surrounding nongaseous formations. This would cause a significant increase in positive amplitude versus angle at the bottom of the gas layer, and a significant increase in negative amplitude versus angle at the top of the gas layer.

4.3.2 The effect of anisotropy

Velocity anisotropy ought to be taken into account when analyzing the amplitude variation with offset (AVO) response of gas sands encased in shales. Although it is generally thought that the anisotropy is weak (10–20%) in most geological settings (Thomsen, 1986), some effects of anisotropy on AVO have been shown to be dramatic using shale/sand models (Wright, 1987). In some cases, the sign of the AVO slope or rate of change of amplitude with offset can be reversed because of anisotropy in the overlying shales (Kim *et al.*, 1993; Blangy, 1994).

The elastic stiffness tensor **C** in transversely isotropic (TI) media can be expressed in compact form as follows:

$$\mathbf{C} = \begin{vmatrix} C_{11} & (C_{11} - 2C_{66}) & C_{13} & 0 & 0 & 0 \\ (C_{11} - 2C_{66}) & C_{11} & C_{13} & 0 & 0 & 0 \\ C_{13} & C_{13} & C_{33} & 0 & 0 & 0 \\ 0 & 0 & 0 & C_{44} & 0 & 0 \\ 0 & 0 & 0 & 0 & C_{44} & 0 \\ 0 & 0 & 0 & 0 & 0 & C_{66} \end{vmatrix}$$

where $C_{66} = \frac{1}{2}(C_{11} - C_{12})$ (4.9)

and where the 3-axis (z-axis) lies along the axis of symmetry.

The above 6 × 6 matrix is symmetric, and has five independent components, C_{11}, C_{13}, C_{33}, C_{44}, and C_{66}. For weak anisotropy, Thomsen (1986) expressed three anisotropic parameters, ε, γ and δ, as a function of the five elastic components, where

$$\varepsilon = \frac{C_{11} - C_{33}}{2C_{33}} \tag{4.10}$$

$$\gamma = \frac{C_{66} - C_{44}}{2C_{44}} \tag{4.11}$$

$$\delta = \frac{(C_{13} + C_{44})^2 - (C_{33} - C_{44})^2}{2C_{33}(C_{33} - C_{44})} \tag{4.12}$$

The constant ε can be seen to describe the fractional difference of the P-wave velocities in the vertical and horizontal directions:

$$\varepsilon = \frac{V_P(90°) - V_P(0°)}{V_P(0°)} \tag{4.13}$$

and therefore best describes what is usually referred to as "P-wave anisotropy."

In the same manner, the constant γ can be seen to describe the fractional difference of SH-wave velocities between vertical and horizontal directions, which is equivalent to the difference between the vertical and horizontal polarizations of the horizontally propagating S-waves:

$$\gamma = \frac{V_{SH}(90°) - V_{SV}(90°)}{V_{SV}(90°)} = \frac{V_{SH}(90°) - V_{SH}(0°)}{V_{SH}(0°)} \tag{4.14}$$

The physical meaning of δ is not as clear as ε and γ, but δ is the most important parameter for normal moveout velocity and reflection amplitude.

Under the plane wave assumption, Daley and Hron (1977) derived theoretical formulas for reflection and transmission coefficients in TI media. The P–P reflectivity in the equation can be decomposed into isotropic and anisotropic terms as follows:

$$R_{PP}(\theta) = R_{IPP}(\theta) + R_{APP}(\theta) \tag{4.15}$$

Assuming weak anisotropy and small offsets, Banik (1987) showed that the anisotropic term can be simply expressed as follows:

$$R_{APP}(\theta) \approx \frac{\sin^2\theta}{2} \Delta\delta \tag{4.16}$$

Blangy (1994) showed the effect of a transversely isotropic shale overlying an isotropic gas sand on offset-dependent reflectivity, for the three different types of gas sands. He found that hard gas sands overlain by a soft TI shale exhibited a larger decrease in positive amplitude with offset than if the shale had been isotropic. Similarly, soft gas sands overlain by a relatively hard TI shale exhibited a larger increase in negative amplitude with offset than if the shale had been isotropic. Furthermore, it is possible for a soft isotropic water sand to exhibit an "unexpectedly" large AVO effect if the overlying shale is sufficiently anisotropic.

4.3.3 The effect of tuning on AVO

As mentioned in the previous section, seismic interference or event tuning can occur as closely spaced reflectors interfere with each other. The relative change in traveltime between the reflectors decreases with increased traveltime and offset. The traveltime hyperbolas of the closely spaced reflectors will therefore become even closer at larger offsets. In fact, the amplitudes may interfere at large offsets even if they do not at small offsets. The effect of tuning on AVO has been demonstrated by Juhlin and Young (1993), Lin and Phair (1993), Bakke and Ursin (1998), and Dong (1998), among others.

Juhlin and Young (1993) showed that thin layers embedded in a homogeneous rock can produce a significantly different AVO response from that of a simple interface of the same lithology. They showed that, for weak contrasts in elastic properties across the layer boundaries, the AVO response of a thin bed may be approximated by modeling it as an interference phenomenon between plane P-waves from a thin layer. In this case thin-bed tuning affects the AVO response of a high-velocity layer embedded in a homogeneous rock more than it affects the response of a low-velocity layer.

Lin and Phair (1993) suggested the following expression for the amplitude variation with angle (AVA) response of a thin layer:

$$R_t(\theta) = \omega_0 \Delta T(0) \cos\theta \cdot R(\theta) \qquad (4.17)$$

where ω_0 is the dominant frequency of the wavelet, $\Delta T(0)$ is the two-way traveltime at normal incidence from the top to the base of the thin layer, and $R(\theta)$ is the reflection coefficient from the top interface.

Bakke and Ursin (1998) extended the work by Lin and Phair by introducing tuning correction factors for a general seismic wavelet as a function of offset. If the seismic response from the top of a thick layer is:

$$d(t, y) = R(y)p(t) \qquad (4.18)$$

where $R(y)$ is the primary reflection as a function of offset y, and $p(t)$ is the seismic pulse as a function of time t, then the response from a thin layer is

$$d(t, y) \approx R(y)\Delta T(0) C(y) p'(t) \qquad (4.19)$$

where $p'(t)$ is the time derivative of the pulse, $\Delta T(0)$ is the traveltime thickness of the thin layer at zero offset, and $C(y)$ is the offset-dependent AVO tuning factor given by

$$C(y) = \frac{T(0)}{T(y)}\left[1 + \frac{V_{RMS}^2 - V^2}{2T(0)^2 V_{RMS}^4} y^2\right] \qquad (4.20)$$

where $T(0)$ and $T(y)$ are the traveltimes at zero offset and at a given nonzero offset, respectively. The root-mean-square velocity V_{RMS}, is defined along a ray path:

$$V_{RMS}^2 = \frac{\int_0^t V^2(t)\,dt}{\int_0^t dt} \qquad (4.21)$$

For small velocity contrasts ($V_{RMS} \approx V$), the last term in equation (4.20) can be ignored, and the AVO tuning factor can be approximated as

$$C(y) \approx \frac{T(0)}{T(y)} \qquad (4.22)$$

For large contrast in elastic properties, one ought to include contributions from P-wave multiples and converted shear waves. The locally converted shear wave is often neglected in ray-tracing modeling when reproduction of the AVO response of potential hydrocarbon reservoirs is attempted. Primaries-only ray-trace modeling in which the Zoeppritz equations describe the reflection amplitudes is most common. But primaries-only Zoeppritz modeling can be very misleading, because the locally converted shear waves often have a first-order effect on the seismic response (Simmons and Backus, 1994). Interference between the converted waves and the primary reflections from the

4.3 AVO analysis

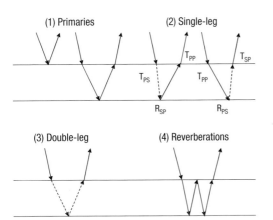

Figure 4.5 Converted S-waves and multiples that must be included in AVO modeling when we have thin layers, causing these modes to interfere with the primaries. (1) Primary reflections; (2) single-leg shear waves; (3) double-leg shear wave; and (4) primary reverberations. (After Simmons and Backus, 1994.) T_{PS} = transmitted S-wave converted from P-wave, R_{SP} = reflected P-wave converted from S-wave, etc.

base of the layers becomes increasingly important as the layer thicknesses decrease. This often produces a seismogram that is different from one produced under the primaries-only Zoeppritz assumption. In this case, one should use full elastic modeling including the converted wave modes and the intrabed multiples. Martinez (1993) showed that surface-related multiples and P-to-SV-mode converted waves can interfere with primary pre-stack amplitudes and cause large distortions in the AVO responses. Figure 4.5 shows the ray images of converted S-waves and multiples within a layer.

4.3.4 Structural complexity, overburden and wave propagation effects on AVO

Structural complexity and heterogeneities at the target level as well as in the overburden can have a great impact on the wave propagation. These effects include focusing and defocusing of the wave field, geometric spreading, transmission losses, interbed and surface multiples, P-wave to vertically polarized S-wave mode conversions, and anelastic attenuation. The offset-dependent reflectivity should be corrected for these wave propagation effects, via robust processing techniques (see Section 4.3.6). Alternatively, these effects should be included in the AVO modeling (see Sections 4.3.7 and 4.5). Chiburis (1993) provided a simple but robust methodology to correct for overburden effects as well as certain acquisition effects (see Section 4.3.5) by normalizing a target horizon amplitude to a reference horizon amplitude. However, in more recent years there have been several more extensive contributions in the literature on amplitude-preserved imaging in complex areas and AVO corrections due to overburden effects, some of which we will summarize below.

AVO in structurally complex areas

The Zoeppritz equations assume a single interface between two semi-infinite layers with infinite lateral extent. In continuously subsiding basins with relatively flat stratigraphy (such as Tertiary sediments in the North Sea), the use of Zoeppritz equations should be valid. However, complex reservoir geology due to thin beds, vertical heterogeneities, faulting and tilting will violate the Zoeppritz assumptions. Resnick *et al.* (1987) discuss the effects of geologic dip on AVO signatures, whereas MacLeod and Martin (1988) discuss the effects of reflector curvature. Structural complexity can be accounted for by doing pre-stack depth migration (PSDM). However, one should be aware that several PSDM routines obtain reliable structural images without preserving the amplitudes. Grubb *et al.* (2001) examined the sensitivity both in structure and amplitude related to velocity uncertainties in PSDM migrated images. They performed an amplitude-preserving (weighted Kirchhoff) PSDM followed by AVO inversion. For the AVO signatures they evaluated both the uncertainty in AVO cross-plots and uncertainty of AVO attribute values along given structural horizons.

AVO effects due to scattering attenuation in heterogeneous overburden

Widmaier *et al.* (1996) showed how to correct a target AVO response for a thinly layered overburden. A thin-bedded overburden will generate velocity anisotropy and transmission losses due to scattering attenuation, and these effects should be taken into account when analyzing a target seismic reflector. They combined the generalized O'Doherty–Anstey formula (Shapiro *et al.*, 1994a) with amplitude-preserving migration/inversion algorithms and AVO analysis to compensate for the influence of thin-bedded layers on traveltimes and amplitudes of seismic data. In particular, they demonstrated how the estimation of zero-offset amplitude and AVO gradient can be improved by correcting for scattering attenuation due to thin-bed effects. Sick *et al.* (2003) extended Widmaier's work and provided a method of compensating for the scattering attenuation effects of randomly distributed heterogeneities above a target reflector. The generalized O'Doherty–Anstey formula is an approximation of the angle-dependent, time-harmonic effective transmissivity T for scalar waves (P-waves in acoustic 1D medium or SH-waves in elastic 1D medium) and is given by

$$T(f) \propto T_0 e^{-(\alpha(f,\theta)+i\beta(f,\theta))L} \tag{4.23}$$

where f is the frequency and α and β are the angle- and frequency-dependent scattering attenuation and phase shift coefficients, respectively. The angle θ is the initial angle of an incident plane wave at the top surface of a thinly layered composite stack; L is the thickness of the thinly layered stack; T_0 denotes the transmissivity for a homogeneous isotropic reference medium that causes a phase shift. Hence, the equation above represents the relative amplitude and phase distortions caused by the thin layers with regard to the reference medium. Neglecting the quantity T_0 which describes the transmission response for a homogeneous isotropic reference medium (that is, a pure phase shift), a phase-reduced transmissivity is defined:

$$\tilde{T}(f) \propto e^{-(\alpha(f,\theta)+i\beta(f,\theta))L} \tag{4.24}$$

For a P-wave in an acoustic 1D medium, the scattering attenuation, α, and the phase coefficient, β, were derived from Shapiro et al. (1994b) by Widmaier et al. (1996):

$$\alpha(f,\theta) = \frac{1}{\cos^2\theta} \frac{4\pi^2 a\sigma^2 f^2}{V_0^2 + 16\pi^2 a^2 f^2 \cos^2\theta} \tag{4.25}$$

and

$$\beta(f,\theta) = \frac{\pi f \sigma^2}{V_0 \cos\theta} \left[1 - \frac{8\pi^2 a^2 f^2}{V_0^2 + 16\pi^2 a^2 f^2 \cos^2\theta}\right] \tag{4.26}$$

where the statistical parameters of the reference medium include spatial correlation length a, standard deviation σ, and mean velocity V_0. The medium is modeled as a 1D random medium with fluctuating P-wave velocities that are characterized by an exponential correlation function. The transmissivity (absolute value) of the P-wave decreases with increasing angle of incidence.

If the uncorrected seismic amplitude (i.e., the analytical P-wave particle displacement) is defined according to ray theory by:

$$U(S, G, t) = R_C \frac{1}{\gamma} W(t - \tau_M) \tag{4.27}$$

where U is the seismic trace, S denotes the source, G denotes the receiver, t is the varying traveltime along the ray path, R_C is the reflection coefficient at the reflection point M, γ is the spherical divergence factor, W is the source wavelet, and τ_M is the traveltime for the ray between source S, via reflection point M, and back to the receiver G.

A reflector beneath a thin-bedded overburden will have the following compensated seismic amplitude:

$$U^T(S, G, t) = \tilde{T}_{tw}(t) * R_C \frac{1}{\gamma} W(t - \tau_M) \tag{4.28}$$

where the two-way, time-reduced transmissivity is given by:

$$\tilde{T}_{tw}(t) = \tilde{T}_{MG}(t) * \tilde{T}_{SM}(t) \tag{4.29}$$

The superscript T of $U^T(S, G, t)$ indicates that thin-bed effects have been accounted for. Moreover, equation (4.28) indicates that the source wavelet, W(t), is convolved with the transient transmissivity both for the downgoing ($\tilde{T}_{SM}$) and the upgoing raypaths ($\tilde{T}_{MG}$) between source (S), reflection point (M), and receiver (G).

In conclusion, equation (4.28) represents the angle-dependent time shift caused by transverse isotropic velocity behavior of the thinly layered overburden. Furthermore, it describes the decrease of the AVO response resulting from multiple scattering additional to the amplitude decay related to spherical divergence.

Widmaier et al. (1995) presented similar formulations for elastic P-wave AVO, where the elastic correction formula depends not only on variances and covariances of P-wave velocity, but also on S-wave velocity and density, and their correlation and cross-correlation functions.

Ursin and Stovas (2002) further extended on the O'Doherty–Anstey formula and calculated scattering attenuation for a thin-bedded, viscoelastic medium. They found that in the seismic frequency range, the intrinsic attenuation dominates over the scattering attenuation.

AVO and intrinsic attenuation (absorption)

Intrinsic attenuation, also referred to as anelastic absorption, is caused by the fact that even homogeneous sedimentary rocks are not perfectly elastic. This effect can complicate the AVO response (e.g., Martinez, 1993). Intrinsic attenuation can be described in terms of a transfer function $\hat{G}(\omega, t)$ for a plane wave of angular frequency ω and propagation time t (Luh, 1993):

$$\hat{G}(\omega, t) = \exp(\omega t / 2 Q_e + i(\omega t / \pi Q_e) \ln \omega / \omega_0) \qquad (4.30)$$

where Q_e is the effective quality factor of the overburden along the wave propagation path and ω_0 is an angular reference frequency.

Luh demonstrated how to correct for horizontal, vertical and offset-dependent wavelet attenuation. He suggested an approximate, "rule of thumb" equation to calculate the relative change in AVO gradient, δG, due to absorption in the overburden:

$$\delta G \approx \frac{f_1 \tau}{Q_e} \qquad (4.31)$$

where f_1 is the peak frequency of the wavelet, and τ is the zero-offset two-way travel time at the studied reflector.

Carcione et al. (1998) showed that the presence of intrinsic attenuation affects the P-wave reflection coefficient near the critical angle and beyond it. They also found that the combined effect of attenuation and anisotropy affects the reflection coefficients at non-normal incidence, but that the intrinsic attenuation in some cases can actually compensate the anisotropic effects. In most cases, however, anisotropic effects are dominant over attenuation effects. Carcione (1999) furthermore showed that the unconsolidated sediments near the sea bottom in offshore environments can be highly attenuating, and that these waves will for any incidence angle have a vector attenuation perpendicular to the sea-floor interface. This vector attenuation will affect AVO responses of deeper reflectors.

4.3.5 Acquisition effects on AVO

The most important acquisition effects on AVO measurements include source directivity, and source and receiver coupling (Martinez, 1993). In particular, acquisition footprint is a large problem for 3D AVO (Castagna, 2001). Irregular coverage at the surface will cause uneven illumination of the subsurface. These effects can be corrected for using inverse operations. Different methods for this have been presented in the literature (e.g., Gassaway *et al.*, 1986; Krail and Shin, 1990; Chemingui and Biondi, 2002). Chiburis' (1993) method for normalization of target amplitudes with a reference amplitude provided a fast and simple way of correcting for certain data collection factors including source and receiver characteristics and instrumentation.

4.3.6 Pre-processing of seismic data for AVO analysis

AVO processing should preserve or restore relative trace amplitudes within CMP gathers. This implies two goals: (1) reflections must be correctly positioned in the subsurface; and (2) data quality should be sufficient to ensure that reflection amplitudes contain information about reflection coefficients.

> **AVO processing**
>
> Even though the unique goal in AVO processing is to preserve the true relative amplitudes, there is no unique processing sequence. It depends on the complexity of the geology, whether it is land or marine seismic data, and whether the data will be used to extract regression-based AVO attributes or more sophisticated elastic inversion attributes.
>
> Cambois (2001) defines AVO processing as any processing sequence that makes the data compatible with Shuey's equation, if that is the model used for the AVO inversion. Cambois emphasizes that this can be a very complicated task.

Factors that change the amplitudes of seismic traces can be grouped into Earth effects, acquisition-related effects, and noise (Dey-Sarkar and Suatek, 1993). Earth effects include spherical divergence, absorption, transmission losses, interbed multiples, converted phases, tuning, anisotropy, and structure. Acquisition-related effects include source and receiver arrays and receiver sensitivity. Noise can be ambient or source-generated, coherent or random. Processing attempts to compensate for or remove these effects, but can in the process change or distort relative trace amplitudes. This is an important trade-off we need to consider in pre-processing for AVO. We therefore need to select a basic but robust processing scheme (e.g., Ostrander, 1984; Chiburis, 1984; Fouquet, 1990; Castagna and Backus, 1993; Yilmaz, 2001).

Common pre-processing steps before AVO analysis

Spiking deconvolution and wavelet processing

In AVO analysis we normally want zero-phase data. However, the original seismic pulse is causal, usually some sort of minimum phase wavelet with noise. Deconvolution is defined as convolving the seismic trace with an inverse filter in order to extract the impulse response from the seismic trace. This procedure will restore high frequencies and therefore improve the vertical resolution and recognition of events. One can make two-sided, non-causal filters, or shaping filters, to produce a zero-phase wavelet (e.g., Leinbach, 1995; Berkhout, 1977).

The wavelet shape can vary vertically (with time), laterally (spatially), and with offset. The vertical variations can be handled with deterministic Q-compensation (see Section 4.3.4). However, AVO analysis is normally carried out within a limited time window where one can assume stationarity. Lateral changes in the wavelet shape can be handled with surface-consistent amplitude balancing (e.g., Cambois and Magesan, 1997). Offset-dependent variations are often more complicated to correct for, and are attributed to both offset-dependent absorption (see Section 4.3.4), tuning effects (see Section 4.3.3), and NMO stretching. NMO stretching acts like a low-pass, mixed-phase, nonstationary filter, and the effects are very difficult to eliminate fully (Cambois, 2001). Dong (1999) examined how AVO detectability of lithology and fluids was affected by tuning and NMO stretching, and suggested a procedure for removing the tuning and stretching effects in order to improve AVO detectability. Cambois recommended picking the reflections of interest prior to NMO corrections, and flattening them for AVO analysis.

Spherical divergence correction

Spherical divergence, or geometrical spreading, causes the intensity and energy of spherical waves to decrease inversely as the square of the distance from the source (Newman, 1973). For a comprehensive review on offset-dependent geometrical spreading, see the study by Ursin (1990).

Surface-consistent amplitude balancing

Source and receiver effects as well as water depth variation can produce large deviations in amplitude that do not correspond to target reflector properties. Commonly, statistical amplitude balancing is carried out both for time and offset. However, this procedure can have a dramatic effect on the AVO parameters. It easily contributes to intercept leakage and consequently erroneous gradient estimates (Cambois, 2000). Cambois (2001) suggested modeling the expected average amplitude variation with offset following Shuey's equation, and then using this behavior as a reference for the statistical amplitude balancing.

Multiple removal
One of the most deteriorating effects on pre-stack amplitudes is the presence of multiples. There are several methods of filtering away multiple energy, but not all of these are adequate for AVO pre-processing. The method known as f–k multiple filtering, done in the frequency–wavenumber domain, is very efficient at removing multiples, but the dip in the f–k domain is very similar for near-offset primary energy and near-offset multiple energy. Hence, primary energy can easily be removed from near traces and not from far traces, resulting in an artificial AVO effect. More robust demultiple techniques include linear and parabolic Radon transform multiple removal (Hampson, 1986; Herrmann et al., 2000).

NMO (normal moveout) correction
A potential problem during AVO analysis is error in the velocity moveout correction (Spratt, 1987). When extracting AVO attributes, one assumes that primaries have been completely flattened to a constant traveltime. This is rarely the case, as there will always be residual moveout. The reason for residual moveout is almost always associated with erroneous velocity picking, and great efforts should be put into optimizing the estimated velocity field (e.g., Adler, 1999; Le Meur and Magneron, 2000). However, anisotropy and nonhyperbolic moveouts due to complex overburden may also cause misalignments between near and far offsets (an excellent practical example on AVO and nonhyperbolic moveout was published by Ross, 1997). Ursin and Ekren (1994) presented a method for analyzing AVO effects in the offset domain using time windows. This technique reduces moveout errors and creates improved estimates of AVO parameters. One should be aware of AVO anomalies with polarity shifts (class IIp, see definition below) during NMO corrections, as these can easily be misinterpreted as residual moveouts (Ratcliffe and Adler, 2000).

DMO correction
DMO (dip moveout) processing generates common-reflection-point gathers. It moves the reflection observed on an offset trace to the location of the coincident source–receiver trace that would have the same reflecting point. Therefore, it involves shifting both time and location. As a result, the reflection moveout no longer depends on dip, reflection-point smear of dipping reflections is eliminated, and events with various dips have the same stacking velocity (Sheriff and Geldhart, 1995). Shang et al. (1993) published a technique on how to extract reliable AVA (amplitude variation with angle) gathers in the presence of dip, using partial pre-stack migration.

Pre-stack migration

Pre-stack migration might be thought to be unnecessary in areas where the sedimentary section is relatively flat, but it is an important component of all AVO processing. Pre-stack migration should be used on data for AVO analysis whenever possible, because it will collapse the diffractions at the target depth to be smaller than the Fresnel zone and therefore increase the lateral resolution (see Section 4.2.3; Berkhout, 1985; Mosher *et al.*, 1996). Normally, pre-stack time migration (PSTM) is preferred to pre-stack depth migration (PSDM), because the former tends to preserve amplitudes better. However, in areas with highly structured geology, PSDM will be the most accurate tool (Cambois, 2001). An amplitude-preserving PSDM routine should then be applied (Bleistein, 1987; Schleicher *et al.*, 1993; Hanitzsch, 1997).

Migration for AVO analysis can be implemented in many different ways. Resnick *et al.* (1987) and Allen and Peddy (1993) among others have recommended Kirchhoff migration together with AVO analysis. An alternative approach is to apply wave-equation-based migration algorithms. Mosher *et al.* (1996) derived a wave equation for common-angle time migration and used inverse scattering theory (see also Weglein, 1992) for integration of migration and AVO analysis (i.e., migration-inversion). Mosher *et al.* (1996) used a finite-difference approach for the pre-stack migrations and illustrated the value of pre-stack migration for improving the stratigraphic resolution, data quality, and location accuracy of AVO targets.

Example of pre-processing scheme for AVO analysis of a 2D seismic line

(Yilmaz, 2001.)
 (1) Pre-stack signal processing (source signature processing, geometric scaling, spiking deconvolution and spectral whitening).
 (2) Sort to CMP and do sparse interval velocity analysis.
 (3) NMO using velocity field from step 2.
 (4) Demultiple using discrete Radon transform.
 (5) Sort to common-offset and do DMO correction.
 (6) Zero-offset FK time migration.
 (7) Sort data to common-reflection-point (CRP) and do inverse NMO using the velocity field from step 2.
 (8) Detailed velocity analysis associated with the migrated data.
 (9) NMO correction using velocity field from step 8.
 (10) Stack CRP gathers to obtain image of pre-stack migrated data. Remove residual multiples revealed by the stacking.
 (11) Unmigrate using same velocity field as in step 6.
 (12) Post-stack spiking deconvolution.
 (13) Remigrate using migration velocity field from step 8.

Some pitfalls in AVO interpretation due to processing effects

- Wavelet phase. The phase of a seismic section can be significantly altered during processing. If the phase of a section is not established by the interpreter, then AVO anomalies that would be interpreted as indicative of decreasing impedance, for example, can be produced at interfaces where the impedance increases (e.g., Allen and Peddy, 1993).
- Multiple filtering. Not all demultiple techniques are adequate for AVO pre-processing. Multiple filtering, done in the frequency–wavenumber domain, is very efficient at removing multiples, but the dip in the f–k domain is very similar for near-offset primary energy and near-offset multiple energy. Hence, primary energy can easily be removed from the near-offset traces, resulting in an artificial AVO effect.
- NMO correction. A potential problem during AVO analysis is errors in the velocity moveout correction (Spratt, 1987). When extracting AVO attributes, one assumes that primaries have been completely flattened to a constant traveltime. This is rarely the case, as there will always be residual moveout. Ursin and Ekren (1994) presented a method for analyzing AVO effects in the offset domain using time windows. This technique reduces moveout errors and creates improved estimates of AVO parameters. NMO stretch is another problem in AVO analysis. Because the amount of normal moveout varies with arrival time, frequencies are lowered at large offsets compared with short offsets. Large offsets, where the stretching effect is significant, should be muted before AVO analysis. Swan (1991), Dong (1998) and Dong (1999) examine the effect of NMO stretch on offset-dependent reflectivity.
- AGC amplitude correction. Automatic gain control must be avoided in pre-processing of pre-stack data before doing AVO analysis.

Pre-processing for elastic impedance inversion

Several of the pre-processing steps necessary for AVO analysis are not required when preparing data for elastic impedance inversion (see Section 4.4 for details on the methodology). First of all, the elastic impedance approach allows for wavelet variations with offset (Cambois, 2000). NMO stretch corrections can be skipped, because each limited-range sub-stack (in which the wavelet can be assumed to be stationary) is matched to its associated synthetic seismogram, and this will remove the wavelet variations with angle. It is, however, desirable to obtain similar bandwidth for each inverted sub-stack cube, since these should be comparable. Furthermore, the data used for elastic impedance inversion are calibrated to well logs before stack, which means that average amplitude variations with offset are automatically accounted for. Hence, the complicated procedure of reliable amplitude corrections becomes much less labor-intensive than for

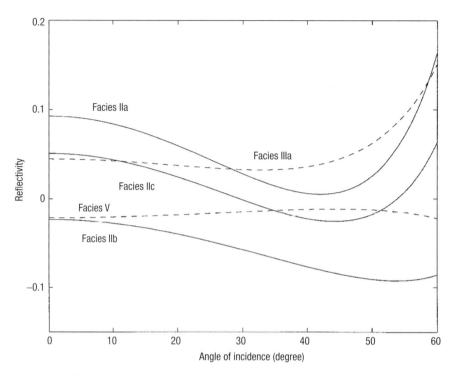

Figure 4.6 AVO curves for different half-space models (i.e., two layers – one interface). Facies IV is cap-rock. Input rock physics properties represent mean values for each facies.

standard AVO analysis. Finally, residual NMO and multiples still must be accounted for (Cambois, 2001). Misalignments do not cause intercept leakage as for standard AVO analysis, but near- and far-angle reflections must still be in phase.

4.3.7 AVO modeling and seismic detectability

AVO analysis is normally carried out in a deterministic way to predict lithology and fluids from seismic data (e.g., Smith and Gidlow, 1987; Rutherford and Williams, 1989; Hilterman, 1990; Castagna and Smith, 1994; Castagna *et al.*, 1998).

Forward modeling of AVO responses is normally the best way to start an AVO analysis, as a feasibility study before pre-processing, inversion and interpretation of real pre-stack data. We show an example in Figure 4.6 where we do AVO modeling of different lithofacies defined in Section 2.5. The figure shows the AVO curves for different half-space models, where a silty shale is taken as the cap-rock with different underlying lithofacies. For each facies, V_P, V_S, and ρ are extracted from well-log data and used in the modeling. We observe a clean sand/pure shale ambiguity (facies IIb and facies V) at near offsets, whereas clean sands and shales are distinguishable at far offsets. This example depicts how AVO is necessary to discriminate different lithofacies in this case.

4.3 AVO analysis

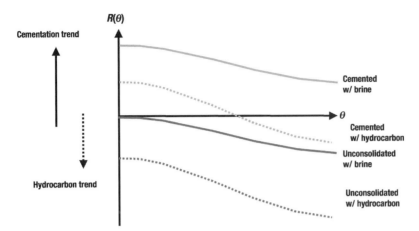

Figure 4.7 Schematic AVO curves for cemented sandstone and unconsolidated sands capped by shale, for brine-saturated and oil-saturated cases.

Figure 4.7 shows another example, where we consider two types of clean sands, cemented and unconsolidated, with brine versus hydrocarbon saturation. We see that a cemented sandstone with hydrocarbon saturation can have similar AVO response to a brine-saturated, unconsolidated sand.

The examples in Figures 4.6 and 4.7 indicate how important it is to understand the local geology during AVO analysis. It is necessary to know what type of sand is expected for a given prospect, and how much one expects the sands to change locally owing to textural changes, before interpreting fluid content. It is therefore equally important to conduct realistic lithology substitutions in addition to fluid substitution during AVO modeling studies. The examples in Figures 4.6 and 4.7 also demonstrate the importance of the link between rock physics and geology (Chapter 2) during AVO analysis.

When is AVO analysis the appropriate technique?

It is well known that AVO analysis does not always work. Owing to the many cases where AVO has been applied without success, the technique has received a bad reputation as an unreliable tool. However, part of the AVO analysis is to find out if the technique is appropriate in the first place. It will work only if the rock physics and fluid characteristics of the target reservoir are expected to give a good AVO response. This must be clarified before the AVO analysis of real data. Without a proper feasibility study, one can easily misinterpret AVO signatures in the real data. A good feasibility study could include both simple reflectivity modeling and more advanced forward seismic modeling (see Section 4.5). Both these techniques should be founded on a thorough understanding of local geology and petrophysical properties. Realistic lithology substitution is as important as fluid substitution during this exercise.

> Often, one will find that there is a certain depth interval where AVO will work, often referred to as the "AVO window." Outside this, AVO will not work well. That is why analysis of rock physics depth trends should be an integral part of AVO analysis (see Sections 2.6 and 4.3.16). However, the "AVO window" is also constrained by data quality. With increasing depth, absorption of primary energy reduces the signal-to-noise ratio, higher frequencies are gradually more attenuated than lower frequencies, the geology usually becomes more complex causing more complex wave propagations, and the angle range reduces for a given streamer length. All these factors make AVO less applicable with increasing depth.

4.3.8 Deterministic AVO analysis of CDP gathers

After simple half-space AVO modeling, the next step in AVO analysis should be deterministic AVO analysis of selected CDP (common-depth-point) gathers, preferably at well locations where synthetic gathers can be generated and compared with the real CDP gathers. In this section, we show an example of how the method can be applied to discriminate lithofacies in real seismic data, by analyzing CDP gathers at well locations in a deterministic way. Figure 4.8 shows the real and synthetic CDP gathers at three adjacent well locations in a North Sea field (the well logs are shown in Figure 5.1, case study 1). The figure also includes the picked amplitudes at a top target horizon superimposed on exact Zoeppritz calculated reflectivity curves derived from the well-log data.

In Well 2, the reservoir sands are unconsolidated, represent oil-saturated sands, and are capped by silty shales. According to the saturation curves derived from deep resistivity measurements, the oil saturation in the reservoir varies from 20–80%, with an average of about 60%. The sonic and density logs are found to measure the mud filtrate invaded zone (0–10% oil). Hence, we do fluid substitution to calculate the seismic properties of the reservoir from the Biot–Gassmann theory assuming a uniform saturation model (the process of fluid substitution is described in Chapter 1). Before we do the fluid substitution, we need to know the acoustic properties of the oil and the mud filtrate. These are calculated from Batzle and Wang's relations (see Chapter 1). For this case, the input parameters for the fluid substitution are as follows.

Oil GOR	64 l/l
Oil relative density	32 API
Mud-filtrate density	1.09 g/cm^3
Pore pressure at reservoir level	20 MPa
Temperature at reservoir level	77.2 °C

4.3 AVO analysis

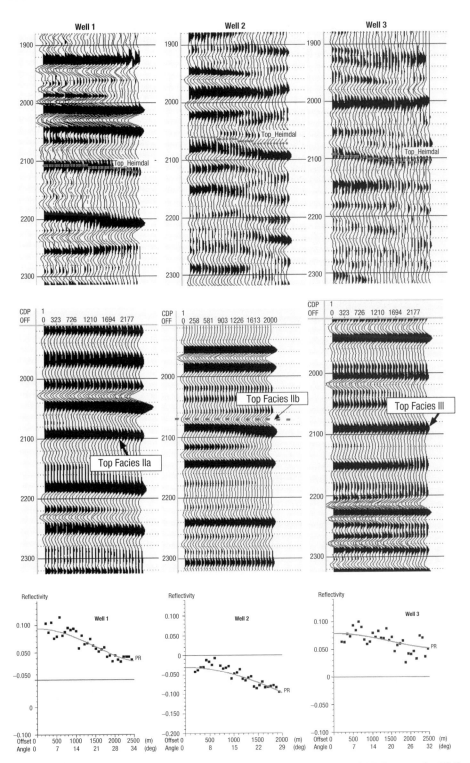

Figure 4.8 Real CDP gathers (upper), synthetic CDP gathers (middle), and AVO curves for Wells 1–3 (lower).

The corresponding AVO response shows a negative zero-offset reflectivity and a negative AVO gradient. In Well 1, we have a water-saturated cemented sand below a silty shale. The corresponding AVO response in this well shows a strong positive zero-offset reflectivity and a relatively strong negative gradient. Finally, in Well 3 we observe a strong positive zero-offset reflectivity and a moderate negative gradient, corresponding to interbedded sand/shale facies capped by silty shales. Hence, we observe three distinct AVO responses in the three different wells. The changes are related to both lithology and pore-fluid variations within the turbidite system. For more detailed information about this system, see case study 1 in Chapter 5.

Avseth *et al.* (2000) demonstrated the effect of cementation on the AVO response in real CDP gathers around two wells, one where the reservoir sands are friable, and the other where the reservoir sands are cemented. They found that if the textural effects of the sands were ignored, the corresponding changes in AVO response could be interpreted as pore-fluid changes, just as depicted in the reflectivity modeling example in Figure 4.7.

> **Importance of AVO analysis of individual CDP gathers**
>
> Investigations of CDP gathers are important in order to confirm AVO anomalies seen in weighted stack sections (Shuey's intercept and gradient, Smith and Gidlow's fluid factor, etc.). The weighted stacks can contain anomalies not related to true offset-dependent amplitude variations.

4.3.9 Estimation of AVO parameters

Estimating intercept and gradient

The next step in an AVO analysis should be to extract AVO attributes and do multivariate analysis of these. Several different AVO attributes can be extracted, mapped and analyzed. The two most important ones are zero-offset reflectivity ($R(0)$) and AVO gradient (G) based on Shuey's approximation. These seismic parameters can be extracted, via a least-squares seismic inversion, for each sample in a CDP gather over a selected portion of a 3D seismic volume.

For a given NMO-corrected CDP gather, $R(t, x)$, it is assumed that for each time sample, t, the reflectivity data can be expressed as Shuey's formula (equation (4.8)):

$$R(t, x) = R(t, 0) + G(t) \sin^2 \theta(t, x) \tag{4.32}$$

where $\theta(t, x)$ is the incident angle corresponding to the data sample recorded at (t, x).

4.3 AVO analysis

For a layered Earth, the relationship between offset (x) and angle (θ) is given approximately by:

$$\sin \theta(t, x) \approx \frac{x}{\left(t_0^2 + x^2/V_{\text{RMS}}^2\right)^{1/2}} \frac{V_{\text{INT}}}{V_{\text{RMS}}^2} \quad (4.33)$$

where V_{INT} is the interval velocity and V_{RMS} is the average root-mean-square velocity, as calculated from an input velocity profile (for example obtained from sonic log).

For any given value of zero-offset time, t_0, we assume that R is measured at N offsets (x_i, $i = 1, N$). Hence, we can rewrite the defining equation for this time as (Hampson and Russell, 1995):

$$\begin{bmatrix} R(x_1) \\ R(x_2) \\ . \\ . \\ . \\ R(x_N) \end{bmatrix} = \begin{bmatrix} 1 & \sin^2\theta(t, x_1) \\ 1 & \sin^2\theta(t, x_2) \\ . & . \\ . & . \\ . & . \\ 1 & \sin^2\theta(t, x_N) \end{bmatrix} \begin{bmatrix} R(t, 0) \\ G(t) \end{bmatrix} \quad (4.34)$$

This matrix equation is in the form of $\mathbf{b} = \mathbf{Ac}$ and represents N equations in the two unknowns, $R(t, 0)$ and $G(t)$. The least-squares solution to this equation is obtained by solving the so-called "normal equation":

$$\mathbf{c} = (\mathbf{A}^T\mathbf{A})^{-1}(\mathbf{A}^T\mathbf{b}) \quad (4.35)$$

This gives us the least-squares solution for $R(0)$ and G at time t.

Inversion for elastic parameters

Going beyond the estimation of intercept and gradient, one can invert pre-stack seismic amplitudes for elastic parameters, including V_P, V_S and density. This is commonly referred to as AVO inversion, and can be performed via nonlinear methods (e.g., Dahl and Ursin, 1992; Buland et al., 1996; Gouveia and Scales, 1998) or linearized inversion methods (e.g., Smith and Gidlow, 1987; Loertzer and Berkhout, 1993). Gouveia and Scales (1998) defined a Bayesian nonlinear model and estimated, via a nonlinear conjugate gradient method, the maximum a-posteriori (MAP) distributions of the elastic parameters. However, the nonlinearity of the inversion problem makes their method very computer intensive. Loertzer and Berkhout (1993) performed linearized Bayesian inversion based on single interface theory on a sample-by-sample basis. Buland and Omre (2003) extended the work of Loertzer and Berkhout and developed a linearized Bayesian AVO inversion method where the wavelet is accounted for by convolution. The inversion is performed simultaneously for all times in a given time window, which makes it possible to obtain temporal correlation between model parameters close in time. Furthermore, they solved the AVO inversion problem via Gaussian priors and obtained an explicit analytical form for the posterior density, providing a computationally fast estimation of the elastic parameters.

> **Pitfalls of AVO inversion**
>
> - A linear approximation of the Zoeppritz equations is commonly used in the calculation of $R(0)$ and G. The two-term Shuey approximation is known to be accurate for angles of incidence up to approximately 30°. Make sure that the data inverted do not exceed this range, so the approximation is valid.
> - The Zoeppritz equations are only valid for single interfaces. Inversion algorithms that are based on these equations will not be valid for thin-bedded geology.
> - The linear AVO inversion is sensitive to uncharacteristic amplitudes caused by noise (including multiples) or processing and acquisition effects. A few outlying values present in the pre-stack amplitudes are enough to cause erroneous estimates of $R(0)$ and G. Most commercial software packages for estimation of $R(0)$ and G apply robust estimation techniques (e.g., Walden, 1991) to limit the damage of outlying amplitudes.
> - Another potential problem during sample-by-sample AVO inversion is errors in the moveout correction (Spratt, 1987). Ursin and Ekren (1994) presented a method for analyzing AVO effects in the offset domain using time windows. This technique reduces moveout errors and creates improved estimates of AVO parameters.

4.3.10 AVO cross-plot analysis

A very helpful way to interpret AVO attributes is to make cross-plots of intercept ($R(0)$) versus gradient (G). These plots are a very helpful and intuitive way of presenting AVO data, and can give a better understanding of the rock properties than by analyzing the standard AVO curves.

AVO classes

Rutherford and Williams (1989) suggested a classification scheme of AVO responses for different types of gas sands (see Figure 4.9). They defined three AVO classes based on where the top of the gas sands will be located in an $R(0)$ versus G cross-plot. The cross-plot is split up into four quadrants. In a cross-plot with $R(0)$ along x-axis and G along y-axis, the 1st quadrant is where $R(0)$ and G are both positive values (upper right quadrant). The 2nd is where $R(0)$ is negative and G is positive (upper left quadrant). The 3rd is where both $R(0)$ and G are negative (lower left quadrant). Finally, the 4th quadrant is where $R(0)$ is positive and G is negative (lower right quadrant). The AVO classes

4.3 AVO analysis

Table 4.1 *AVO classes, after Rutherford and Williams (1989), extended by Castagna and Smith (1994), and Ross and Kinman (1995)*

Class	Relative impedance	Quadrant	R(0)	G	AVO product
I	High-impedance sand	4th	+	−	Negative
IIp	No or low contrast	4th	+	−	Negative
II		3rd	−	−	Positive
III	Low impedance	3rd	−	−	Positive
IV	Low impedance	2nd	−	+	Negative

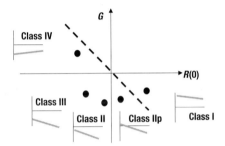

Figure 4.9 Rutherford and Williams AVO classes, originally defined for gas sands (classes I, II and III), along with the added classes IV (Castagna and Smith, 1994) and IIp (Ross and Kinman, 1995). Figure is adapted from Castagna *et al.* (1998).

must not be confused with the quadrant numbers. Class I plots in the 4th quadrant with positive $R(0)$ and negative gradients. These represent hard events with relatively high impedance and low V_P/V_S ratio compared with the cap-rock. Class II represents sands with weak intercept but strong negative gradient. These can be hard to see on the seismic data, because they often yield dim spots on stacked sections. Class III is the AVO category that is normally associated with bright spots. These plot in the 3rd quadrant in $R(0)$–G cross-plots, and are associated with soft sands saturated with hydrocarbons (see Plate 4.10).

Ross and Kinman (1995) distinguished between a class IIp and class II anomaly. Class IIp has a weak but positive intercept and a negative gradient, causing a polarity change with offset. This class will disappear on full stack sections. Class II has a weak but negative intercept and negative gradient, hence no polarity change. This class may be observed as a negative amplitude on a full-offset stack.

Castagna and Swan (1997) extended the classification scheme of Rutherford and Williams to include a 4th class, plotting in the 2nd quadrant. These are relatively rare, but occur when soft sands with gas are capped by relatively stiff shales characterized by V_P/V_S ratios slightly higher than in the sands (i.e., very compacted or silty shales).

> **Summary of AVO classes**
>
> - AVO class I represents relatively hard sands with hydrocarbons. These sands tend to plot along the background trend in intercept–gradient cross-plots. Moreover, very hard sands can have little sensitivity to fluids, so there may not be an associated flat spot. Hence, these sands can be hard to discover from seismic data.
> - AVO class II, representing transparent sands with hydrocarbons, often show up as dim spots or weak negative reflectors on the seismic. However, because of relatively large gradients, they should show up as anomalies in an $R(0)$–G cross-plot, and plot off the background trend.
> - AVO class III is the "classical" AVO anomaly with negative intercept and negative gradient. This class represents relatively soft sands with high fluid sensitivity, located far away from the background trend. Hence, they should be easy to detect on seismic data.
> - AVO class IV are sands with negative intercept but positive gradient. The reflection coefficient becomes less negative with increasing offset, and amplitude decreases versus offset, even though these sands may be bright spots (Castagna and Swan, 1997). Class IV anomalies are relatively rare, but occur when soft sands with gas are capped by relatively stiff cap-rock shales characterized by V_P/V_S ratios slightly higher than in the sands (i.e., very compacted or silty shales).
>
> The AVO classes were originally defined for gas sands. However, today the AVO class system is used for descriptive classification of observed anomalies that are not necessarily gas sands. An AVO class II that is drilled can turn out to be brine sands. It does not mean that the AVO anomaly was not a class II anomaly. We therefore suggest applying the classification only as descriptive terms for observed AVO anomalies, without automatically inferring that we are dealing with gas sands.

AVO trends and the effects of porosity, lithology and compaction

When we plot $R(0)$ and G as cross-plots, we can analyze the trends that occur in terms of changes in rock physics properties, including fluid trends, porosity trends and lithology trends, as these will have different directions in the cross-plot (Figure 4.11). Using rock physics models and then calculating the corresponding intercept and gradients, we can study various *"What if"* scenarios, and then compare the modeled trends with the inverted data.

Brine-saturated sands interbedded with shales, situated within a limited depth range and at a particular locality, normally follow a well-defined "background trend" in AVO cross-plot (Castagna and Swan, 1997). A common and recommended approach in qualitative AVO cross-plot analysis is to recognize the "background" trend and then look for data points that deviate from this trend.

4.3 AVO analysis

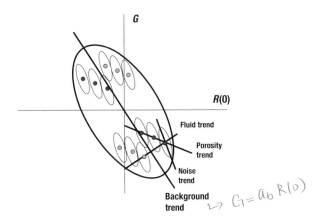

Figure 4.11 Different trends occurring in an intercept–gradient cross-plot. (Adapted from Simm et al., 2000.)

Castagna et al. (1998) presented an excellent overview and a framework for AVO gradient and intercept interpretation. The top of the sands will normally plot in the 4th quadrant, with positive $R(0)$ and negative G. The base of the sands will normally plot in the 2nd quadrant, with negative $R(0)$ and positive G. The top and base of sands, together with shale–shale interfaces, will create a nice trend or ellipse with center in the origin of the $R(0)$–G coordinate system. This trend will rotate with contrast in V_P/V_S ratio between a shaly cap-rock and a sandy reservoir (Castagna et al., 1998; Sams, 1998). We can extract the relationship between V_P/V_S ratio and the slope of the background trend (a_b) by dividing the gradient, G, by the intercept, $R(0)$:

$$a_b = \frac{G}{R(0)} \tag{4.36}$$

Assuming the density contrast between shale and wet sand to be zero, we can study how changing V_P/V_S ratio affects the background trend. The density contrast between sand and shale at a given depth is normally relatively small compared with the velocity contrasts (Foster et al., 1997). Then the background slope is given by:

$$a_b = 1 - 8\left[\frac{(V_{S1} + V_{S2})}{(V_{P1} + V_{P2})}\frac{\Delta V_S}{\Delta V_P}\right] \tag{4.37}$$

where V_{P1} and V_{P2} are the P-wave velocities in the cap-rock and in the reservoir, respectively; V_{S1} and V_{S2} are the corresponding S-wave velocities, whereas ΔV_P and ΔV_S are the velocity differences between reservoir and cap-rock. If the V_P/V_S ratio is 2 in the cap-rock and 2 in the reservoir, the slope of the background trend is -1, that is a 45° slope diagonal to the gradient and intercept axes. Figure 4.12 shows different lines corresponding to varying V_P/V_S ratio in the reservoir and the cap-rock.

The rotation of the line denoting the background trend will be an implicit function of rock physics properties such as clay content and porosity. Increasing clay content

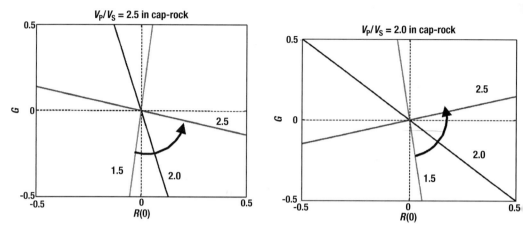

Figure 4.12 Background trends in AVO cross-plots as a function of varying V_P/V_S ratio in cap-rock and reservoir. (We assume no density contrast.) Notice that a V_P/V_S ratio of 1.5 in the reservoir can have different locations in the AVO cross-plot depending on the cap-rock V_P/V_S ratio. If the V_P/V_S ratio of the cap-rock is 2.5, the sand will exhibit AVO class II to III behavior (left), whereas if the cap-rock V_P/V_S ratio is 2.0, the sand will exhibit class I to IIp behavior (right).

at a reservoir level will cause a counter-clockwise rotation, as the V_P/V_S ratio will increase. Increasing porosity related to less compaction will also cause a counter-clockwise rotation, as less-compacted sediments tend to have relatively high V_P/V_S ratio. However, increasing porosity related to less clay content or improved sorting will normally cause a clockwise rotation, as clean sands tend to have lower V_P/V_S ratio than shaly sands. Hence, it can be a pitfall to relate porosity to AVO response without identifying the cause of the porosity change.

The background trend will change with depth, but the way it changes can be complex. Intrinsic attenuation, discussed in Section 4.3.4 (Luh, 1993), will affect the background trend as a function of depth, but correction should be made for this before rock physics analysis of the AVO cross-plot (see Section 4.3.6). Nevertheless, the rotation due to depth trends in the elastic contrasts between sands and shales is not straightforward, because the V_P/V_S in the cap-rock as well as the reservoir will decrease with depth. These two effects will counteract each other in terms of rotational direction, as seen in Figure 4.12. Thus, the rotation with depth must be analyzed locally. Also, the contrasts between cap-rock and reservoir will change as a function of lithology, clay content, sorting, and diagenesis, all geologic factors that can be unrelated to depth. That being said, we should not include too large a depth interval when we extract background trends (Castagna and Swan, 1997). That would cause several slopes to be superimposed and result in a less defined background trend. For instance, note that the top of a soft sand will plot in the 3rd quadrant, while the base of a soft sand will plot in the 1st quadrant, giving a background trend rotated in the opposite direction to the trend for hard sands.

Fluid effects and AVO anomalies

As mentioned above, deviations from the background trend may be indicative of hydrocarbons, or some local lithology or diagenesis effect with anomalous elastic properties (Castagna et al., 1998). Foster et al. (1997) mathematically derived hydrocarbon trends that would be nearly parallel to the background trend, but would not pass through the origin in $R(0)$ versus G cross-plots. For both hard and soft sands we expect the top of hydrocarbon-filled rocks to plot to the left of the background trend, with lower $R(0)$ and G values compared with the brine-saturated case. However, Castagna et al. (1998) found that, in particular, gas-saturated sands could exhibit a variety of AVO behaviors.

As listed in Table 4.1, AVO class III anomalies (Rutherford and Williams, 1989), representing soft sands with gas, will fall in the 3rd quadrant (the lower left quadrant) and have negative $R(0)$ and G. These anomalies are the easiest to detect from seismic data (see Section 4.3.11).

Hard sands with gas, representing AVO class I anomalies, will plot in the 4th quadrant (lower right) and have positive $R(0)$ and negative G. Consequently, these sands tend to show polarity reversals at some offset. If the sands are very stiff (i.e., cemented), they will not show a large change in seismic response when we go from brine to gas (cf. Chapter 1). This type of AVO anomaly will not show up as an anomaly in a product stack. In fact, they can plot on top of the background trend of some softer, brine-saturated sands. Hence, very stiff sands with hydrocarbons can be hard to discriminate with AVO analysis.

AVO class II anomalies, representing sands saturated with hydrocarbons that have very weak zero-offset contrast compared with the cap-rock, can show great overlap with the background trend, especially if the sands are relatively deep. However, class II type oil sands can occur very shallow, causing dim spots that stick out compared with a bright background response (i.e., when heterolithics and brine-saturated sands are relatively stiff compared with overlying shales). However, because they are dim they are easy to miss in near- or full-stack seismic sections, and AVO analysis can therefore be a very helpful tool in areas with class II anomalies.

Castagna and Swan (1997) discovered a different type of AVO response for some gas sands, which they referred to as class IV AVO anomalies (see Table 4.1), or a "false negative." They found that in some rare cases, gas sands could have negative $R(0)$ and positive G, hence plotting in the 2nd quadrant (upper left quadrant). They showed that this may occur if the gas-sand shear-wave velocity is lower than that of the overlying formation. The most likely geologic scenario for such an AVO anomaly is in unconsolidated sands with relatively large V_P/V_S ratio (Foster et al., 1997). That means that if the cap-rock is a shale, it must be a relatively stiff and rigid shale, normally a very silt-rich shale. This AVO response can confuse the interpreter. First, the gradients of sands plotting in the 2nd quadrant tend to be relatively small, and less sensitive to fluid type than the gradients for sands plotting in the 3rd quadrant. Second, these AVO anomalies will actually show up as dim spots in a gradient stack. However, they should stand out in an $R(0)$–G cross-plot, with lower $R(0)$ values than the background trend. Seismically, they should stand out as negative bright spots.

> **Pitfalls**
>
> - Different rock physics trends in AVO cross-plots can be ambiguous. The interpretation of AVO trends should be based on locally constrained rock physics modeling, not on naive rules of thumb.
> - Trends within individual clusters that do not project through the origin on an AVO cross-plot cannot always be interpreted as a hydrocarbon indicator or unusual lithology. Sams (1998) showed that it is possible for trends to have large offsets from the origin even when no hydrocarbons are present and the lithology is not unusual. Only where the rocks on either side of the reflecting surface have the same V_P/V_S ratio will the trends (not to be confused with background trends as shown in Figure 4.12) project through the origin. Sams showed an example of a brine sand that appeared more anomalous than a less porous hydrocarbon-bearing sand.
> - Residual gas saturation can cause similar AVO effects to high saturations of gas or light oil. Three-term AVO where reliable estimates of density are obtained, or attenuation attributes, can potentially discriminate residual gas saturations from commercial amounts of hydrocarbons (see Sections 4.3.12 and 4.3.15 for further discussions).

Noise trends

A cross-plot between $R(0)$ and G will also include a noise trend, because of the correlation between $R(0)$ and G. Because $R(0)$ and G are obtained from least-square fitting, there is a negative correlation between $R(0)$ and G. Larger intercepts are correlated with smaller slopes for a given data set. Hence, uncorrelated random noise will show an oval, correlated distribution in the cross-plot as seen in Figure 4.13 (Cambois, 2000).

Furthermore, Cambois (2001) formulated the influence of noise on $R(0)$, G and a range-limited stack (i.e., sub-stack) in terms of approximate equations of standard deviations:

$$\sigma_{R(0)} = \frac{3}{2}\sigma_s \tag{4.38}$$

$$\sigma_G = \frac{3\sqrt{5}}{2}\frac{\sigma_s}{\sin^2\theta_{max}} \tag{4.39}$$

$$\sigma_G = \sqrt{5}\frac{\sigma_{R(0)}}{\sin^2\theta_{max}} \tag{4.40}$$

4.3 AVO analysis

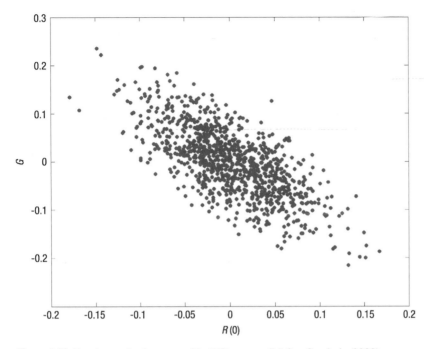

Figure 4.13 Random noise has a trend in $R(0)$ versus G (after Cambois, 2000).

and

$$\sigma_n = \sqrt{n} \cdot \sigma_s \tag{4.41}$$

where σ_s is the standard deviation of the full-stack response, σ_n is the standard deviation of the sub-stack, and n is the number of sub-stacks of the full fold data. As we see, the stack reduces the noise in proportion to the square root of the fold. These equations indicate that the intercept is less robust than a half-fold sub-stack, but more robust than a third-fold sub-stack. The gradient is much more unreliable, since the standard deviation of the gradient is inversely proportional to the sine squared of the maximum angle of incidence. Eventually, the intercept uncertainty related to noise is more or less insensitive to the maximum incidence angle, whereas the gradient uncertainty will decrease with increasing aperture (Cambois, 2001).

Simm *et al.* (2000) claimed that while rock property information is contained in AVO cross-plots, it is not usually detectable in terms of distinct trends, owing to the effect of noise. The fact that the slope estimation is more uncertain than the intercept during a least-square inversion makes the AVO gradient more uncertain than the zero-offset reflectivity (e.g., Houck, 2002). Hence, the extension of a trend parallel to the gradient axis is an indication of the amount of noise in the data.

> **Fluid versus noise trends**
>
> In areas where fluid changes in sands cause large impedance changes, we tend to see a right-to-left lateral shift along the intercept direction. This direction is almost opposite to the noise direction, which is predominantly in the vertical/gradient direction. In these cases there should be a fair chance of discriminating hydrocarbon-saturated sands from brine-saturated sands, even in relatively noisy data.

Simm *et al.* (2000) furthermore stressed that one should create AVO cross-plots around horizons, not from time windows. Horizon cross-plot clearly targets the reservoir of interest and helps determine the noise trend while revealing the more subtle AVO responses. Moreover, only samples of the maximum amplitudes should be included. Sampling parts of the waveforms other than the maxima will infill the area between separate clusters, and a lot of samples with no physical significance would scatter around the origin in an $R(0)$–G cross-plot. However, picking only peaks and troughs raises a delicate question: what about transparent sands with low or no impedance contrast with overlying shales? These are significant reflections with very small $R(0)$ values that could be missed if we invert the waveform only at absolute maxima (in commercial software packages for AVO inversion, the absolute maxima are commonly defined from $R(0)$ sections). Another issue is shale–shale interfaces. These are usually very weak reflections that would be located close to the origin in an AVO cross-plot, but they are still important for assessment of a local background trend.

There are also other types of noise affecting the AVO cross-plot data, such as residual moveout. It is essential to try to reduce the noise trend in the data before analyzing the cross-plot in terms of rock physics properties. A good pre-processing scheme is essential in order to achieve this (see Section 4.3.6).

Cambois (2000) is doubtful that AVO cross-plots can be used quantitatively, because of the noise effect. With that in mind, it should still be possible to separate the real rock physics trends from the noise trends. One way to distinguish the noise trend is to cross-plot a limited number of samples from the same horizon from a seismic section. The extension of the trend along the gradient axis indicates the amount of noise in the data (Simm *et al.*, 2000). Another way to investigate noise versus rock physics trends is to plot the anomaly cluster seen in the AVO cross-plot as color-coded samples onto the seismic section. If the cluster is mainly due to random noise, it should be scattered randomly around in a seismic section. However, if the anomaly corresponds with a geologic structure and closure, it may represent hydrocarbons (see Plate 4.10).

Finally, we claim that via statistical rock physics we can estimate the most likely fluid and lithology from AVO cross-plots even in the presence of some noise. This is referred to as probabilistic AVO analysis, and was first introduced by Avseth *et al.* (1998b). This method works by estimating probability distribution functions of $R(0)$

and *G* that include the variability and background trends. Houck (2002) presented a methodology for quantifying and combining the geologic or rock physics uncertainties with uncertainties related to noise and measurement, to obtain a full characterization of the uncertainty associated with an AVO-based lithologic interpretation. These methodologies for quantification of AVO uncertainties are explained in Section 4.3.12.

> **How to assess the noise content in AVO cross-plots**
>
> - Make cross-plots of full stack versus gradient, in addition to $R(0)$ versus G. The stack should have no correlation with the gradient, so if trends in $R(0)$–G plots are still observed in stack vs. G, these trends should be real and not random noise (Cambois, 2000).
> - Identify the location of AVO anomalies in seismic sections. Color-code AVO anomalies in $R(0)$–G plots and then superimpose them onto your seismic sections. Do the anomalies make geologic sense (shape, location), or do they spread out randomly?
> - Plot the regression coefficient of $R(0)$ and G inversion onto the seismic to identify the areas where $R(0)$ and G are less reliable.
> - Cross-plot a limited number of samples from the same horizon from a seismic section. The extension of the trend along the gradient axis indicates the amount of noise in the data (Simm *et al.*, 2000).

4.3.11 AVO attributes for hydrocarbon detection

The information in the AVO cross-plots can be reduced to one-dimensional parameters based on linear combinations of AVO parameters. This will make the AVO information easier to interpret. Various attributes have been suggested in the literature, and we summarize the most common below (AVO inversion-based attributes are discussed in Section 4.4).

Far- versus near-stack attributes

One can create several AVO attributes from limited-range stack sections. The far stack minus the near stack (FN) is a "rough" estimate of an AVO gradient, and in particular it is found to be a good attribute from which to detect class II AVO anomalies (Ross and Kinman, 1995). For class II type prospects, the far stack alone can be a good attribute for improved delineation. However, for class IIp anomalies, both the near and the far stack can be relatively dim, but with opposite polarities. Then the difference between far and near will manifest the significant negative gradient that is present. In contrast, a conventional full stack will completely zero-out a class IIp anomaly. Ross and Kinman (1995) suggested the following equation for the FN attribute depending on whether

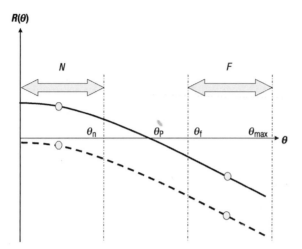

Figure 4.14 Schematic illustration of class II and class IIp and definitions of parameters used in equation (4.42). The angle θ_p is the angle of polarity change. Separation of near and far stack away from the angle of phase reversal will increase the dynamic range of the FN attribute. (Modified from Ross and Kinman, 1995.)

there is a class II or class IIp anomaly:

$$\text{FN} = \sum_{\theta=\theta_f}^{\theta_{\max}} F - c \sum_{\theta=0}^{\theta_n} N \qquad (4.42)$$

where $c = 0$ if class II, and $c = 1$ if class IIp. The variables F and N are the far- and near-stack amplitudes, respectively. The angles θ_n, θ_f, and $\theta_{\max}$ are the limitations of angle ranges applied to calculate average amplitudes of N and F, as depicted in Figure 4.14. The mid-offset range where the phase reversal occurs is avoided. This gives larger average amplitudes of N and F, and hence increases the dynamic range of equation (4.42).

Cross-plots of near (N) versus far minus near (FN) will indicate similar trends to a cross-plot of intercept versus gradient. However, one should make sure that the near stack and the far stack are balanced correctly, so the amplitude changes from near stack to far stack are representative for true AVO effects. Also, one should make sure the horizons are time-aligned correctly during the subtraction procedure.

Other attributes include far minus near times far (FNXF) and far minus near times near (FNXN). The first is a good attribute to enhance class II AVO anomalies, where the near stack is weak and the far is a strong negative. The second is a good attribute to enhance hydrocarbon-related class III AVO anomalies, and at the same time reduce the corresponding brine-saturated class II AVO response.

AVO attributes combining intercept and gradient

The estimated intercept, $R(0)$, and AVO gradient, G, can be plotted as colored section displays or visualized as 3D cubes. We can also plot combinations of these, such as the

4.3 AVO analysis

AVO product:

$$\text{PR} = R(0)G \tag{4.43}$$

This is the product of $R(0)$ and G and is a very helpful parameter in areas where we expect soft sands with hydrocarbons, that is AVO class III (according to the Rutherford and Williams nomenclature), or a classic bright spot. Soft sands with hydrocarbons will have a strong negative intercept and a strong negative gradient. The product will be a strong positive. Nonhydrocarbon reflectors will be weak or have negative products. Thus, the product stack is a nice attribute to distinguish hydrocarbon-related bright spots from "false" bright spots (bright spots with no gradient, or opposite polarity associated with anomalous lithologies). Another advantage of the AVO product is obtained if it is calculated from the complex analytical $R(0)$ and G values (see Swan, 1993). Then this attribute will be independent of the wavelet phase and it will be less sensitive to small stacking velocity errors than is the case for the AVO gradient. Furthermore, in the case of thin beds with opposite reflectivities at the top and the base, the respective AVO products will reinforce each other and the correct sign will be preserved. Nevertheless, the product stack will not be able to recognize relatively hard (class I) or transparent sands (class II) with hydrocarbons, and should only be applied in areas where AVO class III is expected.

Castagna and Smith (1994) compared different AVO attributes, and found that the reflection coefficient difference, $R_P - R_S$, is a better gas-sand discriminator than the AVO product, because it will work for any type of sand, whether these are AVO class I, II or III. They verified that $R_P - R_S$ can be expressed in terms of intercept and gradient as follows:

$$R_P - R_S \approx (R(0) + G)/2 \tag{4.44}$$

This relation is exact when $V_P/V_S = 2$. Castagna and Smith therefore concluded that $(R(0) + G)/2$ should be an excellent hydrocarbon indicator in siliciclastic environments. They also demonstrated that this attribute is physically more intuitive to use than the product stack, as shown by the following derivations:

$$R_P = (\Delta V_P/V_P + \Delta \rho/\rho)/2 \tag{4.45}$$
$$R_S = (\Delta V_S/V_S + \Delta \rho/\rho)/2 \tag{4.46}$$

and

$$R_P - R_S = (\Delta V_P/V_P - \Delta V_S/V_S)/2 \tag{4.47}$$

The equations above reveal that pore-fluid changes will affect R_P much more than R_S, and therefore $R_P - R_S$ will have a large fluid sensitivity. On the other hand, both lithology and porosity changes will affect R_P and R_S similarly. Hence, the reflection difference tends to cancel out lithology and porosity variations, while fluid changes are enhanced. Finally, this attribute will always be negative for gas sands, and always more negative than brine sands, if these are also negative.

Castagna and Smith (1994), as well as Swan (1993), suggested a local calibration of this attribute, later referred to as *scaled Poisson's reflectivity* (e.g., Ross, 2002):

$$\text{SPR} = \alpha R(0) + \beta G \quad (4.48)$$

where α and β are empirical constants extracted during the calibration procedure. These could be obtained from local well-log information or background rock physics trends extracted by other means (e.g., Chapter 2).

> The AVO product will highlight class III AVO anomalies as positive values, compared with the background trend and other type of anomalies which will show up as negative values. However, it should only be used in areas where feasibility studies show that AVO class III is expected for hydrocarbon-saturated sands, and at the same time brine sands will have a positive or transparent impedance contrast to surrounding shales. The AVO product is normally negative for brine-saturated sands. However, relatively soft sands with brine can exhibit a positive AVO product. Moreover, the AVO product may be positive, close to zero, or negative for gas sands, depending on the acoustic impedance contrast with the overlying shale (Castagna and Smith, 1994).
>
> The reflection difference, $R_P - R_S$, is found to be a more universal AVO attribute in siliciclastic environments (Castagna and Smith, 1994). Brine sands capped by shales tend to have reflection differences close to zero. Moreover, the reflection difference is always negative for gas sands, regardless of the impedance contrast with the overlying shale. This is because R_P is fluid sensitive while R_S is not, and at the same time R_P and R_S are more or less equally affected by lithology and porosity changes.

Poisson reflectivity

Verm and Hilterman (1995) suggested AVO attributes based on further approximations of Shuey's (1985) equations:

$$R(\theta) \approx \text{NI} \cos^2\theta + \text{PR} \sin^2\theta \quad (4.49)$$

where NI is the normal incidence reflectivity, and PR is the so-called Poisson reflectivity defined as

$$\text{PR} = \left[\frac{(\sigma_2 - \sigma_1)}{(1 - \sigma_{\text{avg}})^2} \right] \quad (4.50)$$

where σ_1 and σ_2 are the Poisson's ratio above and below the interface, respectively, and σ_{avg} is the average Poisson's ratio for the two layers above and below the interface.

4.3 AVO analysis

Verm and Hilterman's approximation is based on Shuey's assumptions along with the assumption that the background V_P/V_S ratio is 2. Also, the higher-order terms that describe the angle-dependent reflectivity beyond 30° have been dropped. The attributes $R(0)$ and NI are equivalent, but unlike the gradient G, PR includes only Poisson's ratio but not density.

The fluid factor

Smith and Gidlow (1987) introduced the concept of "weighted stacking." This is a technique where the principle is to create a "difference stack" relative to a wet background trend. They referred to this attribute as the *fluid factor*, and AVO anomalies related to hydrocarbons would be enhanced in these attributes.

First, $R(0)$ and G are calculated via least-square inversion as explained in Section 4.3.9. Then, following Smith and Gidlow, the difference between $R(0)$ and G can be approximated by the change in V_S at an interface, normalized by the average V_S in the layers above and below the interface (Wiggins et al., 1983).

$$\frac{\Delta V_S}{V_S} \approx R(0) - G \tag{4.51}$$

where $\Delta V_S = V_{S2} - V_{S1}$, and $V_S = (V_{S2} + V_{S1})/2$.

Using *Gardner's relation* (Gardner et al., 1974) for sandstones allows us to replace densities with P-wave velocities:

$$\Delta \rho / \rho \approx 0.25 \Delta V_P / V_P \tag{4.52}$$

where $\Delta \rho = \rho_2 - \rho_1$, and $\rho = (\rho_2 + \rho_1)/2$; and $\Delta V_P = V_{P2} - V_{P1}$, and $V_P = (V_{P2} + V_{P1})/2$.

Combining Gardner's relation with

$$R(0) = \frac{1}{2} \left(\frac{\Delta V_P}{V_P} + \frac{\Delta \rho}{\rho} \right)$$

the following approximation is obtained:

$$\frac{\Delta V_P}{V_P} \approx \frac{8 R(0)}{5} \tag{4.53}$$

Using the Mudrock Line (Castagna et al., 1985) we obtain $\Delta V_P = 1.16 \Delta V_S$, or $\Delta V_P / V_P = 1.16 (V_S/V_P)(\Delta V_S/V_S)$. This only holds for brine-saturated siliciclastics. Therefore, for hydrocarbon-saturated rocks, a residual called the *fluid factor*, ΔF, is defined as the difference between observed $\Delta V_P/V_P$ (derived from equation (4.53)) and $\Delta V_P/V_P$ predicted from $\Delta V_S/V_S$ (the latter is derived from equation (4.51)):

$$\Delta F = \frac{\Delta V_P}{V_P} - 1.16 \left(\nu \frac{\Delta V_S}{V_S} \right) \tag{4.54}$$

where ν is the background V_S/V_P ratio which can be predicted by application of the Mudrock Line to interval velocities obtained from conventional velocity analysis.

Fatti *et al.* (1994) redefined the fluid factor in terms of P-wave reflectivity and S-wave reflectivity:

$$\Delta F = R_P - 1.16 \left(\frac{V_S}{V_P} \right) R_S \qquad (4.55)$$

They also suggested an alternative way of looking at equation (4.55), in which the fluid factor is the difference between the real P-wave reflection coefficient R_P and the calculated R_P for the same sandstone in a water-saturated state. The calculated R_P is derived from the S-wave reflection coefficient, R_S, using the local Mudrock-Line relationship. Equation (4.55) then takes the form

$$\Delta F(t) = R_P(t) - g(t) R_S(t) \qquad (4.56)$$

where t is two-way traveltime, $R_P(t)$ is the P-wave reflectivity trace, $R_S(t)$ is the S-wave reflectivity trace, and $g(t)$ is a slowly time-varying gain function. The gain function is expressed as:

$$g(t) = M(V_S/V_P) \qquad (4.57)$$

where M is the slope of the Mudrock Line. Fatti *et al.* (1994) suggested that this should be a value extracted locally rather than that of Castagna *et al.* (1985).

Smith and Sutherland (1996) introduced a quality factor (not to be confused with absorption-related quality factor) to find the optimal gain function, g:

$$Q = \frac{\sum [(R_P - g R_S)_{GB} - (R_P - g R_S)_{SG}]}{\sum [R_P - g R_S]_{SB}} \qquad (4.58)$$

where GB indicates gas sand over brine sand, SG indicates shale over gas sand, and SB indicates shale over brine sand. A highest possible Q value is desirable.

Ross (2000) demonstrated that for time windows where ν is constant, this attribute is equivalent to Castagna and Smith's SPR attribute.

Pitfalls using the fluid factor and SPR attributes

The fluid factor trace (Fatti *et al.*, 1994; Smith and Sutherland, 1996) and the reflection difference (Castagna and Smith, 1994) are both constructed so that all reflectors associated with brine-saturated siliciclastics have a low amplitude, whereas rocks that lie off the Mudrock Line (Castagna *et al.*, 1985), or a local version of this equation, will show bright amplitudes. In particular, gas sands will brighten up on the fluid factor trace. However, there are several lithologies that do not follow the Mudrock Line which will also brighten up on these attributes, including carbonates and igneous rocks. These are rock types that may appear locally within a siliciclastic environment.

AVO polarization attributes

Recently, Mahob and Castagna (2003) introduced some new AVO attributes that take into account wavelet characteristics. One of these attributes is the so-called polarization angle, which is defined as follows. For a time window about a single reflection from a given interface, the AVO intercept and gradient have a preferred orientation in the $R(0)$–G plane. The angle defining the preferred orientation in the intercept–gradient space is called the polarization angle. This angle can be found by eigenvector analysis (Keho, 2000; Keho et al., 2001):

$$\phi = \arctan\left(\frac{P_y}{P_x}\right) \quad (4.59)$$

where P_x and P_y are the components of the eigenvector of the correlation matrix (Mahob and Castagna, 2003).

Another attribute created by Mahob and Castagna is the polarization angle difference, which is the difference between the polarization angle and the background trend angle:

$$\Delta\phi = \phi - \phi_{\text{trend}} \quad (4.60)$$

Furthermore, they defined the AVO strength which is the total length of a cloud of points in an $R(0)$–G cross-plot (representing one event). The strength is defined as

$$L = L_{\min} + L_{\max} \quad (4.61)$$

where

$$L_{\min} = \sqrt{R(0)_{\min}^2 + G_{\min}^2} \quad (4.62)$$

and

$$L_{\max} = \sqrt{R(0)_{\max}^2 + G_{\max}^2} \quad (4.63)$$

where $R(0)_{\min}$ is the minimum signed value within the time window of the analysis of $R(0)$ and $G_{\min}$ is the corresponding G at $R(0)_{\min}$. Similarly, $R(0)_{\max}$ and $G_{\max}$ are the corresponding maximum values within the same time window.

The polarization product is the multiplication of polarization angle and AVO strength, and is a measure of the magnitude of the AVO response along the trace. Finally, the linear-correlation coefficient, r, is a measure of how well defined the polarization spread is. The square of this coefficient tells us about the reliability of the polarization attributes, and is defined as:

$$r^2 = \frac{(\text{cov}(R(0), G))^2}{\text{var}(R(0)) \times \text{var}(G)} \quad (4.64)$$

Applying these polarization attributes, anomalies that fall on the background trend can be discriminated from the true background trend. Mahob and Castagna concluded that these attributes better discriminate gas sands and brine sands from background shales than conventional AVO attributes. However, these attributes will not work well if the signal-to-noise ratio of the data is very poor, or if the data are characterized by very low frequencies.

4.3.12 AVO anomalies caused by residual gas saturation

One of the most notorious pitfalls of AVO analysis is related to residual gas saturation (low gas saturation (<30%) due to leakage of a reservoir unit previously characterized by high gas saturation (>60%)), or low gas saturation due to gas coming out of solution from water or oil, caused by a drop in pore pressure. It is well known that just a small amount of gas in the pore space of a rock will cause a dramatic decrease in the bulk modulus of the rock. This effect is described by the Gassmann theory, assuming a uniform saturation distribution (see Chapter 1). Then the lower bound Wood's equation (or Reuss average) will apply, where just a few percent gas will cause a significant drop in the effective fluid modulus, and consequently a significant drop in the saturated bulk modulus of the rock. The problem in AVO analysis is that residual gas saturations (fizz-water) will yield similar seismic properties to commercial gas saturations. If we are dealing with a light oil, there may also be similar ambiguities between residual gas and commercial oil, or even residual oil and commercial oil.

Figure 4.15 shows an example of calculated P-wave velocity (V_P), acoustic impedance (AI), and V_P/V_S ratio as a function of oil or gas saturation versus brine saturation (i.e., two-phase fluid mixtures) for an unconsolidated sand with a porosity of 30%. In this example, representative of an offshore West African reservoir of Oligocene age, we assume the following reservoir and fluid properties: brine salinity = 250 000 ppm, oil relative density = 29 API, gas gravity = 0.7, reservoir temperature $T = 70\,^\circ C$, and pore pressure $P = 33$ MPa. The resulting curves show that the rock with just a few percent of gas will have the same V_P as with commercial amounts of oil. Because a very low gas saturation has little effect on the bulk density and almost no effect on the shear-wave velocity, the same ambiguity that is observed in V_P will also be seen in acoustic impedance and V_P/V_S. In conclusion, two-term AVO will not be able to discriminate between a seismic anomaly caused by a few percent gas and an anomaly caused by commercial amounts of oil, in this case. This is found to be a universal problem, and many wells have been drilled on AVO-driven prospects that indicated hydrocarbons, but proved to be residual amounts of gas. These were scientifically correct but commercial failures.

Han and Batzle (2002) pointed out that fizz-water is an ill-defined and misapplied concept. They found that dissolved gas or gas coming out of solution from water or oil at pressures higher than 20 MPa has little effect on effective fluid properties. This conclusion was based on experiments showing that at pressures over 20 MPa gas coming out of solution has a negligible effect on total gas–water mixture compressibility because

4.3 AVO analysis

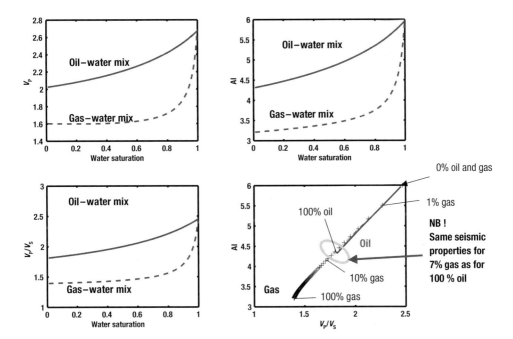

Figure 4.15 Rock physics modeling of V_P, AI, and V_P/V_S versus saturation for oil–water and gas–water mixtures. There is an ambiguity between low gas saturation and commercial oil saturation in all parameters. Hence, AVO analysis will not be able to discriminate between the two scenarios.

the exsolved gas has very low volume and high density at those high pressures. Hence, low gas saturation should have large effects on seismic properties only in shallow formations with low pore pressures. It is important, however, to bear in mind that significant residual gas may occur at pore pressures greater than 20 MPa. If a gas reservoir formerly filled with high gas saturation has leaked, leaving only a few percent of gas, we could still observe a significant drop in P-wave velocity, even at pore pressures greater than 20 MPa.

Density is the only elastic seismic parameter that can discriminate residual gas saturation from commercial hydrocarbon saturation, because low gas saturation should imply bulk densities similar to 100% brine saturation, whereas commercial gas saturation should result in a significant drop in bulk density. Density can be derived seismically from three-term AVO (see Section 4.3.15).

Residual gas can also be discriminated from commercial gas using converted P-to-S elastic impedance calculated from multicomponent seismic data (Zhu *et al.*, 2000; Gonzalez *et al.*, 2003a; see Section 4.4 for further discussion).

Alternatively, residual gas saturation can be discriminated using attenuation attributes, since it has been shown that partial gas saturation will give larger attenuation than either commercial gas saturation or oil saturation. However, this technology is still

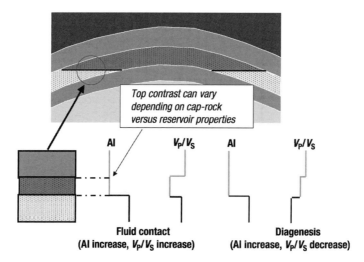

Figure 4.16 Flat spots are normally caused by fluid contacts. However, false flat spots may occur because of other geologic factors. The most common pitfall in flat-spot driven prospects is diagenetic horizons. AVO should be an efficient tool to discriminate qualitatively between fluid contacts and diagenetic horizons.

immature, and few examples have been published in the literature showing successful applications of attenuation attributes.

4.3.13 Flat-spot analysis using AVO

The most trusted and successful seismic hydrocarbon indicator is the flat spot. However, as mentioned in Section 4.2.4, there are numerous examples of seismic flat spots that turned out not to be associated with fluid contacts. Some of the most common pitfalls related to flat spots are diagenetic horizons, which often occur as abrupt, horizontal contacts (e.g., opal-A to opal-CT transitions in diatomaceous ooze deposits). These often cut across dipping sedimentary strata, just like fluid contacts. However, AVO should be an efficient tool to discriminate diagenetic flat spots from hydrocarbon-related flat spots. A schematic example is shown in Figure 4.16, where an anticline structure contains a sandy reservoir filled with oil. Alternatively, the contact indicated by the arrow could be a diagenetic horizon. Seismically, these two scenarios will give similar near-stack responses. In both cases there will be an increase in impedance across the flat spot. Any oil–water, gas–water or gas–oil contact will cause an increase in impedance, as will a diagenetic contact going from uncemented rock (e.g., opal A) to cemented rock (opal CT). The V_P/V_S ratio should increase across a fluid contact, since V_S is more or less fluid-insensitive, while V_P will be lower in gas than in oil, and lower in oil than in brine. In contrast, a diagenetic contact should imply a decrease in V_P/V_S ratio.

4.3.14 AVO detection of overpressure

AVO can be applied to discriminate between shallow water flows (SWF) and overpressured zones (e.g., Dutta et al., 2002; Mukerji et al., 2002; Carcione, 2001). It is important to discover these in order to prevent drilling hazards. Mallick and Dutta (2002) found that the V_P/V_S ratio was an excellent parameter for overpressure detection. Overpressure zones are normally very soft impedance events, and can be erroneously interpreted as gas sands if only P-wave impedance is considered. However, the V_P/V_S ratio in overpressured zones tends to be abnormally high compared with a hydrostatic background trend (see Chapter 2), whereas gas sands have abnormally low V_P/V_S ratios compared with such a background trend. Hence, overpressure can be detected from velocity–depth trend analysis (e.g., from well logs or P and S traveltime inversion) as well as from AVO analysis. Carcione et al. (2003) estimated pore pressure in reservoir rocks in the Tune field, North Sea, using P-wave velocities estimated from reflection tomography. However, they suggested the use of AVO to verify overpressure, to avoid ambiguities with pore fluids and lithology.

4.3.15 Wide-angle AVO analysis

Three-term AVO for density estimation

Three-term AVO analysis can be used to estimate density from pre-stack seismic amplitudes. Shuey's three-term approximation to the Zoeppritz equations for P-wave reflectivity is given by equation (4.7): $R(\theta) \approx R(0) + G \sin^2\theta + F(\tan^2\theta - \sin^2\theta)$. Theoretically, the density contrast at an interface can be calculated by subtracting the third coefficient from the intercept:

$$R(0) - F = \frac{1}{2}\left(\frac{\Delta V_P}{V_P} + \frac{\Delta \rho}{\rho}\right) - \frac{1}{2}\frac{\Delta V_P}{V_P} = \frac{1}{2}\frac{\Delta \rho}{\rho} \tag{4.65}$$

The advantage of extracting density from seismic amplitudes, in addition to V_P and V_S, is the fact that residual gas can be discriminated from commercial gas (see Section 4.3.12). Examples of the use of three-term AVO to calculate density include the papers by Kabir et al. (2000), Roberts et al. (2002), and Buland and Omre (2003), among others. However, Hilterman (2003) demonstrated that for density to be used to discriminate residual gas saturation, the porosity of the reservoir has to be known. A sandstone unit with relatively high porosity and residual gas saturation will have similar density to a sandstone unit with relatively low porosity and commercial amounts of gas. One point to be made here is that residual gas saturation often occurs in the more shaly sands, as leakage of gas from reservoir sands is more efficient from high-permeability clean sandstones.

Chen et al. (2001) pointed out, as demonstrated by Swan (1993), that the method of calculating density from three-term AVO is very difficult because of the poor signal-to-noise ratio of the third-term coefficient F, which they referred to as the curvature term. A complicating factor that adds uncertainty to this procedure is the effect of anisotropy, which starts to dominate the reflection coefficient at mid- to far-offset ranges. Considering elliptical anisotropy, they demonstrated how to correct AVO parameters, including the gradient G and the curvature F, for this anisotropy effect. Based on Rüger's (1997) approximation to the offset-dependent reflection coefficient in transverse isotropic media, they performed empirical corrections to improve the correction of anisotropy for three-term AVO responses. Rüger's approximation includes modifications of the gradient and curvature as follows:

$$G_{rug} = G_{iso} + \Delta G_{rug} \tag{4.66}$$

and

$$F_{rug} = F_{iso} + \Delta F_{rug} \tag{4.67}$$

where G_{iso} and F_{iso} are Shuey's isotropic coefficients, and ΔG_{rug} and ΔF_{rug} are anisotropic corrections, which can be expressed as follows (Rüger, 1997):

$$\Delta G_{rug} = \Delta \delta / 2 \tag{4.68}$$

and

$$\Delta F_{rug} = \Delta \varepsilon / 2 \tag{4.69}$$

where $\Delta \delta$ and $\Delta \varepsilon$ are the changes in Thomsen's anisotropy parameters across the interface (average value of top medium minus average value of bottom medium).

Chen *et al.* (2001) found that even for angles less than 30°, there may be large errors generated by Rüger's approximation. They introduced empirical corrections to Rüger's equations:

$$G_{emp} = G_{iso} + \Delta G_{rug} + \Delta G_{emp} \tag{4.70}$$

and

$$F_{emp} = F_{iso} + \Delta F_{rug} + \Delta F_{emp} \tag{4.71}$$

where ΔG_{emp} and ΔF_{emp} are empirical anisotropic corrections. By comparing Rüger's approximation and exact Daley and Hron (1977) reflection coefficients, they found the empirical relationship between ΔG_{emp} and δ, or ΔF_{emp} and ε, using trial and error. The resulting relationships are expressed as follows:

$$\Delta G_{emp} = R(0)\left(g_1 \delta^{1/2} + g_2 \Delta \delta\right) \tag{4.72}$$

and

$$\Delta F_{emp} = f_0 R(0)^2 + f_1 R(0)\varepsilon + f_2 R(0)^2 \varepsilon + f_3 (\Delta \varepsilon)^2 \tag{4.73}$$

where g_1, g_2, f_0, f_1, f_2, and f_3 are regression coefficients that are functions of the V_P/V_S ratio, and δ and ε are the average Thomsen parameters across the interface. Hence, the empirical corrections depend on both the average anisotropy and the change in anisotropy across the interface.

> **Limitations of Rüger's approximation**
>
> Chen et al. (2001) found that empirical corrections to the Rüger's equations are necessary before three-term AVO can be used to interpret or correct for anisotropic AVO effects.

Chen et al. (2001) found that anisotropy may cause large changes in the position, and minor changes in the slope, of data clouds in an $R(0)$ versus G cross-plot. Background trends may be shifted from a diagonal line going through the origin, to a line intercepting the gradient axis at nonzero values. This can easily be misinterpreted as fluid anomalies in AVO cross-plots, if assuming isotropic media. Similarly, in G versus F cross-plot, trends due to anisotropy can be misinterpreted as changes in density across an interface.

Cambois (2001) argued that if the effects of anisotropy are mild enough to be neglected, the aperture is sufficient, and the signal-to-noise ratio is exceptional, reliable estimates of density contrasts may be obtained from the three-term Shuey equation. However, there are additional sources of errors beyond anisotropy that can make density calculations unreliable, including acquisition (source directivity and array responses become more significant) and processing effects (the parabolic assumption for multiples is not valid) on wide-angle data. Also, anisotropy can cause nonhyperbolic moveouts, or "hockey-stick" signatures in pre-stack CDP gathers (Hilterman, 2001). This could be corrected for using higher-order or anisotropic moveout.

Ultra-far AVO analysis

In the last few years, there has also been increasing focus on extracting elastic properties from wide-angle AVO beyond critical angle, that is, normally angles beyond 50–60° (this procedure is sometimes referred to as ultra-far AVO analysis). Linearized two-term or three-term approximations to Zoeppritz equations will break down, and the exact Zoeppritz equations should be applied in modeling and inversion of ultra-far-offset reflectivities. Roberts (2000) demonstrated the potential of ultra-far AVO analysis by using the exact Zoeppritz equations to obtain improved estimates of S-wave velocities from amplitudes close to and beyond critical angle. Hawkins et al. (2001) extended Roberts' study, and obtained estimates of V_P, V_S and density from wide-angle AVO including post-critical-offset ranges. Simmons and Backus (1994), however, have warned that primaries-only Zoeppritz modeling can be very misleading, because thin-bed effects will imply interference between converted waves and primaries at far offsets. In fact, they found that for primaries-only modeling of thin-bedded media, synthetic seismograms obtained by using a linearized approximation to the Zoeppritz equations to describe the reflection coefficients are more accurate than those obtained via the exact Zoeppritz reflection coefficients. This also indicates that AVO parameter estimation from long-offset AVO based on exact Zoeppritz equations is a very unreliable procedure. Hence, full elastic waveform inversion should be applied to invert ultra-far seismic amplitudes for elastic properties. However, such an inversion is highly non-unique, and very computer-intensive, making the procedure not yet practical for full 3D inversions.

Qualitative assessment of ultra-far reflectivity still has some promising aspects. Wide-angle AVO recorded from conventional streamers can be used to extract information about converted wave energy. In an exploration stage, this could save the cost of acquiring ocean-bottom multicomponent data. Furthermore, the AVO class II and III type hydrocarbon anomalies defined by Rutherford and Williams (1989), which often are difficult to discriminate from brine-saturated sands, can be more easily discriminated on ultra-far, pre-critical offsets. Class II anomalies with transparent near-offset reflectivities will further enhance their energy in ultra-far, pre-critical-offset ranges (Hilterman *et al.*, 1998). Class I anomalies, which are the most difficult anomalies to detect (because the sands are usually stiff with low fluid sensitivity; see Chapter 1) may attain class II characteristics with polarity change at very far, but pre-critical, offsets (Hilterman *et al.*, 2000).

Pros and cons of wide-angle AVO analysis

Pros:
- Can potentially estimate density from three-term linearized AVO or nonlinear inversion of exact Zoeppritz equations (e.g., Roberts *et al.*, 2002).
- Can obtain information about converted waves, hence the extra cost of Ocean Bottom Cable (OBC) surveying may be avoided.
- AVO class I in two-term AVO may turn into AVO class IIp, with polarity change at very far offsets. This will improve the ability to discriminate hydrocarbons from brine sands in relatively stiff sands (Hilterman *et al.*, 2000).
- AVO class IIp with weak stack response based on near- to mid-offset ranges (i.e., dim spots) can be enhanced with further brightening of the amplitudes on very far, pre-critical-offset ranges (Hilterman *et al.*, 1998).
- With ultra-far AVO analysis, one may be able to improve the imaging of sub-salt reservoirs (e.g., Towner and Lindsey, 2000).

Cons:
- The Zoeppritz equation assumes a single layer interface. With the decreased resolution on ultra-far offsets, this assumption is less likely to be valid with increasing offsets. Simmons and Backus (1994) found that modeling of thin-bedded media based on a linearized approximation to the Zoeppritz equations was more accurate than modeling based on the exact Zoeppritz equations.
- There are complications with nonhyperbolic moveouts, NMO stretch and anisotropy effects that tend to worsen with increasing offset (Cambois, 2001).
- The oil industry has little experience with processing and interpretation of very-far-offset data, because traditionally these have been muted away during processing (Castagna, 2001).

Conclusion
The use of wide-angle AVO is still an immature technology that requires extended future research, both in processing and interpretation. Quantitative estimates should be used with even more caution than quantitative estimates from two-term AVO analysis.

4.3.16 Probabilistic AVO analysis

The AVO attributes presented in Section 4.3.11 are all one-dimensional parameters calculated from two-dimensional cross-plots. This procedure is convenient for qualitative interpretation, but will actually reduce the information content in the cross-plots. But quantitatively it is desirable not to reduce the two-dimensional information to one-dimensional parameters. Using statistical techniques, one can classify the different characteristic zones in a cross-plot, and then display the classification result onto the seismic sections.

Another problem with quantitative interpretation of AVO cross-plots, which is not accounted for in the conventional AVO attributes, is that a given point in the cross-plot does not correspond to a unique combination of rock physics properties. Many combinations of rock properties will yield the same $R(0)$ and G (e.g., Sams, 1998). Moreover, owing to natural variability in geologic and fluid parameters, one given geologic scenario may span a relatively large possible outcome area in the AVO cross-plot, not just a discrete point. Hence, a hydrocarbon-like AVO response might occasionally result from a brine-associated reflection, and hydrocarbon-saturated sands might not always produce an anomalous AVO response (Houck, 2002).

Quantifying AVO uncertainties related to variability in rock properties

In this section we show how we can do probabilistic AVO analysis taking into account the natural variability and uncertainties in rock properties. As explained in Chapter 3,

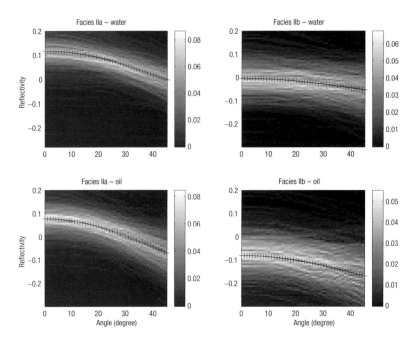

Figure 4.17 AVO pdfs for cemented sandstone and unconsolidated sands with brine and oil. The cap-rock is represented by a silty shale. There are relatively large uncertainties in AVO response related to the variability within each facies and there are overlaps between different facies and pore-fluid scenarios. However, the most likely AVO responses are distinct for each facies and pore-fluid scenario. The superimposed black ticked lines are the deterministic AVO responses calculated from the median values of the cdfs. Equation (4.8) is used to calculate these pdfs. The results from this equation start to deviate away from the exact Zoeppritz solution beyond 30°.

from well-log analysis combined with rock physics modeling, we first extract cumulative density functions (cdfs) of seismic properties for different lithofacies and fluid scenarios. Based on the cdfs of velocities and density, we create probability density functions (pdfs) of AVO response for different lithofacies combinations, and assess uncertainties in seismic signatures related to the natural variability within each facies. Figure 4.17 shows examples of AVO pdfs derived from well-log data from the Glitne field, North Sea. The plots have been generated from Monte Carlo simulated seismic properties drawn randomly from lithofacies cdfs, one for the cap-rock shale and one for the underlying sandy facies. First we simulated V_P and then V_S followed by density. It is important to make sure that the simulation honors the correlation among the three parameters. The procedure of correlated Monte Carlo simulations of rock properties is explained in Chapter 3. The corresponding reflectivity simulations are calculated using equation (4.6). The workflow of this methodology is described in Chapter 6.

Next, we can generate bivariate probability density functions of zero-offset reflectivity versus the AVO gradient (Figure 4.18). The center or peak of each contour plot represents the most likely set of $R(0)$ and G for each facies. These pdfs show how $R(0)$

4.3 AVO analysis

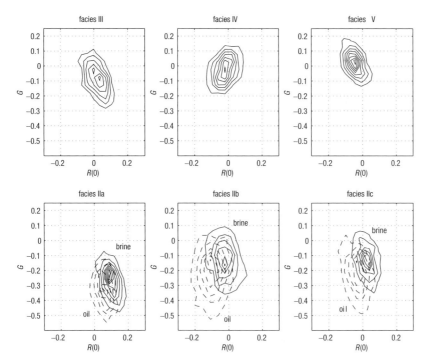

Figure 4.18 Bivariate distribution of the different seismic lithofacies in the $R(0)$–G plane, assuming facies IV as cap-rock. The center of each contour plot represents the most likely set of $R(0)$ and G for each facies. The contours represent iso-probability values, decreasing away from the innermost contour.

and G can vary for a given facies combination, and that different facies combinations can have overlaps. However, the most likely set of $R(0)$ and G is a unique characteristic of a given facies combination, corresponding to the modeled AVO curves in Figure 4.17. In general, these pdfs create a probabilistic link between facies and seismic properties that can be used to predict the most likely facies, and the conditional probability of a given facies, from seismic data. The pdfs are used to statistically classify the seismically derived $R(0)$ and G into the most likely facies class. Chapter 3 explains some common statistical classification techniques, such as discriminant classification, Bayesian classification and classification with neural nets. Case studies 1 and 3 in Chapter 5 describe two examples of probabilistic AVO analysis.

Statistical AVO constrained by rock physics depth trends

The seismic signature of hydrocarbons can be very different from one depth to another owing to different compaction trends for different lithologies. It is therefore necessary to include depth as a parameter when we use AVO analysis to predict lithology and pore fluids from seismic data. The statistical AVO classification technique

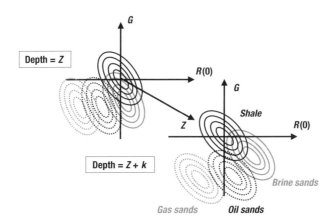

Figure 4.19 Schematic illustration of depth-dependent AVO probability density functions.

presented above was extended by Avseth *et al.* (2001c, 2003) to account for burial depth (Figure 4.19).

First, expected depth trends in rock physics properties are calculated for different lithologies and pore fluids (see Chapter 2). These trends are calculated from empirical calibrated porosity–depth models representing the local burial and compaction history.

Next, the corresponding AVO depth trends are derived from the depth trends in rock properties. Estimated acoustic impedance and V_P/V_S trends can be used to calculate the expected AVO response with depth, for various sand/shale interfaces. Different models (i.e., interface categories) can be generated from the knowledge of local geology and depositional environment. These are based on realistic layer configurations in a given depositional system, but will of course be simplified compared with the true sedimentologic observations. The number of interface categories should be kept as low as possible while still honoring geologic variations that may be seismically significant. Including too many interface categories may introduce too much overlap between individual classes in a binary AVO cross-plot. Values of V_P, V_S, and density for the different facies included in the interface categories are derived from the rock physics depth trends (see Chapter 2). These are assumed to be the mean values for the different facies at the target level. Assuming multi-Gaussian distributions, the variances can be selected on the basis of information from analog areas, or from nearby wells. If necessary, covariances between the different parameters may be defined via regression analysis. Normally, there is a higher correlation between V_P and V_S than between V_P and ρ. Moreover, Gassmann theory is used to estimate the rock properties for gas- and oil-saturated sands (see Chapter 1). In this way we can create histograms or cdfs of seismic properties for different facies at the target depth. For each interface category, Monte Carlo simulation is performed (see Chapter 3) by drawing from the cdfs of V_P, V_S, and ρ, and inserting into Shuey's equation, $R(\theta) \approx R(0) + G \sin^2\theta$, valid for angles less than 30°. The Monte Carlo simulations give distributions of $R(0)$ versus G, based on the mean and covariances in V_P, V_S, and ρ for the different interface categories. In this way AVO probability density functions can be obtained for any given depth of burial.

The final step is to apply the modeled AVO pdfs to predict the most likely facies and pore fluid from seismic data. Values of $R(0)$ and G estimated from pre-stack gathers along the line are calibrated to the modeled AVO pdfs. A background "window" should be identified in the seismic section near or around the target interval. The covariance matrix of $R(0)$ and G for the background trend in the seismic data is calibrated with the covariance matrix of $R(0)$ and G for the background trend in the model, either by matching the covariances or by univariate variance matching. This calibration is then applied to seismic data in the target area. After calibrating the seismic data with the modeled AVO pdfs, the AVO classification can be performed. Using for instance the Mahalanobis distance (see Chapter 3) one can estimate the most likely layer category for each sample in the seismic data.

This technique has been applied to an unconsolidated, mud-rich deep-water turbidite system offshore West Africa; see case study 4 in Chapter 5.

Limitations of AVO classification constrained by rock physics depth trends:
- The AVO classification technique described here will give classification of interfaces, not layers. However, the methodology can also be applied to layer inversion results (Section 4.4). For instance, elastic inversion could be classified using pdfs of AI versus V_P/V_S.
- Remember that fluid properties will be depth-dependent. Pressure and temperature control the compressibility of fluids, but the chemical properties of fluids can also change with depth. In particular, oil reference density (API gravity) tends to be depth-dependent, where biodegradation of oil decreases with depth. Hence, shallow reservoirs will normally contain thicker oil than deeper reservoirs. Trend lines of oil reference density versus depth would be valuable information to be included in this AVO classification technique.
- A future extension of this methodology will be to include facies transition probabilities and spatial statistics to improve the constraints on the classification of vertical and lateral geologic variations from seismic data.

Combining geologic uncertainties and measurement uncertainties

Lithology and pore-fluid interpretations from AVO cross-plots are ambiguous, for two reasons. Geologic uncertainty arises because different lithologies will occupy overlapping ranges of elastic properties. Measurement uncertainty arises because an observed AVO response is imperfectly related to rock properties. Almost always, AVO inversion results will be influenced and deteriorated by processing artifacts, noise, or tuning effects. Houck (2002) presented a technique to quantify and combine these two components of uncertainty to obtain a full characterization of the uncertainty associated with an AVO-based lithology interpretation.

Houck defined the following model for the offset dependence of real seismic amplitudes:

$$d_i = s_i(R(0) + G\sin^2\theta_i) + n_i \qquad (4.74)$$

where s_i is an offset-dependent scale factor and n_i is offset-dependent noise. The scale factor includes uncorrected acquisition and propagation effects, reflector geometry, processing artifacts, and where we happened to sample the seismic wavelet. The noise includes ambient noise, source-generated scattered noise, multiples, tuning effects, and higher-order terms in the reflection coefficient expansion. He assumed that the scale factor does not change with offset, and that the additive noise on each trace is uncorrelated and Gaussian with zero mean and constant variance. Uncertainty in the scale factor means that the true locations of the seismic points in the $R(0)$ and G plane are unknown. In reality, we do not know the correct scaling between seismic amplitudes and true reflectivity values. Not all scale factors are possible, though. Scale factors that produce reflectivities outside the expected range should be excluded, and, even within the allowable range, some scale factors may be more likely than others. Bachrach *et al.* (2003) addressed issues of uncertainty related to resolution and accuracy of seismic inversions applied to reservoir property estimation. They used the methodology described in this book, and extended it to take into account the scale differences between well-log data and seismic inversion results.

The use of spatial statistics to improve reservoir characterization from AVO

The probabilistic AVO methodologies summarized above perform prediction on a trace-by-trace basis. For a given horizon, that also implies sample-by-sample basis. Spatial statistical methods can, however, be applied to account for spatial correlations. Knowing that geologic features, like lithofacies, are characterized by systematic lateral variations, we can account for this in the seismic reservoir characterization. Recent work by Caers *et al.* (2001) and Eidsvik *et al.* (2002, 2004) demonstrates methodologies for integrating rock physics models, AVO attributes and spatial statistical techniques for improved reservoir mapping of the same turbidite field that is shown in case study 1 in Chapter 5. Also, spatial correlations are accounted for in the reservoir mapping from near and far impedances carried out in case study 2 in Chapter 5. (Also see Section 3.3.4.)

4.4 Impedance inversion

Ambiguities in lithologic and fluid identification based only on normal-incidence reflection amplitudes and impedance (ρV) can often be effectively removed by adding information about V_P/V_S-related attributes, for example from non-normal incidence data. This provides the incentive for AVO analysis described in the previous sections. However, synthetic seismic modeling has shown that sometimes it can be difficult to use the seismic amplitudes quantitatively owing to practicalities of picking, resolution problems, and thin-layer effects. Hence another approach to lithofacies identification is based on seismic impedance inversions. Impedance inversions take into account the full waveform of the seismic trace, not just the amplitudes. In the overall scheme of integrated reservoir characterization, impedance inversion is a tool to derive seismic attributes (P-impedance, Poisson's ratio, etc.) that can be linked to rock properties (lithology, porosity, pore fluids, etc.) using rock physics models and statistical techniques. Inversion-derived attributes incorporate the underlying physics of wave propagation using models of different levels of approximation. This is in contrast to purely statistical or mathematically defined attributes derived directly from the seismic traces, without explicitly using any physical model. The purely statistical attributes are harder to relate to rock properties, and require more comprehensive calibration and training data sets.

The goal of geophysical inversion is to estimate model parameters from observed data. So, for example, one might want to invert measured seismic traces to estimate the P-wave impedance of the subsurface layers. The parameters that can be estimated depend on the data and the assumed model. With pre-stack seismic data one might invert for the P-wave impedance and Poisson's ratio (or S-wave impedance) of the subsurface model. If we have only stacked P-wave data, however, we can invert for P-wave impedance alone, and information about Poisson's ratio or S-wave impedance is lost.

General inverse theory is a mathematically rich discipline, and many excellent books on geophysical inverse theory exist that the reader may wish to consult (e.g., Menke, 1989; Parker, 1994; Tarantola, 1987; Sen and Stoffa, 1995). Our goal in this section is just to give a brief overview of seismic impedance inversion methods as applicable for reservoir characterization. The typical set-up of inverse problems is as follows. We have a theoretical model **A** that relates (linearly or nonlinearly) the model parameters **m** to the data **d**. The actual observed data are denoted by $\mathbf{d}_{obs}$ while the calculated data from the forward model are denoted by $\mathbf{d}_{cal}$. The goal is to find those model parameters that minimize some function (called the objective function) of the misfit between $\mathbf{d}_{obs}$ and $\mathbf{d}_{cal}$. The misfit arises not only because we do not know the correct model parameters, but also because the model itself is imperfect, and the observed data are not noise-free. Often we minimize the squared error (i.e., the L_2 norm of the misfit) between observation and model predictions. This gives the least-squares solution. Robust approaches involve minimizing the L_1 norm or the absolute value of the misfit. The objective function can include not only the misfit between $\mathbf{d}_{obs}$ and $\mathbf{d}_{cal}$ but also other constraints derived from prior models or smoothness requirements. In a general probabilistic framework of the inverse problem (e.g. Tarantola, 1987; Gouviea, 1996) the goal is to obtain the posterior distribution of the model parameters given the prior distribution and the likelihood of observing the data. The likelihood may be estimated from the forward model, or from exhaustive training data if such data are available. Following Sen and Stoffa (1995), model-based inversion methods may be categorized as follows.

Linear methods
In these methods, data and model parameters are linearly related and can be expressed in a matrix equation as:

$$\mathbf{Am} = \mathbf{d}$$

The least-square solution $\hat{\mathbf{m}}$ is given by the well-known normal equations of linear algebra,

$$\hat{\mathbf{m}} = (\mathbf{A}^T\mathbf{A})^{-1}\mathbf{A}^T\mathbf{d}_{obs}$$

and the covariance of the estimate (assuming unbiased model and uncorrelated errors) is

$$C_{\hat{m}\hat{m}} = \sigma^2(\mathbf{A}^T\mathbf{A})^{-1}$$

where σ^2 is the error variance. Uncertainty in the estimated model parameters may be given in terms of $\sqrt{C_{\hat{m}\hat{m}}}$ if we assume a Gaussian distribution of the parameters. A particular case of this linear model fitting was described in the earlier section where seismic amplitudes R measured at different offsets (or angles) were fit to the simple straight-line model $R(\theta) = R(0) + G\sin^2\theta$ to obtain the least-squares estimate of the AVO intercept, $R(0)$, and gradient, G.

Iterative gradient-based methods
These methods attempt to solve nonlinear problems by linearizing around an initial solution. Iterative linear steps are taken to update the current model on the basis of gradient information. The iteration is stopped when the updates are below some tolerance. Gradient descent methods such as Newton's method, steepest descent, and conjugate gradient can be used to minimize the objective function. Gradient descent methods are susceptible to the choice of the starting point, and can easily get trapped into local mimima.

Exhaustive search methods
This involves computation of synthetic data from the forward model at every point of the model space. Usually for seismic inversions this is not very practical.

Random search methods
The model space is searched randomly using Monte Carlo trials. This is also computationally expensive.

Directed Monte Carlo methods

These are global optimization methods where the random Monte Carlo search is directed using some fitness criteria of the estimate. Methods such as simulated annealing (SA) and genetic algorithms (GA) belong to this category. These are powerful methods and can be very useful for highly nonlinear problems. Probabilistic estimates including estimates of uncertainty can be obtained without the Gaussianity assumption. The monograph by Sen and Stoffa (1995) explains SA and GA as applied to geophysical inverse problems. Mallick (1999) describes some practical aspects of using GA for pre-stack waveform inversion.

Geostatistical sequential simulation methods

Geostatistical simulations are often used in reservoir characterization to integrate different kinds of data while at the same time incorporating the spatial correlation of reservoir heterogeneities. One approach is to first derive seismic impedances from conventional (gradient-based) inversion techniques and then perform geostatistical co-kriging or indicator simulation of reservoir properties using the impedance as secondary data (e.g. Doyen, 1988; Doyen and Guidish, 1992; Zhu and Journel, 1993; Mukerji *et al.*, 2001). However, in the geostatistical inversion methodology (Bortoli *et al.*, 1993; Haas and Dubrule, 1994) geostatistical simulations are more closely integrated with the seismic inversion at the initial stage itself. So far the geostatistical simulation methodology has been used mostly for post-stack seismic inversion. In principle it can be used for pre-stack inversion, though in practice computation time poses a limitation. The methodology consists of local trace-by-trace optimization combined with sequential geostatistical sampling based on the horizontal and vertical variogram (Rowbotham *et al.*, 1998). The variogram statistically quantifies the spatial correlation. Each trace location is visited in a random path. At each location, a number of possible vertical seismic impedance logs are simulated using sequential Gaussian simulation (Deutsch and Journel, 1996). The simulation is constrained by the existing impedances at the well locations, and by the vertical and horizontal variograms. The synthetic seismograms computed from the simulated impedance logs using a 1D convolution model are compared with the actual seismic data. The simulated log that gives the best fit to the seismic data is retained and used as a constraint for simulating vertical logs at the next random location. The seismic data constrain the inversion within the seismic bandwidth while the higher spatial frequencies are stochastically constrained by the variograms obtained from well logs. A Bayesian framework for stochastic impedance inversion is described by Eide *et al.* (1997).

In all of the stochastic inversion methods multiple realizations of seismic impedance are obtained. These multiple realizations can be statistically analyzed to estimate probabilities and uncertainties of the results.

The simplest seismic impedance inversion is trace-by-trace post-stack inversion based on a 1D convolutional model of the seismic trace. Many vendors offer user-friendly GUI-driven commercial software that do trace-based impedance inversions. Since this is a post-stack 1D inversion, the output is P-impedance when P-wave data are used. At the time of writing, very few vendors offer commercial software to do full-waveform pre-stack impedance inversions. Pre-stack inversion of P data can be used to estimate not only the P-impedance but also shear-wave-related attributes such as Poisson's ratio or S-impedance. Of course one can also estimate S-impedance directly from post-stack inversion of shear-wave data when available. As described in Chapter 1, having both P- and S-related attributes can be very useful in discriminating lithologies and pore-fluid saturations. An alternative to full pre-stack impedance inversion is offered by the concept of offset impedance or angle-dependent impedance, also called "elastic impedance." As explained below, this generalized impedance allows us to use trace-based 1D inversion algorithms and existing software on far-offset partial stacks to invert for elastic impedance that carries information about shear-wave attributes in the form of V_P/V_S ratio. But before we describe offset impedances, let us look at the steps required for a standard post-stack, trace-based inversion for P impedance.

4.4.1 Post-stack 1D impedance inversion

In 1D inversions the seismic trace $S(t)$ is modeled as a convolution of the normal-incidence reflectivity series $r(t)$ with the wavelet $w(t)$:

$$S(t) = w(t) * r(t) \tag{4.75}$$

The normal-incidence reflectivity is defined in terms of the contrast in the seismic impedance ($I = \rho V$) as

$$r = \frac{I_{j+1} - I_j}{I_{j+1} + I_j} \approx \frac{1}{2} \mathrm{d}(\log I) \tag{4.76}$$

where the approximation holds for small impedance contrasts. Since the seismic trace is a bandlimited version of the derivative of log I, a simple inversion consists of just reversing the process by integrating the seismic trace, after scaling the amplitudes. This gives us a bandlimited estimate of log I up to an integration constant. A simple Hilbert transform of the trace is a "poor man's" impedance inversion since the Hilbert transform is equivalent to bandlimited integration. The low-frequency trend has to be supplied separately. Most robust approaches use constrained minimization to combine a prior model with the observed seismic data as outlined below.

We start with migrated post-stack seismic data, well logs (sonic and density), and interpreted horizon picks. Well logs are tied to the seismic data by comparison of the nearest traces to the 1D synthetic computed from the sonic log. This usually involves iteration with wavelet estimation. Extracting a reliable wavelet is an important step in the inversion process. Wavelet extraction is a complex processing issue, but most software packages include a few different methods to extract a wavelet. A direct deterministic method of course is to measure the source wavelet using nearby surface receivers. Purely statistical methods use the autocorrelation of the seismic traces over a small window to estimate the amplitude spectrum of the wavelet, while the phase spectrum is user-defined. An inverse Fourier transform then gives the time-domain wavelet. Sonic log information can be used in addition to the seismic data to extract the wavelet. This depends on a good tie between log and seismic and good depth-to-time conversion. A good seismic-to-log tie of course depends on having a good wavelet, so there is some iteration between tying the logs and extracting the wavelet. A robust method is to extract the amplitude spectrum from the seismic autocorrelation and use the well log to estimate an average phase. In reality, wavelets are not constant from trace to trace over a section, or from one traveltime to another. One could, in theory, extract a spatially and temporally varying wavelet. In practice, usually a single average wavelet is extracted and used for the impedance inversion.

> In summary, from the end-user point of view, these are the steps involved in a trace-based seismic impedance inversion.
> - Import migrated post-stack seismic data, well logs, and horizon picks.
> - Tie well logs to seismic and extract a reliable average wavelet.
> - Build a background initial impedance model using logs (maybe along with seismic RMS velocity estimates), horizon picks and other geological and structural information.
> - Carry out the inversion. In commercial software this involves selecting amongst the choice of available algorithms and setting the algorithm parameters.
> - Examine the residuals to identify zones that show anomalously high residual values. Go back to seismic data, log ties and prior model to see if the anomalous residuals can be explained and a better inversion obtained.
> - Export the inverted impedance values and when possible the uncertainty associated with the estimated impedances.

Another important step in impedance inversion is building the background model or prior model. This supplies information about the low-frequency (spatial frequency) component of the impedance. The seismic trace, being a bandlimited version of the Earth's reflectivity, lacks any information about the low-frequency trend. This bandlimited nature also gives rise to the inherent nonuniqueness of the seismic inversion problem. There may be many combinations of impedance models that fit the data equally well, but differ only in the low-frequency trend. The low-frequency background model may be estimated from sonic logs or RMS velocity estimates from seismic data. Simple model builders in commercial software use the well-log impedance values and interpolate them along horizon picks. More sophisticated model builders allow placement of faults and unconformities and other complex structures that may be available from prior geologic and seismic interpretations. The prior model and the seismic data are then combined in a constrained least-squares inversion to obtain the inverted impedance estimates. The prior model may be used in different ways: to estimate depth trends of upper and lower bounds to constrain the inversion, or in a weighted objective function that includes not only the misfit with the data but also the deviations from the prior model. The weights depend on our relative confidence in the prior model versus the seismic traces. The constrained minimization problem can then be solved using gradient-based optimization techniques. In some software programs the objective function is based not on a trace-by-trace misfit but on a global misfit between forward model predictions and observed data for multiple traces in the section.

4.4.2 Far-offset elastic impedances

As described in the earlier section, usually impedance inversion is applied to zero-offset, or near-offset, stacked sections to estimate the acoustic impedance ρV, and therefore does not give us any V_P/V_S information. A direct method of obtaining S-wave impedance is, of course, from inversion of post-stack S-wave data. When only P-wave data are available, one way to estimate S-wave-related attributes (e.g., Poisson's ratio, or V_P/V_S) is by pre-stack inversion of multi-offset data. While full pre-stack inversion is not yet common in commercial software, it will almost surely become the trend in future. Here we describe an alternative approach based on a *pseudo-impedance attribute* (Connolly, 1998; Mukerji *et al.*, 1998b) which is a far-offset equivalent of the more conventional zero-offset impedance. This far-offset impedance has been called the "elastic impedance" (EI) as it contains information about the V_P/V_S ratio. This approach allows us to use the same trace-based 1D algorithm for inversion of the far-offset stack as for the near-offset stack, to get an elastic impedance cube. Although only approximate, the inversion for this pseudo-impedance parameter is economical and simple compared with full pre-stack inversion. The key to using this extracted attribute effectively for quantitative reservoir characterization is calibration with log data.

The acoustic impedance, $I_a = \rho V$, can be expressed as

$$I_a = e^{2 \int R(0)\,dt} \qquad (4.77)$$

where $R(0)$ is the normal-incidence reflection coefficient. Similarly, the elastic impedance may be defined in terms of the elastic P–P reflection coefficient at θ, $R(\theta)$, as:

$$I_e(\theta) = e^{2 \int R(\theta)\,dt} \qquad (4.78)$$

Substituting in this equation one of the well-known approximations for $R(\theta)$ (see for example Aki and Richards, 1980) in terms of V_P, V_S, and density contrasts:

$$R(\theta) = R(0) + A \sin^2\theta + B \tan^2\theta \qquad (4.79)$$

4.4 Impedance inversion

where

$$R(0) = \frac{1}{2}\left(\frac{\Delta V_P}{V_P} + \frac{\Delta \rho}{\rho}\right)$$

$$A = -2\left(\frac{V_S}{V_P}\right)^2 \left(\frac{2\Delta V_S}{V_S} + \frac{\Delta \rho}{\rho}\right)$$

$$B = \frac{1}{2}\frac{\Delta V_P}{V_P}$$

We can express I_e as:

$$I_e(\theta) = \rho V_P \cdot e^{\tan^2\theta \int d(\ln V_P)} \cdot e^{-4\sin^2\theta (V_S/V_P)^2 \int 2d(\ln V_S)} \cdot e^{-4\sin^2\theta (V_S/V_P)^2 \int d(\ln \rho)} \quad (4.80)$$

or

$$I_e(\theta) = V_P^{(1+\tan^2\theta)} \rho^{(1-4K\sin^2\theta)} V_S^{(-8K\sin^2\theta)} \quad (4.81)$$

where $K = (V_S/V_P)^2$ is taken to be a constant. In deriving this expression we have used the fact that $e^{\ln x} = x$. The elastic impedance reduces to the usual acoustic impedance, $I_a = \rho V$, when $\theta = 0$. Unlike the acoustic impedance, the elastic impedance is not a function of the rock properties alone but depends on the angle. Using only the first two terms in the approximation for $R(\theta)$ gives a similar expression for I_e with the $\tan^2\theta$ terms replaced by $\sin^2\theta$:

$$I_e(\theta) = V_P^{(1+\sin^2\theta)} \rho^{(1-4K\sin^2\theta)} V_S^{(-8K\sin^2\theta)} \quad (4.82)$$

This has been termed the first-order elastic impedance (Connolly, 1999), and it goes to $(V_P/V_S)^2$ at $\theta = 90°$ assuming K to be $1/4$. Note that the elastic impedance has strange units and dimensions, and they change with angle. Whitcombe (2002) defines a useful normalization for the elastic impedance:

$$I_e(\theta) = [V_{P0}\rho_0]\left(\frac{V_P}{V_{P0}}\right)^{(1+\tan^2\theta)} \left(\frac{\rho}{\rho_0}\right)^{(1-4K\sin^2\theta)} \left(\frac{V_S}{V_{S0}}\right)^{(-8K\sin^2\theta)} \quad (4.83)$$

where the normalizing constants V_{P0}, V_{S0}, and ρ_0 may be taken to be either the average values of velocities and densities over the zone of interest, or the values at the top of the target zone. Now the elastic impedance has the same dimensionality as the acoustic impedance. Mention must be made also of the extended elastic impedance (EEI) of Whitcombe et al. (2002), which is defined over angle χ ranging from $-90°$ to $+90°$. It should not be interpreted as the actual reflection angle, but rather as the independent input variable in the definition of EEI. The EEI is expressed as:

$$I_e(\chi) = [V_{P0}\rho_0]\left(\frac{V_P}{V_{P0}}\right)^{(\cos\chi+\sin\chi)} \left(\frac{\rho}{\rho_0}\right)^{(\cos\chi-4K\sin\chi)} \left(\frac{V_S}{V_{S0}}\right)^{(-8K\sin\chi)} \quad (4.84)$$

Under certain approximations, the EEI for specific values of the independent variable χ becomes proportional to rock elastic parameters such as bulk modulus and shear modulus.

For impedance inversion using 1D trace-based algorithms, the elastic impedance may be written as:

$$I_e(\theta) = V_P \rho^* \tag{4.85}$$

$$\rho^* = V_P^{(\tan^2\theta)} \rho^{(1-4K\sin^2\theta)} V_S^{(-8K\sin^2\theta)} \tag{4.86}$$

where ρ^* is a pseudo-density, and 1D convolutional models can now be used for forward modeling. In actual inversion of field seismic data, the far-offset traces are stacked over some appropriate small angle range $\Delta\theta = \theta_2 - \theta_1$, instead of using just a single reflection angle θ. The stacked elastic impedance attribute $\bar{I}_e(\theta_1, \theta_2)$ can be obtained by first integrating the reflection coefficient over the angle range, and then using equation (4.78) to give:

$$\bar{I}_e(\theta_1, \theta_2) = V_P^{1+T_\theta} \rho^{1-4(V_S/V_P)^2 S_\theta} V_S^{-8(V_S/V_P)^2 S_\theta} \tag{4.87}$$

where

$$T_\theta = \frac{\tan\theta_2 - \tan\theta_1}{\Delta\theta} - 1$$

$$S_\theta = \frac{1}{2} - \frac{1}{4}\frac{(\sin 2\theta_2 - \sin 2\theta_1)}{\Delta\theta}$$

Equation (4.87) reduces to equation (4.81) in the limit when $\theta_2 \to \theta_1$.

The elastic and acoustic impedance attributes derived from well logs can be used to test for the distinguishability of the facies based on these attributes. Log data are first used to build a bivariate calibration pdf for I_e and I_a. A key step, as mentioned in Chapter 3, is to extend the log derived data, using Gassmann's equations, to incorporate velocity and impedance attributes for pore fluids not encountered in the well. Lithology substitution should also be done if necessary, using for example the cementation models or texture models described in Chapter 2. This helps to incorporate velocity and impedance values corresponding to lithology variations not encountered in the well. Figure 4.20 shows different seismic lithofacies defined from wells in a North Sea reservoir, in an I_e–I_a cross-plot. Facies that overlap in acoustic impedance can be discriminated by their elastic impedance and vice versa. Different pore fluids can also be discriminated on the I_e–I_a cross-plot.

Classification success rate can then be tested statistically. One simple method is by omitting one sample at a time from the training data, and using the rest of the data to classify the omitted sample (jackknife validation). This is done for all the samples in the training data. Figure 4.21 shows the classification success rate using a

4.4 Impedance inversion

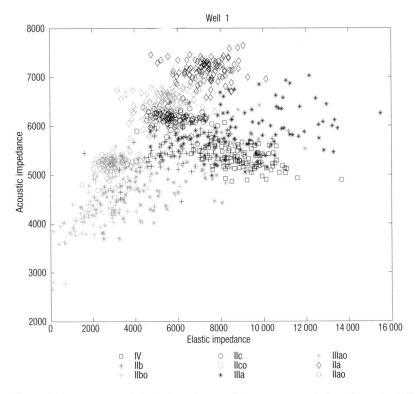

Figure 4.20 Cross-plot of elastic impedance at 30° versus acoustic impedance for different lithofacies. Light symbols indicate oil-saturated facies.

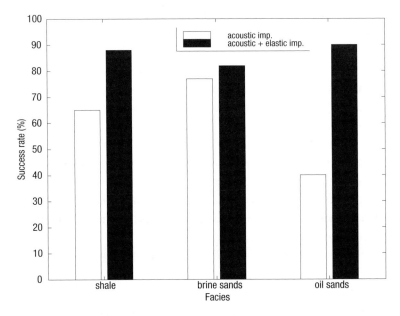

Figure 4.21 Classification success rate at the wells.

discriminant analysis (see Chapter 3) to classify the samples. Classification was done using zero-offset impedance (I_a) alone, and then using both I_a and I_e together. Figure 4.21 clearly shows the value (in this particular case) of the far-offset impedance attribute in increasing the average success rate for all facies from about 60% (I_a only) to about 85% (I_a and I_e).

The following section describes two other data examples, one from the Australian shelf, and another from the Gulf of Mexico. Figure 4.22A shows a cross-plot of acoustic impedance versus elastic impedance from logs in a well on the Australian shelf. The sands are fairly incompressible, with a large acoustic impedance compared with the shales. Sensitivity to fluid type is small, and oil-sand impedance (not shown) is very close to brine-sand impedance. In this case lithologies can be separated more easily than fluid type. Changing the pore fluid to gas (using Gassmann's equation) shifts the sand points slightly to the lower left. While brine sands are well separated from shales by their acoustic impedance, the acoustic impedance of gas sands overlaps that of shales. It would be difficult to separate gas sands from shales from just their zero-offset impedance. Figure 4.22B shows classification success rates for the three different facies. The histograms again show the value of the far-offset impedance attribute. The average success rate increases from about 68% (I_a only) to about 78% (I_a and I_e).

Figure 4.23A shows a cross-plot of acoustic impedance versus elastic impedance from a Gulf of Mexico well. Here the sands are not very consolidated, and show a strong sensitivity to pore fluids. Gas, oil, and brine sands are well separated on the cross-plot both in acoustic impedance and elastic impedance. In this instance, acoustic impedance by itself is a good discriminator of the sands with different pore fluids. Adding elastic impedance does not improve the classification performance (Figure 4.23B) *for this case*.

These examples show how well-log data can be used before doing inversions to decide whether it is worthwhile to do far-offset inversions, and how much improvement might be expected by including far-offset impedance in the reservoir characterization strategy.

The convolutional model does not handle properly all the reflections at far offsets as the primary reflections get mixed with other events. The approximations used to derive the expressions for elastic impedance get worse at larger angles. The first-order two-term elastic impedance has been found to give more stable results than the three-term elastic impedance (Mallick, 2001). Mallick compares pre-stack inversion with partial stack elastic impedance inversions and recommends a hybrid approach. In this approach full pre-stack inversion is done at a few control points to get reliable estimates of P and S impedance. These pre-stack inversions are used as anchors for cheaper 1D trace-based inversions over large data volumes. On the basis of these results, small zones may be selected for detailed analysis by pre-stack inversions. Bachrach and Dutta (2004) used

4.4 Impedance inversion

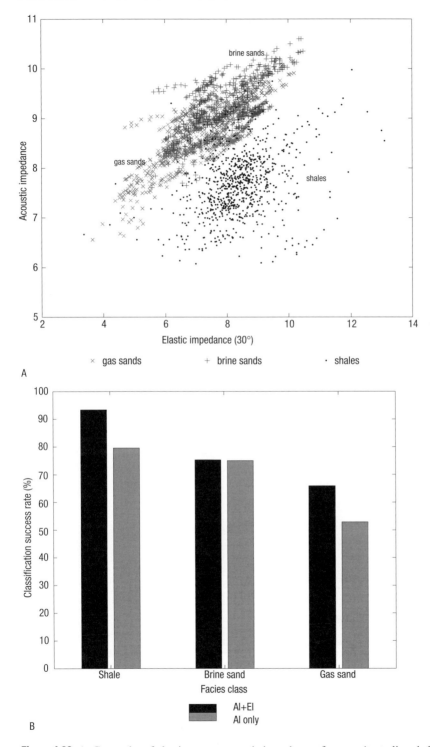

Figure 4.22 A, Cross-plot of elastic versus acoustic impedances from an Australian shelf well log, and B, corresponding classification success rates.

242 Common techniques for quantitative seismic interpretation

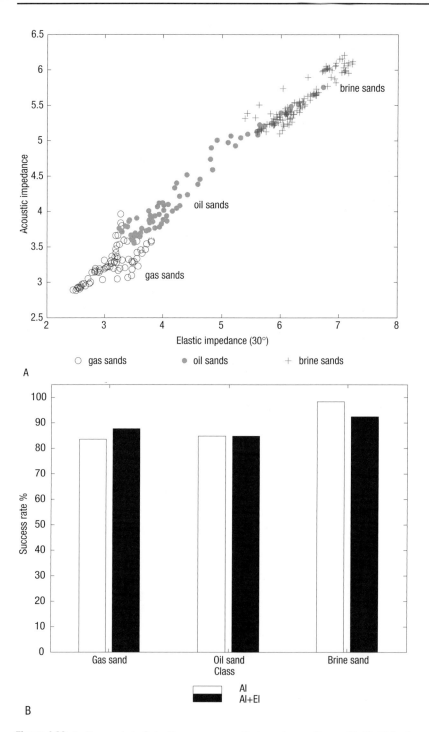

Figure 4.23 A, Cross-plot of elastic versus acoustic impedances from a Gulf of Mexico well log, and B, corresponding classification success rates.

pre-stack inversions and statistical rock physics methodologies described in this book to quantify porosity and saturation distributions from seismic data.

> In applications to reservoir characterization, care must be taken to filter the logs to match the seismic frequencies, as well as to account for the differences in frequency content in near- and far-offset data. A different wavelet has to be extracted for the near- and far-offset angle stacks. The inverted acoustic and elastic impedance co-located at the wells must be calibrated with the known facies and fluid types in the well, before classifying the seismic cube in the interwell region.
>
> The major limitations of using partial stack elastic impedance inversion arise from the assumptions of the 1D convolutional model for far offsets, the assumption of a constant value for K, and errors in estimates of incidence angle.

> As pre-stack inversions become more common, it will be routine to obtain the inverted P and S impedances directly instead of attributes such as the elastic impedance that are indirectly related to shear-wave properties. It should be kept in mind that any other set of attributes that are directly derived from P and S impedances by deterministic functions (e.g., linear combinations, or nonlinear functions) cannot contain any extra information than that originally in P and S impedances. This is a direct result of a fundamental data-processing theorem in information theory (Cover and Thomas, 1991; Takahashi, 2000)

4.4.3 Lambda–mu estimation

Goodway *et al.* (1997) have championed the use of the parameters $\lambda\rho$, $\mu\rho$, and λ/μ obtained from pre-stack seismic data. Here λ and μ are the elastic Lamé parameters and ρ is the density. The Lamé parameter μ is the same as the shear modulus. Goodway *et al.* (1997) use the approximation to the P–P reflectivity in terms of P- and S-wave impedances:

$$R_{\text{PP}}(\theta) = \frac{\Delta I_\text{P}}{2I_\text{P}}(1+\tan^2\theta) - 8\left(\frac{V_\text{S}}{V_\text{P}}\right)^2 \sin^2\theta \frac{\Delta I_\text{S}}{2I_\text{S}}$$
$$- \left[\frac{1}{2}\tan^2\theta - 2\left(\frac{V_\text{S}}{V_\text{P}}\right)^2 \sin^2\theta\right]\frac{\Delta\rho}{\rho} \quad (4.88)$$

where Δ indicates contrast across the reflecting interface, and I_P, I_S, and ρ are the average P-wave impedance, the average S-wave impedance, and the average density over the interface. The average P-wave and S-wave velocities are denoted by V_P and V_S, and they are related to impedances in the usual way: $I_\text{P} = \rho V_\text{P}$, and $I_\text{S} = \rho V_\text{S}$. Ignoring the far-angle third term in density contrast, equation (4.88) can be used to extract P and S reflectivity sections from pre-stack P-wave data, which are then inverted to obtain I_P

and I_S, the P and S impedances. Finally I_P and I_S are used to compute $\lambda\rho$ and $\mu\rho$ from the relations

$$\mu\rho = I_S^2$$
$$\lambda\rho = I_P^2 - 2I_S^2 \qquad (4.89)$$

which follow directly from the equations relating velocities to elastic moduli:

$$V_S = \sqrt{\mu/\rho}$$

$$V_P = \sqrt{(\lambda + 2\mu)/\rho}$$

Since equations (4.89) define a one-to-one mapping from the pair (I_P, I_S) to $(\lambda\rho, \mu\rho)$, the statistical information content stays the same. This equivalence stems from a fundamental theorem in information theory, sometimes termed the "data-processing theorem" (Cover and Thomas, 1991; Takahashi, 2000). Points that overlap in the (I_P, I_S) domain will overlap in the $(\lambda\rho, \mu\rho)$ domain, and the Bayes classification error using the full bivariate probability density functions will be the same. However, the covariances will be different in the two domains. Since the impedances are estimated from seismic inversions, they are subject to errors. Squaring the impedances and taking their linear combinations introduces further errors and bias in the estimates of $\lambda\rho$ and $\mu\rho$. Gray (2002), using assumptions of Gaussian noise distribution and independence of I_P and I_S, showed that the error in $\mu\rho$ is approximately twice the error in I_S, while the error in $\lambda\rho$ is about four times as great as the error associated with I_S. This also assumes that the errors associated with I_P and I_S are about the same.

Instead of first inverting for impedances and then computing $\lambda\rho$ and $\mu\rho$, Gray et al. (1999) and Gray (2002) have advocated directly estimating λ and μ from prestack seismic data by using an approximation for $R_{PP}(\theta)$ expressed directly in terms of contrasts in λ and μ:

$$R_{PP}(\theta) = \frac{\Delta\lambda}{\lambda}\left[\frac{1}{4} - \frac{1}{2}\left(\frac{V_S}{V_P}\right)^2\right](\sec^2\theta) + \frac{\Delta\mu}{\mu}\left(\frac{V_S}{V_P}\right)^2\left(\frac{1}{2}\sec^2\theta - 2\sin^2\theta\right)$$

$$+ \left[\frac{1}{2} - \frac{1}{4}\sec^2\theta\right]\frac{\Delta\rho}{\rho} \qquad (4.90)$$

Here as before, Δ indicates contrast across the reflecting interface, and λ, μ, and ρ are averages over the interface. The reflectivity can also be expressed in terms of bulk modulus, K, and shear modulus contrasts as follows (Gray et al., 1999):

$$R_{PP}(\theta) = \frac{\Delta K}{K}\left[\frac{1}{4} - \frac{1}{2}\left(\frac{V_S}{V_P}\right)^2\right](\sec^2\theta) + \frac{\Delta\mu}{\mu}\left(\frac{V_S}{V_P}\right)^2\left(\frac{1}{3}\sec^2\theta - 2\sin^2\theta\right)$$

$$+ \left[\frac{1}{2} - \frac{1}{4}\sec^2\theta\right]\frac{\Delta\rho}{\rho} \qquad (4.91)$$

These approximations can be used in AVO analysis of pre-stack P-wave data to extract $\Delta\lambda/\lambda$ and $\Delta\mu/\mu$. Usually the far-angle density term is ignored in conventional AVO analysis. The extracted $\Delta\lambda/\lambda$ and $\Delta\mu/\mu$ are then inverted using, for example, standard post-stack amplitude inversion based on convolutional modeling. The direct estimates of λ and μ are less prone to noise than computations of $\lambda\rho$ and $\mu\rho$ from I_P and I_S, and moreover, decouple the elastic parameters from the density.

4.4.4 P-to-S elastic impedance

Non-normal-incidence reflections give rise to converted waves. We now derive expressions for far-offset impedance attributes for converted waves using Aki–Richards approximations for the P-to-S and S-to-P reflectivities, $R_{PS}(\theta)$ and $R_{SP}(\theta)$, respectively (see also Duffaut et al., 2000). The angle-dependent far-offset impedance attributes will be expressed in terms of integrated reflectivities as

$$I_{PS}(\theta) = e^{2\int R_{PS}(\theta)\,dt} \qquad (4.92)$$

$$I_{SP}(\theta) = e^{2\int R_{SP}(\theta)\,dt} \qquad (4.93)$$

For converted waves, it is important to distinguish between the incidence angle and the reflection angle as they are not the same. The P-to-S reflectivity is given by

$$R_{PS}(\theta) = -\frac{\sin\theta_P}{2\cos\theta_S}\left[\left(1 - 2\frac{V_S^2}{V_P^2}\sin^2\theta_P + 2\frac{V_S}{V_P}\cos\theta_P\cos\theta_S\right)\frac{\Delta\rho}{\rho}\right.$$
$$\left. - \left(4\frac{V_S^2}{V_P^2}\sin^2\theta_P - 4\frac{V_S}{V_P}\cos\theta_P\cos\theta_S\right)\frac{\Delta V_S}{V_S}\right] \qquad (4.94)$$

where θ_P and θ_S are the angles made by the incident P-wave and reflected S-wave with the normal to the plane interface. Using the relation $\sin\theta_P = (V_P/V_S)\sin\theta_S$ to express the reflectivity in terms of just the reflected wave angle θ_S we get

$$R_{PS}(\theta_S) = -\frac{\tan\theta_S}{2(V_S/V_P)}\left(1 - 2\sin^2\theta_S + 2\cos\theta_S\sqrt{(V_S/V_P)^2 - \sin^2\theta_S}\right)\frac{\Delta\rho}{\rho}$$
$$+ \frac{\tan\theta_S}{2(V_S/V_P)}\left(4\sin^2\theta_S - 4\cos\theta_S\sqrt{(V_S/V_P)^2 - \sin^2\theta_S}\right)\frac{\Delta V_S}{V_S} \qquad (4.95)$$

Substituting this expression into equation (4.92) and carrying out the integration we get

$$I_{PS}(\theta_S) = \rho^a V_S^b \qquad (4.96)$$

$$a = \frac{\tan\theta_S}{(V_S/V_P)}\left(2\sin^2\theta_S - 1 - 2\cos\theta_S\sqrt{(V_S/V_P)^2 - \sin^2\theta_S}\right)$$

$$b = \frac{4\tan\theta_S}{(V_S/V_P)}\left(\sin^2\theta_S - \cos\theta_S\sqrt{(V_S/V_P)^2 - \sin^2\theta_S}\right)$$

Similar expressions may be obtained for the S-to-P converted waves. The S-to-P reflectivity versus angle is given by

$$R_{SP}(\theta_S) = \frac{\cos\theta_S}{\cos\theta_P}\left(\frac{V_S}{V_P}\right) R_{PS} \qquad (4.97)$$

Expressing all angles in terms of the reflected angle θ_P we get

$$R_{SP}(\theta_S) = -\frac{(V_S/V_P)\tan\theta_P}{2} \times \left(1 - 2(V_S/V_P)^2\sin^2\theta_P \right.$$

$$\left. + 2(V_S/V_P)\cos\theta_P\sqrt{1 - (V_S/V_P)^2\sin^2\theta_P}\right)\frac{\Delta\rho}{\rho} + \frac{(V_S/V_P)\tan\theta_P}{2}$$

$$\times \left(4(V_S/V_P)^2\sin^2\theta_P - 4(V_S/V_P)\cos\theta_P\sqrt{1-(V_S/V_P)^2\sin^2\theta_P}\right)\frac{\Delta V_S}{V_S}$$

$$(4.98)$$

and finally using equation (4.93) we obtain

$$I_{SP}(\theta_P) = \rho^a V_S^b \qquad (4.99)$$

$$a = (V_S/V_P)\tan\theta_P\left(2(V_S/V_P)^2\sin^2\theta_P - 1 - 2(V_S/V_P)\cos\theta_P\sqrt{1-(V_S/V_P)^2\sin^2\theta_P}\right)$$

$$b = 4(V_S/V_P)\tan\theta_P\left((V_S/V_P)^2\sin^2\theta_P - (V_S/V_P)\cos\theta_P\sqrt{1-(V_S/V_P)^2\sin^2\theta_P}\right)$$

Figure 4.24 shows cross-plots of various impedance attributes. They were computed from well-log data using the above equations for the far-offset P-to-P and P-to-S impedances. The log is from the Australian shelf. The points for gas sands were computed from the brine-sand data by fluid substitution using Gassmann's equations. The sands are well consolidated and show little sensitivity to fluid changes. The sands and shales can be separated, provided we have both P and S information. This can be in the form of direct estimates of the P and S impedances (upper left subplot) or it could come from one of the other impedance attributes, which indirectly contain the S-wave information.

Gonzalez et al. (2003) showed how the PS far-offset elastic impedance (PSEI) could be used to discriminate quantitatively not only lithologies (sand/shale) but also "fizz-water" from commercial gas. Homogeneously mixed "fizz-water" with low gas saturation is difficult to differentiate seismically from higher gas saturations. The abrupt reduction in V_P with the first few percent of gas controls the seismic response. Therefore, usually only the presence of gas but not the saturation can be detected with P-to-P seismic. Use of P-to-S converted waves has been suggested as a source of additional information that can help in distinguishing high from low gas saturation (for example, Wu, 2000; Zhu et al., 2000). At near offsets (small angles) V_S and ρ terms contribute

4.4 Impedance inversion

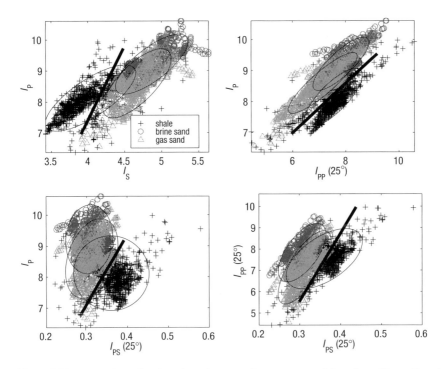

Figure 4.24 Cross-plots of various impedance attributes computed from logs (Australian shelf). The ellipses are contours of Gaussian functions computed from the mean and covariance of each cluster. The thick line (drawn by eye) separates the sands from shales.

equally to PS elastic impedance. For mid-to-large offsets the ρ term dominates. The asymmetric contribution or "decoupling" between V_S and ρ can be exploited in discriminating different reservoir properties (Gonzalez et al., 2003). Unlike the abrupt jump in V_P with the initial presence of a small amount of gas, the density varies more gradually and linearly with gas saturation. The linear behavior of density with saturation makes attributes, that are closely related to density useful proxies for estimating gas saturation. Gonzalez et al. (2003a, 2003b) defined two groups: fizz-water sands with gas saturation between 10% and 20%, and commercial gas sands with gas saturation greater than 50%. Classification success rates for the two groups were then estimated using Monte Carlo simulations and well-log data from Eastern Venezuela. Classification was done with different pairs of attributes: PSEI(10°)–PSEI(50°); I_P–EI(30°); $\rho\lambda$–$\rho\mu$; and λ–μ. Successful classification of commercial gas increased to about 90% with the PSEI attributes, compared with about 75–80% using I_P–EI(30°), and about 70–80% using the various Lamé parameter attributes. Of course it goes without saying that these classification rates only apply to the particular sands and fluids studied by Gonzalez et al. (2003a, 2003b). One advantage of using two PSEI attributes, instead of a combination of P-to-P and P-to-S attributes, is that the time matching of PP and PS data is avoided.

4.4.5 Anisotropic elastic impedance

It is tempting to apply the same simple derivation to the reflectivity equations for anisotropic media. The integration of the anisotropic reflectivity versus angle and azimuth (AVAZ) gives us an impedance-like attribute that depends on both the incidence angle and the azimuth. This anisotropic elastic impedance can be calculated from measured well logs (V_P, V_S, and density) along with some estimates of the Thomsen's anisotropic parameters.

This section details the equations used for deriving anisotropic elastic impedance. We begin with the approximate forms of the reflectivity vs. angle (and azimuth) equations. We will use the notation introduced by Thomsen (1986) for weak transversely isotropic media with density ρ, and stiffness tensor $\mathbf{C}$:

$$\alpha = \sqrt{\frac{C_{33}}{\rho}} \quad \varepsilon = \frac{C_{11} - C_{33}}{2C_{33}}$$

$$\beta = \sqrt{\frac{C_{44}}{\rho}} \quad \gamma = \frac{C_{66} - C_{44}}{2C_{44}}$$

$$\delta = \frac{(C_{13} + C_{44})^2 - (C_{33} - C_{44})^2}{2C_{33}(C_{33} - C_{44})}$$

In the above equations, α and β are the P- and S-wave velocities along the symmetry axis, and ε, γ, and δ are the Thomsen's anisotropic parameters. The P-wave reflection coefficient for weakly anisotropic TIV media (transverse isotropy with vertical symmetry axis) in the limit of small impedance contrast is given by (Thomsen, 1993)

$$R_{\text{PP}}(\theta) = R_{\text{PP-iso}}(\theta) + R_{\text{PP-aniso}}(\theta) \tag{4.100}$$

$$R_{\text{PP-iso}}(\theta) \approx \frac{1}{2}\left(\frac{\Delta Z}{\overline{Z}}\right) + \frac{1}{2}\left[\frac{\Delta \alpha}{\overline{\alpha}} - \left(\frac{2\overline{\beta}}{\overline{\alpha}}\right)^2 \frac{\Delta G}{\overline{G}}\right]\sin^2\theta$$

$$+ \frac{1}{2}\left(\frac{\Delta \alpha}{\overline{\alpha}}\right)\sin^2\theta \tan^2\theta \tag{4.101}$$

$$R_{\text{PP-aniso}}(\theta) \approx \frac{\Delta \delta}{2}\sin^2\theta + \frac{\Delta \varepsilon}{2}\sin^2\theta \tan^2\theta \tag{4.102}$$

where

$\theta = (\theta_2 + \theta_1)/2 \quad \Delta\varepsilon = \varepsilon_2 - \varepsilon_1 \quad \Delta\gamma = \gamma_2 - \gamma_1$

$\overline{\rho} = (\rho_1 + \rho_2)/2 \quad \Delta\rho = \rho_2 - \rho_1 \quad \Delta\delta = \delta_2 - \delta_1$

4.4 Impedance inversion

$\bar{\alpha} = (\alpha_1 + \alpha_2)/2 \qquad \Delta\alpha = \alpha_2 - \alpha_1$
$\bar{\beta} = (\beta_1 + \beta_2)/2 \qquad \Delta\beta = \beta_2 - \beta_1$
$\bar{G} = (G_1 + G_2)/2 \qquad \Delta G = G_2 - G_1 \quad G = \rho\beta^2$
$\bar{Z} = (Z_1 + Z_2)/2 \qquad \Delta Z = Z_2 - Z_1 \quad Z = \rho\alpha$

In the above and following equations, Δ indicates a difference and an overbar indicates an average of the corresponding quantity. All these equations are approximations for the exact reflectivity in anisotropic media, and are valid for "small contrasts," "small anisotropy" and for incidence angle θ up to 30–40°. In TIH media (transverse isotropy with horizontal symmetry axis, e.g., single set of vertical fractures with a horizontal symmetry axis), reflectivity will vary with azimuth, ϕ, as well as offset or incident angle θ. Rüger (1995, 1996) and Chen (1995) derived the P-wave reflection coefficient in the symmetry planes for reflections at the boundary of two TIH media sharing the same symmetry axis. At a horizontal interface between two TIH media with horizontal symmetry axis x_1 and vertical axis x_3, the P-wave reflectivity in the vertical symmetry plane parallel to the x_1 symmetry axis can be written as

$$R_{PP}(\phi = 0°, \theta) \approx R_{PP\text{-iso}}(\theta) + \left[\frac{\Delta\delta^{(V)}}{2} + \left(\frac{2\bar{\beta}}{\bar{\alpha}}\right)^2 \Delta\gamma\right]\sin^2\theta$$

$$+ \frac{\Delta\varepsilon^{(V)}}{2}\sin^2\theta\tan^2\theta \qquad (4.103)$$

where azimuth ϕ is measured from the x_1-axis and incident angle θ is defined with respect to x_3. The isotropic part $R_{PP\text{-iso}}(\theta)$ is the same as before. In the above expression

$\alpha = \sqrt{\dfrac{C_{33}}{\rho}} \qquad \varepsilon^{(V)} = \dfrac{C_{11} - C_{33}}{2C_{33}}$

$\beta = \sqrt{\dfrac{C_{44}}{\rho}} \qquad \delta^{(V)} = \dfrac{(C_{13} + C_{55})^2 - (C_{33} - C_{55})^2}{2C_{33}(C_{33} - C_{55})}$

$\beta^\perp = \sqrt{\dfrac{C_{55}}{\rho}} \qquad \gamma = \dfrac{C_{66} - C_{44}}{2C_{44}}$

In the vertical symmetry plane perpendicular to the symmetry axis, the P-wave reflectivity is the same as the isotropic solution:

$R_{PP}(\phi = 90°, \theta) = R_{PP\text{-iso}}(\theta)$

Common techniques for quantitative seismic interpretation

In nonsymmetry planes, Rüger (1996) derived the P-wave reflectivity $R_{PP}(\phi, \theta)$ using a perturbation technique:

$$R_{PP}(\phi, \theta) \approx R_{PP\text{-}iso}(\theta) + \left\{ \left[\frac{\Delta \delta^{(V)}}{2} + \left(\frac{2\bar{\beta}}{\bar{\alpha}} \right)^2 \Delta \gamma \right] \cos^2\phi \right\} \sin^2\theta$$

$$+ \left\{ \left[\frac{\Delta \varepsilon^{(V)}}{2} \right] \cos^4\phi + \left[\frac{\Delta \delta^{(V)}}{2} \right] \sin^2\phi \cos^2\phi \right\} \sin^2\theta \tan^2\theta \quad (4.104)$$

The impedance-like far-offset attribute may be defined in terms of the elastic P–P reflectivity as

$$I_{aniso}(\theta, \phi) = \exp\left[2 \int R_{PP}(\theta, \phi) \right] \quad (4.105)$$

Using this definition of anisotropic elastic impedance, we integrate the R_{PP} relations for anisotropic media described above to get the following relations. In each case, the anisotropic elastic impedance (I_{aniso}) may be represented in terms of the isotropic elastic impedance, multiplied by a factor due to the anisotropy. For TIV media we have

$$I_{aniso}(\theta, \phi) = I_{iso} A \quad (4.106)$$
$$A = \exp\left[\delta \sin^2\theta + \varepsilon \sin^2\theta \tan^2\theta \right]$$

The isotropic elastic impedance (I_{iso}) is the usual isotropic far-offset impedance that has been described in various forms in the preceding section. One expression for I_{iso} obtained by integrating the Aki–Richards three-term approximation of R_{PP} in isotropic media is given as

$$I_{iso}(\theta) = \alpha^{1+\tan^2\theta} \cdot \rho^{1-4(\beta/\alpha)^2 \sin^2\theta} \cdot \beta^{-8(\beta/\alpha)^2 \sin^2\theta} \quad (4.107)$$

For TIH media, there is a dependence on both incidence angle and the azimuthal angle measured from the horizontal symmetry axis. Integrating the approximate R_{PP} relation we get for TIH media

$$I_{aniso}(\theta, \phi) = I_{iso} A \quad (4.108)$$
$$A = \exp[\delta^{(V)} \cos^2\phi \sin^2\theta (1 + \sin^2\phi \tan^2\theta) $$
$$+ \varepsilon^{(V)} \cos^4\phi \sin^2\theta \tan^2\theta + 2\gamma \cos^2\phi \sin^2\theta (2\beta/\alpha)^2]$$

The isotropic elastic impedance can be estimated from well-log data. To estimate the additional factor A in the anisotropic elastic impedance we need estimates for the Thomsen parameters. These can come from modeling of the anisotropy, for instance using crack models (e.g., Hudson's model; Hudson, 1981), or from a reasonable guess for the percentage P and S anisotropy.

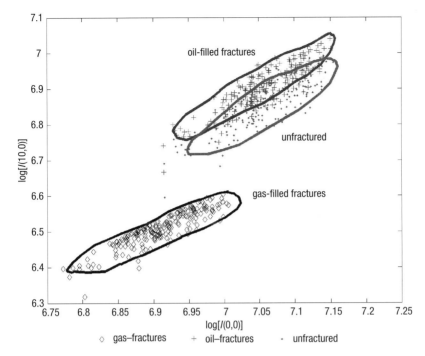

Figure 4.25 Cross-plot of normal-incidence and mid-offset (10°) elastic impedance for fractured and unfractured carbonate reservoir rocks. The azimuth is along the normals of a single set of vertical cracks. The anisotropic offset-impedance attributes may be useful for characterizing fractures and fluids from seismic data. Well-log data and Hudson's model were used to create this cross-plot.

Figure 4.25 shows an example of a cross-plot of near- and mid-offset elastic impedance in anisotropic media. The computations were based on log data taken from a carbonate reservoir. Fractures were modeled using Hudson's formulation to embed a single crack set (crack density 0.07; aspect ratio 0.001) in the background limestone matrix giving an effective TIH fractured reservoir. The cross-plot shows the possibility of distinguishing gas-filled fractures using the anisotropic impedance attribute defined here. Of course, the separability depends on rock and fluid properties, and should be calibrated for the reservoir of interest.

4.4.6 Interpretation of elastic inversion results using rock physics templates

The generation of the rock physics templates (RPTs) was demonstrated in Chapter 2. This is a tool that can be applied by seismic interpreters. Based on a compiled catalog or atlas of RPTs calculated by a rock physics expert, the seismic interpreter can select the appropriate RPT(s) for the zone and area of interest, and interpret elastic inversion results without in-depth knowledge about rock physics theory. If a compilation of relevant RPTs is available for the area under investigation, the ideal interpretation workflow becomes a fairly simple two-step procedure: (1) use well-log data to validate the selected RPT(s) (if no appropriate RPT exists, the user should provide the rock physicist with local geologic input, so that new RPTs can be created for the area under investigation); and (2) use the selected and verified RPT(s) to interpret elastic inversion results.

Plate 4.26 shows an example: a small set of 3D elastic inversion data exist around this well. Plate 4.26 shows the estimated AI and V_P/V_S results for a selected line in a 100-ms time window. Values of V_P/V_S are calculated from elastic impedance according to equation (4.82). The V_P/V_S vs. AI cross-plot of these data is shown in Plate 4.27, with a selected RPT superimposed (the RPT has been verified to well-log data according to the procedure described in Chapter 2).

The rock physics interpretation of Plate 4.27 appears to be straightforward. The population that sits along the theoretical shale trend is interpreted to represent shale. Note that the shale points appear to move closer to the sand trend for the highest AI values. This could reflect shales becoming increasingly more silty, and the points between the shale and brine-sand trends are interpreted to be silty shales and/or shaly sand. The points close to the theoretical brine-sand trend most probably represent clean sand. We do not expect to see a clear oil-sand response, as the oil is fairly heavy in this case, but some of the data that plot significantly below the brine-sand trend may be attributed to oil saturation. The sand appears to have total porosities in the range 22–28%.

4.5 Forward seismic modeling

One common way to do quantitative analysis of seismic amplitudes is to do forward seismic modeling. This is done by creating a synthetic seismic model based on an assumed or interpreted Earth model. The synthetic seismic is then compared to the real seismic data, and if necessary the Earth model must be edited to give a better match. This is the opposite procedure to seismic inversion, where the Earth model is calculated from the seismic data. However, the two procedures are often combined.

4.5.1 "Quick and dirty" 1D acoustic modeling

Plate 4.28 shows a very simple but very illustrative example of how seismic modeling can be applied to verify a geologic model. In a seismic near-stack section (upper left), we observe a channel-levee complex where the levees are relatively bright with a positive impedance contrast (shown in red) at the top surface of the levees. The channel fill is relatively transparent, and it is easy to interpret this as a shale-filled channel. However, in the area of investigation, well-log rock physics analysis shows that the acoustic impedance of clean sands is very similar to the acoustic impedance of shales, while shaly sands are characterized by relatively high impedances (upper right). Hence, if we assume the levees to be mainly represented by shaly sands, the channel to be filled with clean sands, and the cap-rock to be shale (lower left), the resulting zero-offset seismic response (lower right) shows exactly the same seismic response as the real seismic section. Simple zero-offset modeling like this can be very informative and beneficial before or during seismic interpretation, for a qualitative assessment of seismic amplitudes. However, complex geology and AVO effects are not accounted for. For this specific case, it would be of interest to determine whether the channel is filled with transparent shales or transparent sands. More rigorous modeling including AVO effects could reveal the correct lithology.

4.5.2 2D elastic seismic modeling

To show an example of multilayer, offset-dependent seismic modeling, we create a 2D Earth model based on the facies information from a North Sea well (well logs shown in Plate 2.31) combined with stratigraphic information from seismic interpretation of the 2D seismic line intersecting this well. This model is a simplification of the real case. However, it is a realistic model that honors vertical facies variations observed in the type well, and takes into account the interpreted lateral extent and geometry of the observed facies. In Plate 4.29, the seismic section is zero-phase, peak frequency is 30 Hz, and a black peak in the wiggle display represents a positive stacked amplitude. The seismic horizons included in the figure correspond to major lithostratigraphic boundaries.

Top Heimdal is the interface of main interest, representing the top of the reservoir. This horizon changes character laterally, and a polarity change is observed in the stack section. We assume that this lateral variation reflects changes in the reservoir rock properties. We interpret the 2D cross-section to transect laterally from oil-filled lobe sands at the well location, into marginal facies (Facies III) in both directions. Marginal facies are observed conformably underlying the lobe sands, and these are believed to correlate with the marginal facies laterally from the lobe sands. This interpretation is guided by observations made in the seismic amplitude map in Plate 1.1 (see also Figure 5.1), and the fact that Facies III represents the top of the reservoir in one of the marginal wells (see case study 1, Chapter 5). The general conceptual model of turbidite systems (Walker, 1978) shown in Figure 2.30, and application of Walther's law (Middleton, 1973; Chapter 2), also support this interpretation. The resulting Earth model is depicted in Figure 4.30.

Facies and rock physics properties that build up the model are listed in Table 4.2. Seismic forward modeling is conducted using a commercial 2D dynamic ray-tracing package, assuming elastic and isotropic conditions. The seismic pulse used is a zero-phase Ricker wavelet with 30 Hz center frequency. The modeling creates synthetic

Table 4.2 *Rock properties for each facies or layer in the Earth model*

Layer	V_P (m/s)	V_S (m/s)	Density (g/cm³)	V_P/V_S	AI (m/s × g/cm³)
0–150-m water zone	1500	0	1	∞	1500
Overburden (sand and shale)	1850–2390	450–950	1.8–2.2	4–2.5	3300–5260
Tuff (Balder Fm)	2600	1200	2.3	2.17	5980
Facies V (Sele Fm)	2300	950	2.25	2.42	5175
Facies IV (Lista Fm)	2400	1000	2.25	2.4	5400
Facies IIb-oil (Heimdal Fm)	2440	1300	2.02	1.88	4930
Facies IIc-oil (Heimdal Fm)	2630	1400	2.06	1.88	5420
Facies III (Heimdal Fm)	2750	1200	2.2	2.3	6050
Facies IIa (Heimdal Fm)	3100	1600	2.15	1.94	6650
Chalk (Ekofisk Fm)	3500	1700	2.3	1.94	8050

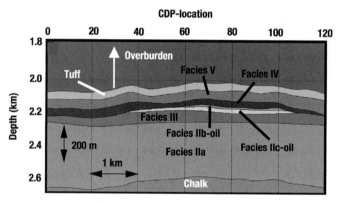

Figure 4.30 The geological model used as input for the seismic modeling. Elastic properties are given in Table 4.2. Note that this figure is not to scale. The lateral extension is 6 km and the vertical/lateral ratio is about 1/6.

pre-stack seismic gathers along the section. Only primary reflectors are included, and the offset-dependent reflectivity is calculated using the Zoeppritz equations (Zoeppritz, 1919) at each interface. These gathers are stacked at limited ranges to create a near-stack, a far-stack and a full-stack seismic section corresponding to our Earth model.

The results from the forward seismic modeling are shown in Figure 4.31. Comparing the synthetic full-stack section with the real stack section, we observe a good fit. This shows that the Earth model can explain the seismic signatures observed in the real data. Considering the Top Heimdal horizon, we clearly observe a phase shift as we go from marginal facies to the lobe on the full stack. Correspondingly, we observe a bright spot on the far stack with large negative amplitudes at the Top Heimdal level, while the near stack shows a much weaker seismic response at this level.

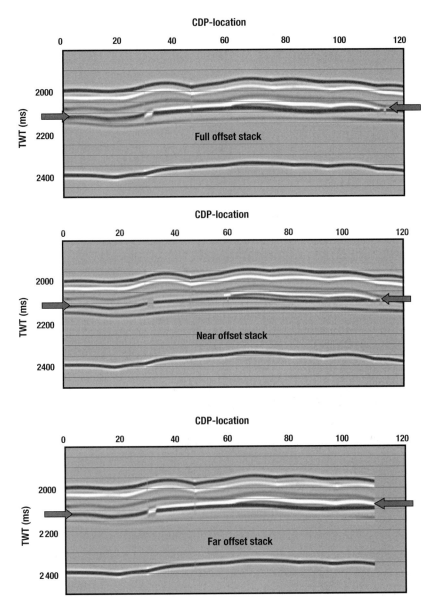

Figure 4.31 Synthetic seismic modeling results, including a full-offset stack section (upper), a near-offset stack (middle) and a far-offset stack (lower). The data are zero-phase, and peak frequency is 30 Hz. White amplitudes represent negative reflectivity. Arrows on the sides indicate the Top Heimdal horizon. Note the much brighter white amplitude on the far-offset stack section compared with the near-offset stack section at the Top Heimdal horizon (CDP range 35–110). Also note the phase change and positive reflectivity along this horizon in the CDP range 0–35. The CDP spacing is 50 m.

Ray theory modeling will break down in very heterogeneous and/or structurally complex media. In these cases, more advanced finite-difference algorithms are normally used for seismic modeling. It is beyond the scope of this book to go into the details of ray theory and finite-difference methods for seismic modeling, but some excellent references on the theory of seismic wave propagation include Cerveny (2001) and Aki and Richards (1980).

A fast convolution-based elastic modeling in complex geologic media was developed by Petersen (1999). His modeling approach takes into account geologic processes (tectonic and sedimentologic), which makes it easy to modify and update the Earth model. It also allows for quick lithology and fluid substitution within complex geologic settings.

4.6 Future directions in quantitative seismic interpretation

We see some clear trends in quantitative seismic interpretation: more rigorous modeling and inversion of the wave propagation phenomena; combining sedimentologic and diagenetic modeling with rock physics modeling to obtain more realistic predictions of seismic properties; probabilistic Monte Carlo simulations to capture uncertainties in both rock physics and inversion results; and incorporation of geostatistical methods to account for spatial correlations in reservoir properties.

Today, two-term AVO analysis is still the most common means to estimate elastic properties from pre-stack seismic data. However, higher-order, ultra-far AVO analysis, although immature, is a technology that can potentially provide us with additional information about reservoir properties from seismic data. Furthermore, full-waveform pre-stack inversions will become more common as computer power increases. Benabentos *et al.* (2002) used hybrid inversions which combined pre-stack inversion with post-stack inversion to quantify lithologies. Bachrach *et al.* (2004) quantified uncertainties in reservoir prediction using full-waveform pre-stack inversion combined with rock physics analysis and mapped the estimated probabilities of different lithologies in deep-water Gulf of Mexico. Not only will we see inversions of the elastic seismic properties, but also increased use of attributes related to attenuation. Attenuation has always been difficult to estimate reliably from seismic data. However, recent techniques give us hope that it will become more common to use Q_P and Q_S in addition to V_P, V_S, and density for reservoir characterization.

Integration of geologic processes, by numerical modeling, will open up new doors in quantitative seismic interpretation. Helseth *et al.* (2004) combined numerical modeling of diagenetic processes with the rock physics models shown in Chapter 2, to predict quantitative depth trends in seismic properties. There is also a trend of using results from quantitative interpretation techniques in virtual reality rooms for improved delineation of geomorphologic elements.

4.6 Future directions

Because reservoir characterization is inherently uncertain and risky, Monte-Carlo-based techniques are essential for managing the uncertainty. Using simple deterministic models without capturing the variability can lead to erroneous decisions. Representation of quantitative seismic interpretation in terms of probabilities allows the results to be more easily incorporated into economic risk analysis. One of the main goals of this book is to show how seismic interpretation can be represented quantitatively in terms of probabilities.

5 Case studies: Lithology and pore-fluid prediction from seismic data

The path of precept is long, that of example short and effectual.

Seneca

The case study examples in this chapter make use of the techniques described in the previous chapters, to estimate the uncertainty and map the probability of occurrence of different facies and fluids away from the well locations by combining attributes from seismic analyses with statistical rock physics.

The first case study uses pre-stack seismic amplitude analyses to delineate reservoir zones in the North Sea. In the second study, again in the North Sea, use is made of seismic impedance inversions, statistical rock physics, and geostatistics to characterize the reservoir by mapping probabilities of occurrence of facies and fluids. In the third case study we show how we can combine statistical rock physics, lithofacies interpretation, and AVO analysis to discriminate between lithologies and thereby improve detectability of hydrocarbons from seismic amplitudes in Grane field, North Sea. The fourth study, from West Africa, shows an example of using seismic amplitude analyses and depth trends in rock properties to classify hydrocarbon zones at different depths. The fifth case study is an example of the full workflow of rock physics template (RPT) analysis, starting with the selection of the most appropriate RPT using well-log cross-plot analysis followed by rock physics interpretation of elastic inversion results using the selected template. The example is from the Grane field in the North Sea, the same field as for case study 3.

5.1 Case 1: Seismic reservoir mapping from 3D AVO in a North Sea turbidite system

5.1.1 Introduction

In this case study (Avseth *et al.*, 2001a, 2001b), we conduct seismic reservoir characterization constrained by well-log rock physics (see Chapter 2) and facies classification (see Chapter 3), and apply it to the Glitne field. The Glitne field is a turbidite system located in South Viking Graben, North Sea, whose reservoir sands represent the

5.1 Case 1: 3D AVO

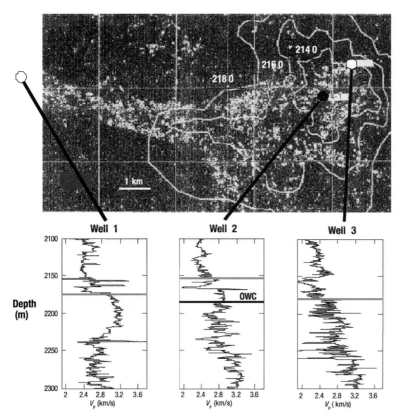

Figure 5.1 Seismic reflectivity map (above) of Top Heimdal Formation, corresponding to the gray lines in the well logs (P-wave velocity) (below). The bright amplitudes reflect relatively strong positive stack responses. The three wells penetrate a submarine fan in the feeder channel (Well 1), in a lobe channel (Well 2), and in the marginal area of the lobe (Well 3). The Heimdal Formation dramatically changes character between the wells. Oil was encountered in Well 2, and the area represents a commercial oil field, the Glitne field. The oil–water contact (OWC) in Well 2 is indicated by the horizontal black line. The contours in the seismic map are in two-way traveltime (ms) and illustrate the structural topography of the lobe. (The reflectivity map is courtesy of Norsk Hydro.)

Heimdal Formation of Late Paleocene age and include an oil field of economic interest (Figure 5.1). By linking lithofacies to rock physics properties, using statistical techniques to account for natural variability within, and overlap between different facies, we obtain a probabilistic link between facies, rock properties, and seismic response. This allows us to predict the most likely lithofacies and conditional probabilities of facies from seismic data. The proposed methodology, including the steps presented in Chapters 1, 2, 3 and 4, ultimately improves the ability to delineate subtle traps and characterize reservoir units in complex depositional systems from seismic data.

We have a comprehensive database available for this study, including thin sections and cores, well-log data from seven wells (five of the wells are located in the field

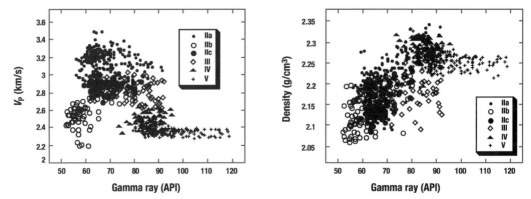

Figure 5.3 P-wave velocity versus gamma ray (left) and density versus gamma ray (right), for different seismic lithofacies in training data (Well 2). Note the ambiguity in P-wave velocity between Facies IIb and IV/V.

of study while two are located in a neighboring field), CDP (common depth point) gathers from selected seismic lines and a 3D seismic cube covering the area of interest. The thin sections and cores are used to guide the facies identification from well-log data. The well-log data available for classification and generation of probability density functions (pdfs) include P-wave velocity (V_P), density and gamma ray for all the seven wells. In addition we have S-wave velocity (V_S) and resistivity data (shallow and deep) from two of the wells. Helium porosity data are available from the cored zone in Well 2. The pre-stack seismic data (CDP gathers) both from the selected 2D lines and from the 3D cube have been pre-processed for true amplitude recovery and AVO analysis. The processing includes spherical divergence correction, pre-stack FK time migration, NMO correction, Radon-transform multiple removal, and surface-consistent offset balancing.

5.1.2 Rock physics and facies analysis of well-log data

As discussed in Chapter 2, relating lithofacies to rock physics properties will improve the ability to use seismic amplitude information for reservoir prediction and characterization in complex depositional systems. There are several reasons why facies are important in seismic interpretion. Firstly, facies occur in predictable patterns in terms of lateral and vertical distribution. Facies can also be linked to sedimentary processes and depositional environments. Moreover, facies have a major control on reservoir geometries and porosity distributions.

The facies definitions used in this case study are listed in Section 2.5.1. The different facies are demarcated on the logs from the type-well in Plate 5.2. Figure 5.3 shows the different seismic lithofacies plotted as P-wave velocity versus gamma ray (left), and density versus gamma ray (right). We observe an overturned V-shape, and an

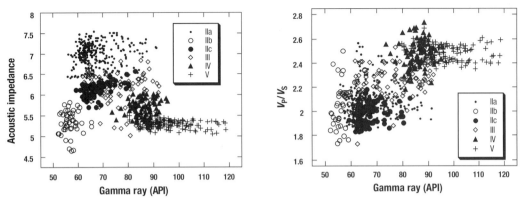

Figure 5.4 Acoustic impedance versus gamma ray (left) and V_P/V_S ratio versus gamma ray (right) in type-well.

ambiguity exists between Facies IIb and IV/V. Cemented sands (IIa) and laminated sands (IIc) as well as interbedded sand–shales have relatively high velocities. The sand–shale ambiguity is not observed in density versus gamma ray. Here we see a more linear trend where density increases with increasing gamma-ray values (i.e., clay content) as we go from clean sands (Facies IIa and IIb) to silty shales (Facies IV). However, we observe that silty shales have higher densities than pure shales. The sand–shale ambiguity observed in terms of velocity is also observed in acoustic impedance, which is the product of V_P and density (Figure 5.4; left). The overturned V-shape we observe can be explained physically: for grain-supported sediments, increasing clay content tends to reduce porosity (i.e., increase density) and therefore stiffen the rock. However, for clay-supported sediments, porosity will increase with increasing clay content because of the intrinsic porosity of clay, and the rock framework will weaken. Hence, velocity will reach a peak when clay content is approximately 40% (see also Section 2.2.3 and Figure 2.8).

The shear-wave sonic log provides us with shear-wave velocity (V_S). Figure 5.4 (right) shows the V_P/V_S ratio versus gamma-ray value. Here we observe that Facies IIb can be distinguished from shales (Facies IV and V), as the V_P/V_S ratio increases with increasing shaliness. Higher V_P/V_S ratios in shales than sands are expected, since the shear strength in shales tends to be relatively low compared with sands, owing to the platy shapes of clay particles.

Potentially, the trends observed in our cross-plots could be influenced by variation in pore fluid, as the thick-bedded sand units identified as IIb and IIc are located within the oil zone. However, the shallow and deep resistivity logs and helium porosity measurements indicate invasion of mud filtrate in the shallow zone (Figure 5.5). The density logs are proven to measure mud filtrate by calibration to the helium porosity measurements (Figure 5.6). We have no direct proof that the velocity logs measure in the invaded zone, but the perfect match between the velocity–porosity data of Facies IIb and the

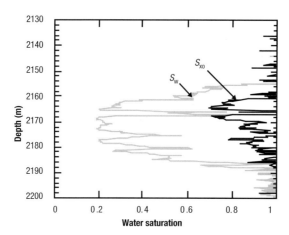

Figure 5.5 Saturation curves derived from resistivity logs in the reservoir zone of Well 2, indicating the effect of mud-filtrate invasion. S_w is water saturation in the reservoir. The oil saturation of the reservoir equals $1 - S_w$. S_{xo} is the water saturation in the invaded zone. The residual oil saturation in the invaded zone equals $1 - S_{xo}$.

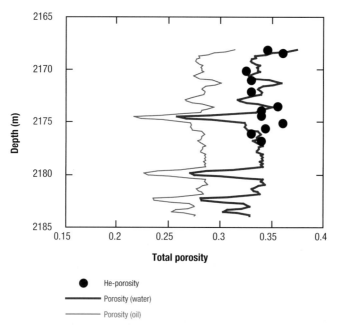

- He-porosity
- Porosity (water)
- Porosity (oil)

Figure 5.6 Porosity logs derived from the density log in the interval 2168–2184 m in Well 2, representing the lower part of the oil zone, where the sands are plane-laminated (Facies IIc). This is the interval where helium porosity data are available. If we assume the pore fluid is oil with a density of 0.78 g/cm^3, the porosity log shows values that are too low compared with the helium porosity measurements. Assuming a saline water density of 1.09 g/cm^3, the porosity log matches with the helium porosities. This proves that the formation is invaded by mud filtrate in the zone measured by the density tool.

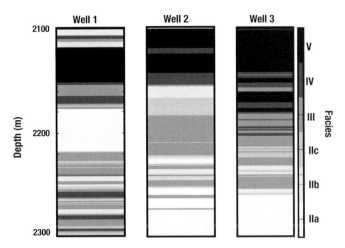

Figure 5.7 Seismic lithofacies classification results in the three wells shown in Figure 5.1. The channel sands in Well 1 (2172–2220 m) are classified as cemented (IIa), while the marginal lobe facies encountered in Well 3 (2180–2240 m) are identified as interbedded sand–shales (III). Well 2 is the type-well and the reservoir facies comprise Facies IIb at the top (2155–2165 m) and Facies IIc beneath (2166–2183 m)

friable-sand model, shown in Plate 2.31 (Section 2.5.3) was obtained assuming mud filtrate as pore fluid. Furthermore, since Facies IIb and IIc in Well 2 have about the same oil saturation, it is clear that the relatively large increase in velocity when we go from Facies IIb to IIc must be related to lithologic and/or textural changes. This indicates that the sonic log is also measuring in the zone invaded by mud filtrate. Hence, the variations in seismic properties observed in Figures 5.3 and 5.4 should only reflect facies variations.

5.1.3 Creating nonparametric facies and pore-fluid pdfs

To correlate and describe the reservoir between the wells is an impossible task without using seismic data, and the goal is therefore to predict from seismic amplitudes the character of the reservoir in the interwell areas. In this section we generate probability density functions (pdfs) of seismic parameters based on the well classification, and these pdfs will then be used to create facies maps from seismic data.

We first do statistical facies classification of all the wells used in the case study. This allows us to create cumulative distribution functions from which we can perform Monte Carlo simulation of seismic parameters and create the AVO pdfs (for details see Chapter 3 and Section 4.3.11). The log data from the type-well (Well 2) are used as training data for a multivariate statistical classification of seismic lithofacies in other wells in the area. Figure 5.7 shows the classification results in Wells 1–3, where only gamma-ray and P-wave velocity logs were used, as the density logs were found to be

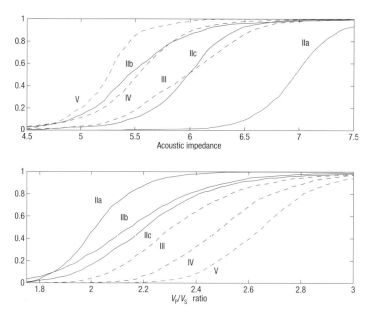

Figure 5.8 Cumulative distribution functions (cdfs) of acoustic impedance and V_P/V_S ratio for each of the brine-saturated facies. We observe a much better discrimination in V_P/V_S ratio than in acoustic impedance.

corrupted by wash-outs and rough borehole surfaces in some intervals. The classification was done in seven wells using Mahalanobis quadratic discriminant analysis (Davis, 2002; Doveton, 1994). This method uses the means and covariances of the training data. Samples are classified according to the minimum of the Mahalanobis distances to each cluster in the training data (Duda and Hart, 1973; Fukunaga, 1990). The Mahalanobis distance classification technique is further explained in Chapter 3.

In Figure 5.7 we observe that the feeder-channel sands in Well 1 (2172–2220 m) are mainly classified as cemented clean sands (Facies IIa), whereas the lobe-channel sands in Well 2 (2155–2165 m) are classified as uncemented sands (Facies IIb). Furthermore, we observe that the Top Heimdal is represented by interbedded shales/sands in the lobe margin area where Well 3 is located (2180–2240 m). As confirmed in the deterministic AVO analysis in Figure 4.7, this dramatic variability in the lateral facies distribution going from a relatively proximal feeder-channel environment to a relatively distal lobe and lobe margin environment has great impact on the seismic signatures in this turbidite system.

Based on the facies classification, we first extract cumulative density functions of seismic properties for each of the lithofacies, and for oil-saturated sand facies (Figures 5.8 and 5.9). The oil-saturated cdfs were calculated from the water-saturated cdfs using the Biot–Gassmann theory (Gassmann, 1951; see Chapter 1). As for the training data, we observe a much better discrimination in V_P/V_S ratio than in acoustic impedance in

5.1 Case 1: 3D AVO

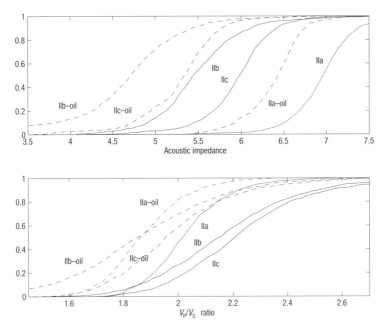

Figure 5.9 Cumulative distribution functions (cdfs) of acoustic impedance and V_P/V_S ratio for oil versus brine saturation in the sandy facies. We observe a much better discrimination in V_P/V_S ratio than in acoustic impedance.

terms of lithofacies. The cdfs in Figure 5.9 show that V_P/V_S ratio also discriminates pore fluids better than acoustic impedance. Hence, as suggested in Section 4.3.7, amplitude versus offset (AVO) analysis must be used to predict lithofacies from seismic data.

From the cdfs of velocities and density, we create probability density functions (pdfs) of AVO response for different lithofacies combinations, and assess uncertainties in seismic signatures related to the natural variability within each facies. The pdfs are generated from Monte-Carlo-simulated seismic properties drawn randomly from the two lithofacies cdfs, one for the cap-rock and one for the underlying facies. First we simulate V_P and then V_S followed by density. We make sure the simulation honors the correlation between the three parameters. The corresponding reflectivity simulations are calculated using Shuey's equation (i.e., equation (4.8)). The AVO pdfs can be plotted either as reflectivity versus offset (cf. Figure 4.14), or as zero-offset reflectivity $R(0)$ versus the AVO gradient G (Figure 5.10). In Figure 5.10 the center or peak of each contour plot represents the most likely set of $R(0)$ and G for each facies. These pdfs show how $R(0)$ and G can vary for a given facies combination, and that different facies combinations can have overlaps. However, the most likely set of $R(0)$ and G is a unique characteristic of a given facies combination. For instance, a cemented sand (Facies IIa) with brine will likely have a relatively large positive $R(0)$ and a relatively large negative G, whereas an oil-saturated cemented sand will more likely have a smaller positive $R(0)$ and larger negative gradient. However, there is a great overlap between the

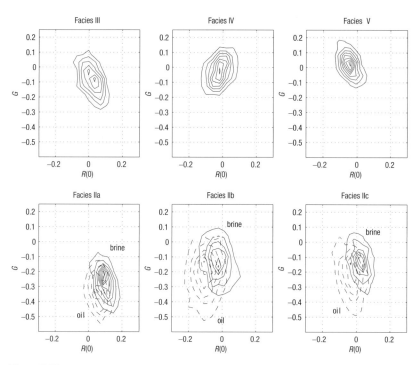

Figure 5.10 Bivariate distribution of the different seismic lithofacies in $R(0)$–G plane, assuming Facies IV as cap-rock. The center of each contour plot represents the most likely set of $R(0)$ and G for each facies.

water-saturated and oil-saturated sands, and oil sands can potentially show larger $R(0)$ and smaller negative G values than water sands. There is also overlap between different types of sands. Another interesting observation is that even a shale–shale interface can cause a significant seismic response. In general, these pdfs create a probabilistic link between facies and seismic properties that can be used to predict the most likely facies and the conditional probability of a given facies, from seismic data.

To better assess important trends in terms of the $R(0)$–G bivariate plots, we lump all oil sands together into one group, all brine sands in another and all shaly facies in a third, and plot them together in the same cross-plot (Figure 5.11). Only the iso-probability contours of 50% and larger are included. In spite of significant overlaps, there is a fairly good separation between shales and sands, and between oil sands and brine sands. From these plots we observe that both $R(0)$ and G are needed to discriminate between facies and pore fluids in our case.

5.1.4 Facies and pore-fluid classification of AVO attributes

We use our AVO pdfs to predict lithofacies and pore fluids from 2D and 3D prestack seismic data. We first conduct a realistic seismic forward modeling along a 2D

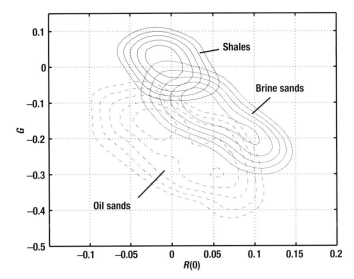

Figure 5.11 AVO pdfs for main facies groups: oil sands, brine sands and shales. Only the iso-probability contours of 50% and larger are included for each group. The $R(0)$ and G pdfs nicely separate the three facies groups, but there are significant overlaps.

cross-section intersecting the type-well (see Section 4.5), and predict most likely lithofacies and pore fluids along the top reservoir horizon from the synthetic data. This becomes a feasibility study on how well the methodology works, as the input Earth model is known. Then we use the same technique to predict facies and pore fluids from the real 2D seismic section intersecting the well. Finally, we characterize facies and pore fluids, and map their occurrence probability, over the whole field using 3D AVO data.

AVO inversion and facies prediction from synthetic seismic data

The next step is to use the offset-dependent reflectivity information in the synthetic seismograms (see Figure 4.31) to see if we are able to predict the correct facies present immediately beneath the Top Heimdal horizon. We extract $R(0)$ and G along this horizon using AVO inversion based on generalized least-squares as available in a commercial AVO package. A common procedure to calculate $R(0)$ and G from pre-stack seismic data is described in Chapter 4.

Combining the inverted AVO parameters, $R(0)$ and G, with the bivariate probability distributions in Figure 5.10, we are able to predict the most likely seismic lithofacies present below the Top Heimdal horizon in the synthetic seismic section. The results are shown in Plate 5.12.

The lithofacies are indicated both in terms of a graph and as a color display. For computational reasons, the facies are given integer numbers 1 through 9, according to the following scheme:

1 = Facies IIa with oil	4 = Facies IIa with brine	7 = Facies III
2 = Facies IIb with oil	5 = Facies IIb with brine	8 = Facies IV
3 = Facies IIc with oil	6 = Facies IIc with brine	9 = Facies V

For convenience, the sandy facies with oil (1 through 3) are red-colored, the sandy facies with brine (4 through 6) are yellow-colored, whereas the shaly facies (7 through 9) are green-colored.

In Plate 5.12, we have superimposed the true $R(0)$ and G values calculated from Table 4.1 with the predicted (inverted) $R(0)$ and G, respectively. There is a relatively nice fit between true and predicted $R(0)$, while true and inverted G show larger discrepancy. The largest discrepancy in $R(0)$ occurs where Facies IIc is the true answer (CDP 36–60). However, Facies IIc is relatively thin ($\sim$10 m) and pinches out laterally. Hence the discrepancy can be related to tuning effects. The total thickness of the Heimdal Formation reservoir sands encountered in Well 2 is about 35 m. This is approximately half a wavelength, and at this location the sands are therefore seismically resolvable. Accordingly, we expect no major tuning at places along the line other than at the pinch-out of Facies IIc. Nevertheless, G shows relatively large discrepancy in several places along the section. This could be due to focusing/defocusing of energy as some of the overlying horizons are rather curved, and this could have caused the nonhyperbolic moveouts that were observed locally. (The synthetic section used for the inversion has not been pre-stack migrated.) The largest discrepancy in G, however, occurs in the pinch-out zone of Facies IIc where we observe tuning of $R(0)$. Consequently, this zone also has substantial error in terms of predicted facies. Shales of type IV and brine-saturated sands of type IIc are predicted where the true answer is oil-saturated sands of type IIc. Elsewhere, the predicted most likely lithofacies underlying the Top Heimdal horizon match very well with the true facies given in the Earth model.

Facies and pore-fluid prediction from real 2D seismic section

Now, we want to use the AVO-pdfs in Figure 5.10 to predict facies and pore fluids from a real 2D seismic section. We select the same 2D line as the one from which we derived our Earth model in the synthetic case (i.e., the seismic line intersecting the type-well, Well 2). Thus, if our Earth model is more or less correct, we should expect the predicted reservoir rocks to be similar in the synthetic and the real cases. The assumption of a consistent cap-rock of Facies IV is reasonable as the Lista Formation which overlies the Heimdal Formation reservoir rocks is normally represented by hemipelagic, silty shales (cf. classification results in Figure 5.7).

Figure 5.13 shows the real 2D seismic stack section (wiggle-trace display, zero-phase wavelet, 30 Hz peak frequency) intersecting the type-well, the same line as shown in Plate 4.29. Plate 5.14 shows the extracted $R(0)$ and G along the picked Top Heimdal horizon, and the predicted most likely seismic lithofacies present below the horizon. We

5.1 Case 1: 3D AVO

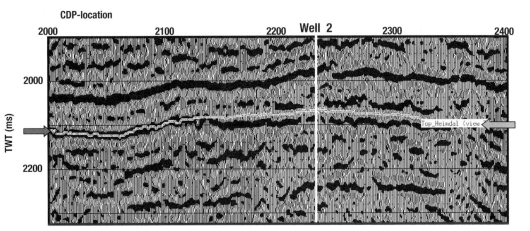

Figure 5.13 Seismic section intersecting the lobe of the submarine fan. The picked horizon and the arrows on the side indicate the top of the Heimdal sands. There is a marked phase-shift along the Top Heimdal horizon. CDP spacing is 18 m.

predict mainly oil sands of type IIb and IIc within the interval where the Top Heimdal horizon has a negative $R(0)$. This is very similar to what is suggested in the Earth model in Figure 4.30, although the subfacies of sands are not always the same. Bear in mind that our oil facies pdfs represent 100% oil saturation, while the true oil saturation in the reservoir is varying between 0.2 and 0.8 (Figure 5.5). This can have an effect on the prediction of sand type (IIb versus IIc).

In the area where the Earth model has shaly sands or interbedded sand-shales (Facies III), the prediction shows a more heterogeneous character. We observe both shaly sands (Facies III) and thick-bedded, cemented sands (Facies IIa) with oil. This indicates that there is probably another lobe-channel intersected by the real 2D line that we did not include in the synthetic modeling. An alternative explanation is that this local oil-saturated sand is a result of tuning effects or noise in the data, as discussed for the synthetic case. A third explanation is lateral facies variations in the Lista Formation above the reservoir, obstructing our assumption of a cap-rock consisting of only Facies IV. These issues are further discussed below.

Facies and pore-fluid prediction and probability maps from 3D AVO data

The next step is to expand on our results from the 2D seismic line and perform facies and pore-fluid prediction from 3D seismic data. Three-dimensional AVO inversion is done on the turbidite system using the same commercial inversion software that was used for the 2D line. Again, we focus only on the horizon representing the top of the system (Top Heimdal). Figure 5.15 shows the 3D topography (in two-way traveltime) of this seismic horizon, where the geometries of the feeder-channel and the lobe structure

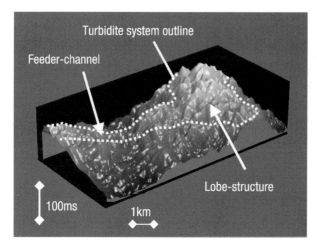

Figure 5.15 Three-dimensional seismic topography of Top Heimdal horizon (traveltime). The depositional geometry of a feeder-channel and fan lobe is outlined (compare to Figure 5.1).

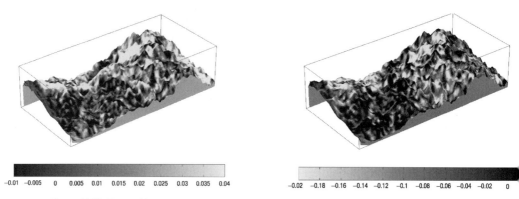

Figure 5.16 Zero-offset reflectivity, $R(0)$ (left), and AVO gradient, G (right), along Top Heimdal horizon.

are outlined. The inversion gives us $R(0)$ and G over the whole area, along this horizon slice. Figure 5.16 shows $R(0)$ (left) and G (right). These plots allow us to predict the most likely seismic lithofacies under this horizon. This is done by combining the $R(0)$ and G inverted from the seismic with the $R(0)$–G bivariate pdfs derived from well-log data. Before we can do this, however, the inverted parameters must be calibrated to the well-log values.

Figure 5.17 shows the comparison between the well-log-derived $R(0)$ and G values and the $R(0)$ and G from the AVO inversion. The upper left subplot shows the global training data from the well logs. In the upper right subplot are the raw unscaled $R(0)$ and G values derived from the least-squares AVO inversion. We calibrate the inverted $R(0)$ and G at Well 3. In this well we observe Facies III beneath the Top Heimdal horizon. We first calculate the mean uncalibrated $R(0)$ and G from a small area around the well

5.1 Case 1: 3D AVO

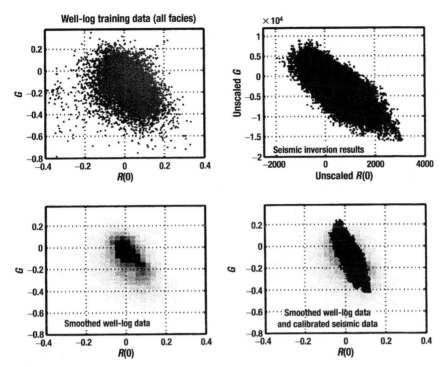

Figure 5.17 Comparing the global training data of $R(0)$ and G derived from well-log data (upper left; Monte Carlo simulated values) to 3D AVO inversion results (upper right). The calibrated AVO parameters show a smaller range than the well-log data, but the scatter matches nicely with the distribution of the well-log pdf (lower left and right). The dark-colored two-dimensional bins in the lower left subplot represent relatively high frequencies of data points within a bin.

(approximately 200 m × 200 m). Then we calibrate these values to the mean values of $R(0)$ and G of Facies III calculated from the well-log data. The calibrated seismic data are shown in the lower right subplot.

The smaller scatter in $R(0)$ and G in the seismic data compared with the well-log data is expected because of the scale difference. We assume that the well-log-derived pdfs can still be used to predict facies and pore fluids from the seismic data. This assumption implies that all the facies present beneath the Top Heimdal horizon are also present in the global well-log training data. In order to compare the well-log $R(0)$ and G with the calibrated $R(0)$ and G from the seismic data, we superimpose the estimated well-log pdf (lower left subplot in Figure 5.17) on the seismic data. The calibrated values match the well-log pdf very nicely.

The next step is to use the well-log-derived AVO pdfs to predict facies and pore fluids from the seismic data. To get a general picture of the reservoir, we first distinguish only between oil versus brine, and sands versus shales. Hence, we group similar facies together. Facies IIa with oil, IIb with oil and IIc with oil are lumped into a facies group referred to as oil sands. Similarly, we have created a brine sands group. Facies III, IV

Case studies

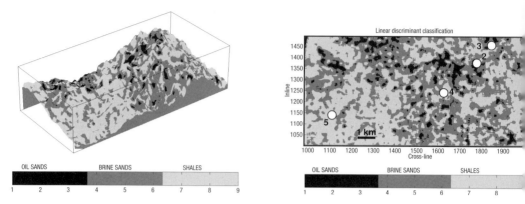

Figure 5.18 Lithofacies prediction beneath a seismic horizon with 3D topography (left) and in map view (right).

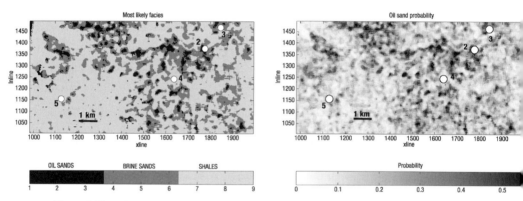

Figure 5.19 Left, most likely facies derived from pdfs; right, oil-sand probability.

and V have been lumped into a facies group of shales. First, we apply the Mahalanobis distance method to calculate the most likely facies group and pore fluid. The results are shown in Figure 5.18 (left: 3D topography; right: map view). We predict oil-saturated sands in the lobe area where the lobe is structurally highest. The rest of the lobe area is most likely water-saturated according to the prediction. Furthermore, we predict oil-filled sands in the upper feeder-channel. Outside the submarine fan, mainly shale is predicted to be the most likely facies. The exception is an area just north of the feeder-channel where oil and brine sands are predicted. If this prediction is correct, it could imply the presence of some overbank sands.

The overall prediction is reasonable in terms of facies and pore-fluid distribution. The sands are mainly predicted in the channel and lobe areas while oil is predicted in the structurally highest areas of the sand deposits.

The next step is to use nonparametric pdfs to calculate the conditional posterior probabilities of the various facies groups and pore fluids. Figure 5.19 shows two map views of the Top Heimdal horizon calculated from the well-log-derived pdfs, one with the nonparametric facies classification results and one with the estimated probability of oil sands given the observed $R(0)$ and G:

$$P\{oil \mid R(0), G\} \tag{5.1}$$

The map to the left shows the most likely facies and pore fluid. The results are very similar to the results from the linear discriminant method (see Figure 5.18). The oil sands are predicted mainly in the feeder-channel and in the central part of the lobe. The map to the right shows the probability of oil sands, and in accordance with the map of most likely facies, we recognize relatively high probabilities in the central lobe, in the upper feeder-channel and in the possible splay deposits north of the feeder-channel. Outside the turbidite system, there are very low probabilities of oil sands.

Plate 5.20 shows the probability of oil sands (upper left), brine sands (upper right), brine and oil sands together (lower left), and shales (lower right). The high probabilities of oil and brine sands together (sand probability map) nicely depict the depositional pattern of a submarine fan. Also note that there are relatively high probabilities of brine sands even where the most likely sands were predicted to be oil sands. This stems from the fact that brine sands and oil sands have a large overlap in terms of $R(0)$ and G (see Figure 5.11). The low probabilities of shales in the lower right map depict the depositional pattern of a submarine fan as shales are found outside the margins of the system and in interchannel areas of the lobe complex.

We want to analyze more detailed probability maps of individual facies to gain a better sedimentologic understanding of the studied turbidite system. Plate 5.21 shows probability maps of the different facies. Since there are nine facies including oil-saturated facies, probabilities larger than 0.11 indicate more likely occurrence than just by random chance. The upper three subplots show the three different subfacies of sand saturated with oil. We observe relatively high probabilities of Facies IIa with oil predominantly in the upper feeder-channel and on the lobe structure, while Facies IIb with oil has relatively low probabilities over the whole system. Facies IIc with oil has relatively high probabilities in scattered areas of the lobe area. Facies IIa with brine shows a very similar probability map to the same facies with oil, with relatively high probabilities in the feeder-channel and proximal parts of the lobe. This could be explained by the fact that Facies IIa is a stiff rock type, resulting in a large overlap between the pdfs of oil and brine (Figure 5.10). Facies IIb with brine shows relatively large probabilities in the southern lobe area, north of the feeder-channel and in a small area just south of the feeder-channel. The two last occurrence probabilities could reflect overbank or splay sands from the feeder-channel. Facies IIc is found to have relatively high probabilities over a large area including the feeder-channel, lobe structure and an area north of the feeder-channel. Facies III shows relatively large probabilities along the feeder-channel and in the distal portions of the lobe. This is in accordance with conceptual models that interbedded sand–shales and shaly sands occur in marginal areas of a turbidite

Table 5.1 *Blind-test results at well locations*

Well	Facies (well-log observation)	P{sand} (only thick-bedded)			P{shale} (includes interbedded sand–shales)	Facies correct?	Fluid correct?
		P{oil}	P{brine}	P{total}			
2	Thick-bedded oil sands (II)	0.25	0.44	0.69	0.31	Yes	No
3	Interbedded sand–shale (III)	0.1	0.31	0.41	0.59	Yes	–
4	Thick-bedded brine sands (II) (thin oil cap)	0.16	0.38	0.54	0.46	Yes	Yes
5	Silty shale (IV)	0.09	0.23	0.32	0.68	Yes	–

$P\{x\}$ indicates the probability that x occurs at a given well location. Shales occur only as brine-saturated, hence we include the dash symbol in the fluid prediction column for Well 3 and Well 5.

system, either as levee deposits associated with channels, or in distal portions of the lobe (see Section 2.5.2). Finally, both Facies IV and V show high probabilities outside the turbidite system.

Blind test at well locations

Four wells penetrate the turbidite system inside the area of seismic inversion (see Figures 5.18 and 5.19). The calibration was done over an area around Well 3, with the mean value of Facies III. The exact $R(0)$ and G values estimated at Well 3 are also classified into the correct Facies III (see Table 5.1). Based on the calibration at Well 3, the three other wells were blind-tested in terms of facies and pore fluids. The results are listed in Table 5.1. Starting from the left, Well 5 encountered only shales at the target level, and the most likely lithofacies according to the seismic prediction is shale. Well 4 is located within lobe sands, but mostly brine-saturated (the oil column is about 10 m out of a total ~45 m of reservoir sands). In contrast, the most likely facies is Facies III, interbedded sand–shales. However, the total probability of thick-bedded sands (oil and brine) is 0.54, which is higher than the probability of shaly facies (0.46). Well 2 (the type-well), located structurally higher on the lobe, encountered 35 m of oil sands. We predict most likely brine sands, but the well is just on the fringe of an area of predicted most likely oil sands.

5.1.5 Discussion

We have shown how we can use statistical rock physics to translate 3D AVO inversion results into lithofacies and pore-fluid probability maps. In our case we have successfully mapped the most likely distribution of good-quality reservoir sands in a North Sea turbidite system and estimated the probability of finding oil within these sands. These maps are ultimate products in the process of geologically characterizing reservoirs from seismic data. They can be used as inputs for various decision and risk analyses during exploration and development, or as constraints for reservoir modeling and flow simulation during production and reservoir forecasting.

Although we have obtained a successful characterization of the turbidite system, it is important to be aware of certain limitations of the methodology proposed in this study. An important factor to be considered is that all the facies in the training data are at a well-log scale, while the prediction is at the seismic scale. This issue could be handled by creating physically upscaled pdfs from the well logs using effective-medium theory, especially for the interbedded sand–shales (Facies III). An upscaling of thin-bedded sequences using effective-medium theory (e.g., Backus, 1962) is needed if the intercalating layers have strong contrasts in elastic properties. However, the interbedded sand–shales in the studied turbidite system consist of thin-bedded sands that have weak contrasts in seismic properties compared with the intercalating shales (cf. small range in V_P within the Facies III cluster in Figure 3.6). In core observations made of Facies III in Well 2, the thin-bedded sands seemed to have a relatively high clay content, while the thin-bedded shales seemed to have a relatively high quartz content (i.e., silt). From Marion's (1992) study of sand–shale mixtures, we know that shaly sands and silty/sandy shales can have similar elastic properties. We therefore do not expect the thin-bed scale effect on rock physics properties to be important for Facies III, nor for the more thick-bedded facies observed in this turbidite system. Another aspect of scale that can cause problems to AVO analysis is tuning effects (see Chapter 4). The AVO inversions employed in this study assume no tuning. As a result the parameter estimates can be wrong in areas where tuning occurs. Consequently, classification and prediction of facies and pore fluids can also be wrong. This was manifested in the synthetic modeling and prediction case (Section 4.4.1). The training pdfs could be recreated to include the uncertainties caused by tuning.

The AVO inversion procedure itself is also a source of error (Chapter 4). We use a linear approximation of the Zoeppritz equations in our calculation of $R(0)$ and G. This approximation is known to be accurate for angles of incidences up to approximately 30° (Shuey, 1985). The data inverted in our case do not exceed this range, so the approximation is valid. The linear AVO inversion is furthermore sensitive to uncharacteristic amplitudes caused by noise (including multiples) or processing and acquisition effects (Chapter 4). A few outlying values present in the pre-stack amplitudes are enough to cause erroneous estimates of $R(0)$ and G. The 3D AVO inversion software used in this study, as opposed to the 2D AVO inversion software, applies a robust estimation technique (Walden, 1991) to limit the damage of outlying amplitudes. Other potential problems in the AVO methodology used here include errors in the moveout correction, cap-rock anisotropy, and focusing and defocusing of wave energy caused by lateral velocity variations in the overburden (see Chapter 4). We have neglected the effect of anisotropy in this study. In particular, some of the shales may be transverse isotropic. We also suspect defocusing and focusing of wave energy to play a role in the Glitne area, based on overburden observations of shale tectonics and deformation at ~1 km depth. The rugged traveltime map in Figure 5.15 can reflect lateral velocity fluctuations related to the shale tectonics. If overburden variation is statistically homogeneous over the area, however, the calibration of the inverted $R(0)$ and G with the well data partly accounts for this uncertainty. Local overburden effects on the other hand (e.g., major faults, shale diapirs, gas pockets), can cause nonlinear moveouts and abrupt changes in the offset-dependent reflectivity. In this case, the straight-line approximation of Shuey (1985) breaks down, and the estimation of $R(0)$ and G will be meaningless.

The pick of seismic horizon also represents an uncertainty. We do not know for sure if the seismic interpretation of the Top Heimdal horizon in our 3D case is correct everywhere. If the horizon is incorrectly picked, the estimated AVO data we use are not representative of our reservoir. As we have observed in the 2D cross-section intersecting Well 2, polarity reversals occur along the Top Heimdal horizon. Picking these can be a very difficult task. In fact the 3D interpretation of the Top Heimdal horizon, conducted prior to this study, was based on the belief that the Heimdal sands always have much higher impedance than overlying shales, resulting in a consistent positive reflector. This study shows that this is not the case, as variation in sand texture has a dramatic impact on the seismic response. However, it was an impossible task to double-check the 3D interpretation at every CDP gather prior to the 3D inversion. Therefore, the predictions from the 3D data can be affected by a subjectively picked horizon that does not necessarily coincide with the true top reservoir horizon. In particular, some of the unconsolidated sands saturated with oil, seen on the 2D data as negative stack amplitudes (Figure 5.13), are not detected in the 3D case (Figure 5.16).

Another important issue is whether the well-log training data are representative of the statistics of the entire reservoir. The well-log pdfs are calculated from vertically stacked facies, whereas the predicted facies are located laterally beside each other. Based on Walther's law of facies, we believe that the different facies observed vertically stacked in the wells are also present laterally over the large area where 3D seismic inversion is done. However, there may be facies observed in the wells that are not present beneath the Top Heimdal horizon. The opposite could also occur, with facies we have not observed in the wells being present beneath the Top Heimdal horizon.

Moreover, it is important to note that in the lithofacies prediction from AVO parameters we assume the cap-rock to be Facies IV (silty shale), which is not necessarily true everywhere. Nonetheless, well-log observations indicate that the Top Heimdal is consistently capped by a silty shale, so the assumption is reasonable. Other cap-rocks could be included in the prediction, but this could on the other hand cause more ambiguities in the results. We are, however, including the variability within silty shales in the calculations of the pdfs.

Finally, regarding the spatial distribution of facies, one expansion on this study has been to include spatial statistics in the facies prediction (Eidsvik et al., 2002). This technique implies a better control on the lateral facies transitions during the prediction.

With all these potential limitations and uncertainties that have not been considered in this case study, we still feel that including the uncertainties related to variability in facies and rock physics properties strengthens the validity of the seismic reservoir characterization of the Glitne field. Also, by linking the rock physics properties and seismic signatures to sedimentary facies, we can determine whether our results are geologically plausible. In a complete assessment of the uncertainties of reservoir characterization from seismic data, however, stochastic models of all the other factors mentioned above should be included.

The blind testing of wells (Table 5.1) represents a means of validating our methodology. The correct facies were predicted in all the wells, whereas pore fluid was incorrect in one well. The match between seismic predictions and well-log observations is not perfect, but that is not expected. Boreholes are "pinpoints" into the underground, while seismic data contain information from a relatively large area given by the Fresnel zone size. This is why we calibrated the seismic data from an area around Well 3. In addition, high-frequency random noise is present in the well-log data, which makes the comparison between seismic and well-log data even more difficult. Nevertheless, the blind-test results indicate that the results in this case study are reliable.

5.1.6 Conclusions

- We have estimated uncertainties and mapped probabilities of occurrence of different lithofacies and pore fluids from AVO attributes in a North Sea turbidite system (the Glitne field).
- We have analyzed real CDP gathers at several well locations, and successfully predicted the seismic lithofacies indicated by the well-log data. This demonstrates the feasibility of using AVO analysis to predict seismic lithofacies.
- Uncertainties in AVO response related to the inherent natural variability of each seismic lithofacies have been quantified using a Monte Carlo technique. The resulting AVO probability plots show that there are overlaps between different facies, but the most likely responses for each facies are nicely separated.
- Bivariate probability plots of zero-offset reflectivity ($R(0)$) versus AVO gradient (G) are created and calibrated to both 2D and 3D AVO attributes. Combining the $R(0)$ and G values estimated from the seismic data with the bivariate probability density functions (pdfs) estimated from well logs, we have used both linear discriminant analysis and Bayesian classification to predict lithofacies and pore fluids from the seismic amplitudes. The linear discriminant analysis is tested out on a synthetic seismic section, and the predicted facies match the "true" facies model very well, except in a zone where wavelet tuning occurs.
- For the 3D real data, the final results are spatial maps of the most likely facies and pore fluids, and their occurrence probabilities. These maps show that the studied turbidite system is a point-sourced submarine fan in which thick-bedded clean sands are present

in the feeder-channel and in the lobe channels, while interbedded sand–shale facies and shaly sands are found in interchannel and marginal areas of the system. Shales are located outside the margins of the turbidite fan. Oil is most likely to be present in the central lobe channel, and in parts of the feeder-channel.

5.2 Case 2: Mapping lithofacies and pore-fluid probabilities in a North Sea reservoir using seismic impedance inversions and statistical rock physics

In this case study (Mukerji *et al.*, 2001) we show how statistical rock physics techniques combined with seismic impedance inversions can be used to classify reservoir lithologies and pore fluids. One of the innovations at that time was to use both the normal-incidence acoustic impedance (ρV_P) and the so-called "elastic" impedance attributes. As described in Chapter 4, the elastic impedance (related to V_P/V_S ratio) incorporates far-offset data, but at the same time can be practically obtained using normal-incidence inversion algorithms. The methods were applied to a North Sea turbidite system (the same field as for case study 1, the Glitne field). We incorporated well-log measurements with calibration from core data to estimate the near- and far-offset reflectivity and impedance attributes. Multivariate probability distributions were estimated from the data to identify the attribute clusters and their separability for different facies and fluid saturations. A training data set was set up using Monte Carlo simulations based on the well-log-derived probability distributions. Fluid substitution by Gassmann's equation was used to extend the training data, thus accounting for pore-fluid conditions not encountered in the well.

Seismic inversion of near-offset and far-offset stacks gave us two 3D cubes of impedance attributes in the interwell region. The near-offset stack approximates a zero-offset section, giving an estimate of the normal-incidence acoustic impedance (ρV). The far-offset stack gives an estimate of a V_P/V_S-related elastic impedance attribute that is equivalent to the acoustic impedance for non-normal incidence. These impedance attributes obtained from seismic inversion were then used with the training pdfs to predict the probability of occurrence of the different lithofacies in the interwell region. Statistical classification techniques (such as those described in Chapter 3) and geostatistical indicator simulations were applied to the 3D seismic data cube. A Markov–Bayes technique was used to update the probabilities obtained from the seismic data by taking into account the spatial correlation as estimated from the facies indicator variograms. The final results are spatial 3D maps of not only the most likely facies and pore fluids, but also their occurrence probabilities. A key ingredient in this case study was the exploitation of physically based seismic-to-reservoir property transforms optimally combined with statistical techniques.

5.2 Case 2: Lithofacies and fluid probabilities

The ultimate goal of this study was to obtain reliable quantitative estimates, with their uncertainties, for relevant reservoir rock and fluid parameters in the area of exploration. The four major components of our study are as follows:

(1) Well-log analysis to define different seismic lithofacies, rock physics analysis including fluid effects and shear-velocity estimation in wells without shear logs, and log-based analysis of near- and far-offset seismic attributes for different lithofacies and pore fluids.

(2) Seismic inversion of near- and far-offset partial stacks to obtain 3D cubes of near- and far-offset impedances.

(3) Nonparametric multivariate probability density estimation of facies and near- and far-offset seismic impedances from wells and co-located seismic inversion to obtain the training pdfs.

(4) Statistical and geostatistical classification of the 3D seismic impedance cubes to predict most likely facies and pore fluids, and obtain the spatial distribution of probabilities of occurrence for different lithofacies and pore fluids.

5.2.1 Defining lithofacies from logs

The geological setting is a Tertiary turbidite system in the North Sea. Deep-water clastic systems and associated turbidite reservoirs are often characterized by very complex, heterogeneous sand distributions. Conventional seismic reservoir characterization may be very uncertain in these depositional environments. Linking 3D seismic imagery with rock physics properties of different facies and pore fluids can provide a powerful strategy for improved quantitative interpretation of seismic data. The North Sea turbidite system was covered by a marine 3D seismic survey which was specially processed for amplitude interpretation. As shown in Plate 5.22, an amplitude anomaly interpreted as a channel-fan submarine system was identified in an early stage, before this study. What does this amplitude anomaly mean *quantitatively* in terms of lithology and pore fluids? How can we link the 3D seismic imagery with rock and fluid properties of the different facies?

The well information is sparse within the coverage (approximately 300 km^2) of the 3D survey. The well data had been subjected to an extensive petrophysical analysis prior to this study, and the results from those analyses were made available to us. Of the five wells drilled in the area of 3D seismic coverage, only one had a shear-wave log. As shear information can be crucial for discriminating lithologic and pore-fluid ambiguities, a calibration well outside the study area was included in the analysis. This well had shear-wave data in the Tertiary interval, and was interpreted to have roughly the same depositional environment as the other wells within the study area. Emphasis was on careful analysis of the well data for calibration and use of the seismic data for lithofacies characterization in the interwell region.

We first defined seismic lithofacies representing seismic-scale sedimentary units with distinguishable characteristic petrophysical properties such as clay content, bedding configuration (massive or interbedded), petrography (grain size, cementation, packing, clay location), and seismic properties (P-wave velocity, S-wave velocity, and density). (See Sections 2.4.2 and 2.5.1.) This was the basis for quantitative facies and fluid estimation from seismic data. Seismic scale here refers to units that can be observed and mapped from seismic data. This depends on vertical and lateral seismic resolution, which is governed by wavelength, Fresnel zone radius and depth to target. In this study the wavelength was about 60 m and the thickness of the seismic lithofacies units about 10 to 15 m.

A key well was identified with a complete suite of good-quality logs that sampled all of the important lithologies in the turbidite system. Well logs play an important role in linking rock parameters to the seismic data. Figure 5.23 shows an example of some of the important logs from the key well. Reasons for choosing this well as the key well are that shear-wave information is available, the important facies of the turbidite system are all encountered in the well, and it is a new well with good-quality modern logs. The total porosity is relatively constant, and there is a marked increase in acoustic impedance at about 2225 m (depths indicated from mean sea level) (Figure 5.23). Cross-plots of acoustic impedance vs. V_P/V_S ratio color-coded to volume shale and porosity (Plate 5.24) provide one way to visualize the lithologic information present in the data. Plate 5.24 shows that the porosity is relatively constant within this depth interval (2000–2400 m) and the shale has both a lower acoustic impedance and higher V_P/V_S ratio than the sands.

Histograms of acoustic impedance and V_P/V_S ratio for sandy (volume shale $V_{sh} <$ 0.35) and shaly (volume shale $V_{sh} > 0.35$) facies are presented in Figure 5.25. This plot helps to show the differences and overlap in the seismic parameters of the two main groups, and also reveals the possible existence of several subgroups within each main group of lithology. The sandy facies shows two subgroups from the acoustic impedance plot. Two shaly subgroups are clearly distinguishable from the V_P/V_S plot. In this case it is the tuffaceous Balder Formation which has the lower V_P/V_S ratio.

Based on the logs, and some core and thin section descriptions, five major facies were identified (see also case study 1). Facies II–V represent a gradual transition from clean sandstones to pure shale (II: thick-bedded sands; III: interbedded sand–shale; IV: silty shales; V: pure shales). Gravels and conglomerates were included as Facies I. Four subfacies of Facies II (thick-bedded sands) were introduced to account for important seismic and petrographic variations within the thick-bedded sands. There is a gradual increase in clay content from Facies IIa to IId, and the cleanest sands (IIa) are slightly cemented. Brine sands and oil sands were grouped as separate categories. The gamma-ray log values and patterns, and velocity and density logs, were used primarily to determine the different facies with contrasting seismic properties.

5.2 Case 2: Lithofacies and fluid probabilities

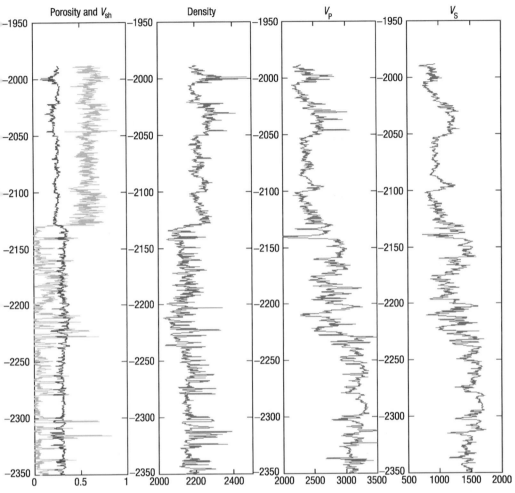

Figure 5.23 Porosity, volume shale (light curve), density, V_P and V_S logs from one of the key wells used in this study.

5.2.2 Pore-fluid effects

An important complicating factor is the presence of flushed zones in hydrocarbon columns at the wells. Because of the water-based drilling fluid, the well logs could be measuring water-saturated rocks from the mud-filtrate invaded zone, instead of measuring the oil-saturated rocks. This was carefully investigated using deep-sounding and shallow-sounding resistivity and Gassmann modeling based on dry ultrasonic data with actual mud-filtrate and reservoir hydrocarbon properties (Avseth, 2000; see also Figures 5.5 and 5.6 in case study 1). The fluid substitution showed that log values in the oil zone were actually very close to the Gassmann estimated water-saturated values. Realistic reservoir fluid properties (oil GOR 64; oil gravity 32 API; oil density 0.78 g/cm^3; oil P-wave velocity 1070 m/s; mud-filtrate density 1.09 gm/cm^3; mud-filtrate P-wave velocity 1700 m/s) were used in the fluid-substitution calcula-

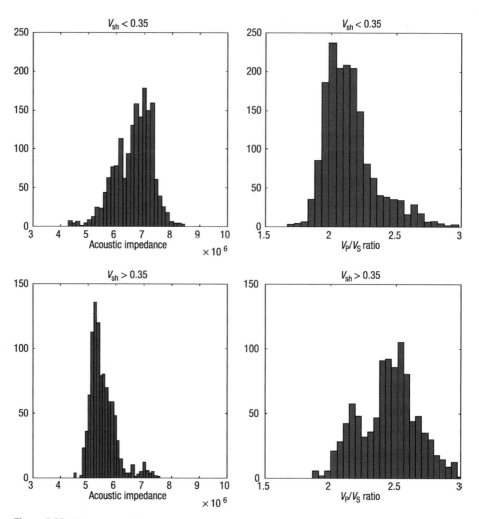

Figure 5.25 Histograms of P-wave acoustic impedance and V_P/V_S ratio grouped as sandy ($V_{sh} < 0.35$) and shaly ($V_{sh} > 0.35$) facies. It can be seen that there are several subgroups within both sands and shales.

tion using Gassmann's equations (Chapter 1). Comparison of log and core porosities, and empirical V_P/V_S relations, also confirmed that the sonic velocities were measures of the water-saturated rocks and not oil-saturated rocks. Using the log data alone to build our calibration pdfs would not capture the properties of the oil-saturated facies. A key step therefore was to extend the log-derived training data, using Gassmann's equations, to derive probability distributions incorporating velocity and impedance attributes for fluid saturations not encountered in the well. The augmented training data were then used to build up the calibration pdfs.

5.2.3 Acoustic and elastic impedance

Ambiguities in lithologic and fluid identification based only on normal-incidence impedance (ρV) can often be effectively removed by adding information about V_P/V_S-related attributes. This provides the incentive for AVO analysis. Synthetic seismic modeling has shown that sometimes it can be difficult to use the seismic amplitudes quantitatively because of practicalities of picking and resolution problems. Another approach to lithofacies identification is based on seismic impedance inversion. Usually this is applied to zero-offset or near-offset sections to estimate the acoustic impedance $I_a = \rho V$, and therefore does not contain V_P/V_S information. Here we used a pseudo-impedance attribute that is a far-offset equivalent of the more conventional zero-offset impedance. The far-offset impedance attribute is sometimes variously termed the angle impedance, or elastic impedance, as it contains information about the V_P/V_S ratio, and depends on the angle of incidence at the target (Mukerji et al., 1998b; Connolly, 1999). This approach allows us to use the same algorithm for inversion of the far-offset stack as for the near-offset stack, and get an elastic impedance cube. The inversion for this pseudo-impedance parameter is therefore economical and simple, with no additional software required. The mathematical details about the definition of the far-offset impedance are described in Chapter 4 and in Mukerji et al. (1998b) and Connolly (1999). In brief, the elastic impedance I_e can be expressed in terms of layer parameters available from logs as:

$$I_e(\theta) = \left(V_P^{1+\tan^2\theta}\right)\left[\rho^{1-4(V_S/V_P)^2 \sin^2\theta}\right]\left[V_S^{-8(V_S/V_P)^2 \sin^2\theta}\right] \quad (5.2)$$

There are other variations of this expression (not used in this study) depending on the form of the approximation used for the angle-dependent reflectivity in deriving the far-offset impedance. Equation (5.2) is based on the Aki–Richards approximation to the full Zoeppritz equations (Aki and Richards, 1980), and is therefore valid for small angles ($<30°$).

Figure 5.26 shows the seismic lithofacies defined from the wells and by fluid substitution in an I_e–I_a cross-plot. These data points are from a depth around 2 km. Facies that overlap in acoustic impedance can be discriminated by their elastic impedance and vice versa. Modeling of far-offset seismic signatures and computation of I_e from logs require knowledge of V_S. Since shear-wave velocity was not available in all wells, we first estimated V_S (where unavailable) using V_P–V_S relations calibrated from wells where shear logs were available. The V_S estimation sub-task involved testing various models for predicting V_S from P-wave velocity, porosity, and shale volume (Jørstad et al., 1999). As described in Jørstad et al. (1999), the V_S prediction error was about 10%. This uncertainty was included in the Monte Carlo simulations for estimating the pdfs of the training data at the wells.

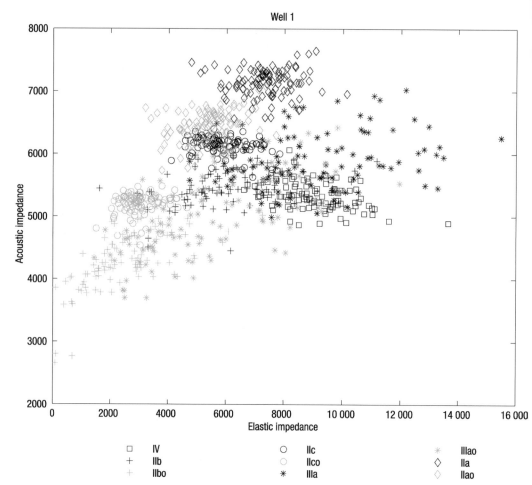

Figure 5.26 Cross-plot of elastic impedance at 30° versus acoustic impedance for different lithofacies. Light markers indicate oil-saturated facies. Facies that overlap in acoustic impedance can be discriminated by their elastic impedance and vice versa.

5.2.4 Seismic inversions

The seismic data used in this study are near-offset and far-offset partial stacks from a marine 3D survey covering approximately 300 km^2 of the North Sea. The survey was processed for true amplitude recovery. The maximum fold is 30, which corresponds to a maximum source–receiver offset of approximately 2500 m. The near-offset stack of 10 traces has an average incidence angle of 8° at the target level, while the far-offset stack of the 10 last traces in each CMP gather has an average incidence angle of 26°.

The post-stack inversion was performed using a commercially available package. The inversion requires as inputs information about the seismic wavelet, the geometric structure from structural seismic interpretation, and a prior model based on well-log impedance. The same method (generalized least-square inversion) is used for both the near- and far-offset stacks, except that the calculated elastic impedance logs for the proper incidence angle were used as the prior model for the far-offset inversions. In other words, for the near-offset (approximately normal incidence) inversion the "usual" sonic V_P and density logs were used as inputs to build the prior model. But for the far-offset inversion, we used the V_P log and a pseudo-density log defined as $\rho = I_e(\theta)/V_P$, so that the product of V_P and the pseudo-density is the desired elastic impedance. Many of the commercial software packages that are geared to handle far-offset impedance inversions now do the calculations automatically, freeing the user from having to define a pseudo-density log.

We used two independent methods to obtain a reliable wavelet estimate. Different wavelets have to be estimated for the near- and far-offset inversions. The first method, provided in the software package, is based on the amplitude spectrum of a selected time window, and a scan of the phases to pick one that best matches synthetic and true seismic traces. Another independent estimate, outside the package, was obtained from the Akaike information criterion (AIC) estimate (e.g. Priestley, 1983) of the filter that best minimizes the error between synthetic and real traces. Both methods indicated that the actual wavelet at the target level was mixed phase with a time shift of ∼20 ms. The prior model was created by extrapolation of well data along the defined structural horizons. We used two well-defined horizons for this purpose: the Top Balder and Top Heimdal horizons. The inversion itself is a 1D trace-by-trace inversion, based on convolution with the wavelet, followed by a minimization of the squared error between the synthetic seismogram and the observed seismogram. A different wavelet is used for the near- and far-offset inversions. Plate 5.27 shows a subset of the near- and far-offset impedance cubes (i.e., acoustic and elastic impedance) from the inversions.

5.2.5 Statistical classification and simulation

The impedance attribute cubes were used to estimate the most likely facies, and the probability of occurrence of each facies at every grid point within the cube. We used two different statistical approaches (see Chapter 3): a quadratic discriminant analysis (e.g. Fukunaga, 1990) based on the Mahalanobis distance, and a Bayes classification and probability estimate based on the complete pdfs for each facies calibrated from the training data. Finally, the probabilities estimated from the seismic classification were used in a geostatistical indicator simulation, thus incorporating the spatial correlation of the facies. The spatial correlation was modeled by variograms estimated from the log data. By simulating multiple realizations, the conditional distributions of the facies were obtained. In general, all three methods gave similar probability maps for the different facies. This is because there was sparse well control to strongly condition the geostatistical simulations, and the effects of the seismic classification dominated the results.

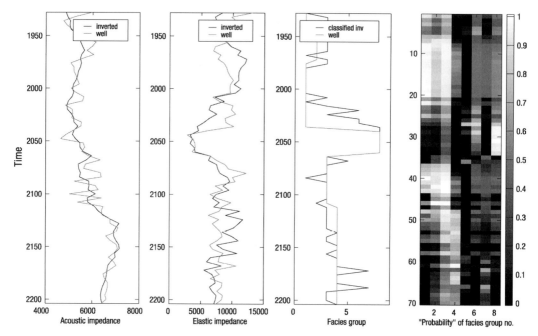

Figure 5.28 Acoustic and elastic impedances from wells and inversion, and facies classification at the well. The first two panels from the left compare the acoustic and elastic impedance traces from the inversion with the same attributes computed from the log values. The third panel compares the quadratic discriminant classification of the facies from the inverted traces with the known facies at the well. The fourth panel displays the normalized Mahalanobis distance (see text) as a measure of the probability for each facies.

We compared the log values and the inverted acoustic and elastic impedance traces co-located with the wells as a check for consistency. The co-located impedance traces were then used to classify the known facies at the wells. This was done with the standard Mahalanobis distance method, which takes into consideration only the means and covariances of the training pdfs. Discriminant analysis is described in Chapter 3, as well as in many standard texts on statistical pattern classification (e.g. Duda and Hart, 1973; Fukunaga, 1990; Doveton, 1994). The attribute space in this case is two-dimensional since we have two attributes – near- and far-offset seismic impedances. For every unknown point, its Mahalanobis distance from each class is computed, and the point is assigned to the class to which it has the smallest Mahalanobis distance.

The inverted seismic impedance traces were calibrated and cross-validated to the well logs. Figure 5.28 shows an example comparing the inverted acoustic and elastic traces with the log values. Panel 3 (from left) in Figure 5.28 shows the facies classification from the inverted traces compared with the known facies at the well. The fourth panel shows the normalized Mahalanobis distance which is a measure of the probability of each facies.

5.2 Case 2: Lithofacies and fluid probabilities

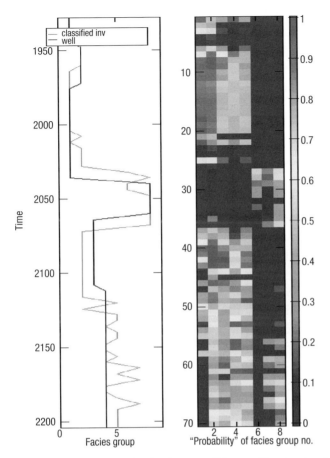

Figure 5.29 Facies classification in one well from calibration data in two other wells. The inverted near- and far-offset impedance traces at the well were classified into facies using discriminant analysis using only training data from two other wells. Facies 1 and 2 are shaly facies, 3 to 5 are brine sands and 6 to 8 are the corresponding oil sands. The average classification success rate in this well increased from 49% (using acoustic impedance alone) to 73% (using both acoustic and elastic impedance).

Another check using the same technique was to classify the facies in one well based on training data and facies definitions in a different well. This is demonstrated in Figure 5.29 where the inverted impedances are classified into facies and compared with the known facies at the well using only training data from two other wells. The classification follows closely the defined facies in the well and picks out the oil-filled sands (around time 2050). The oil-sand facies were not actually present in the two other calibration wells but were added to the training data by fluid substitution of the brine-sand facies. This shows the value of extending the training data using physical models. The average classification success rate was about 73% in this well when using both acoustic impedance and elastic impedance for the classification. Using acoustic impedance alone, the average classification success was 49%. After validation of the inversions at the well locations we were able to go ahead and classify the seismic inversion results. Each sample in the inverted near- and far-stack sub-cube around the reservoir is classified based on the nearest Mahalanobis distance. Example vertical sections from the discriminant analysis are shown in Plate 5.30. Facies coded 1 and 2 are shaly facies, facies 3–5 are water-saturated sands, and facies 6–8 are the corresponding oil sands.

After discriminant classification, we used the complete pdfs to estimate $P(\text{facies} \mid I_a, I_e)$, the conditional probability of occurrence of each facies at each voxel in the 3D cube, given the inverted acoustic and elastic impedances at that voxel. The training pdfs, $P(I_a, I_e \mid \text{facies})$, were obtained from the inverted near- and far-offset traces located at the wells and the known facies at the wells. This is a distribution with two continuous variables (I_a and I_e) for each categorical variable (facies type). The distribution was estimated nonparametrically without assuming any specific form (e.g. Gaussian) for the pdf (Silverman, 1986). Similar nonparametric density estimation techniques were used by Gastaldi et al. (1998) to predict reservoir thickness. The choice of size for the smoothing kernel in our pdf estimation was done by dividing the training data into two subsets, and cross-validating with different smoothing kernel sizes. Obviously, too large a smoothing window smears out the distinction between different facies giving poor validation results, while too small a window makes the pdf estimate very specific to the training subset, with poor ability to generalize to the validation subset. For Bayes classification, the choice of the smoothing window is not too critical. The decision boundary does not change too much with different choices of window. However, estimates of the classification error are more sensitive to choice of window. Figure 5.31 shows a surface plot of the nonparametric calibration pdfs estimated from the training data to be used in the seismic classification. As mentioned earlier, facies codes 1 and 2 represent shales, 3 to 5 are brine sands, while 6 to 8 are the corresponding oil sands obtained by fluid substitution of facies 3 to 5.

All the statistical classifications and geostatistical simulations (described below) were first carried out on a pilot subset of the cube ($251 \times 100 \times 100 \sim 2.5$ million cells) before a final production run for the whole cube ($475 \times 250 \times 100 \sim 11.9$ million cells). Plate 5.32 displays time slices of the estimated conditional probability maps for three different facies groups, while Plate 5.33 shows the spatial distribution of iso-probability surfaces (surface of 80% probability) for shales and oil sands. The surfaces in Plate 5.33 represent probabilities of occurrence of sands and shales, and are not the sand and shale bodies themselves. For example, at the location of the iso-probability surface for oil sands there still exists a 20% probability that it is not oil sand.

To take into account the spatial correlation of the different facies we used geostatistical techniques of indicator kriging and simulation. We describe below one possible method for geostatistical simulation incorporating seismic information and well data. Certainly this is not the only one. Since the geostatistical simulations take into

5.2 Case 2: Lithofacies and fluid probabilities

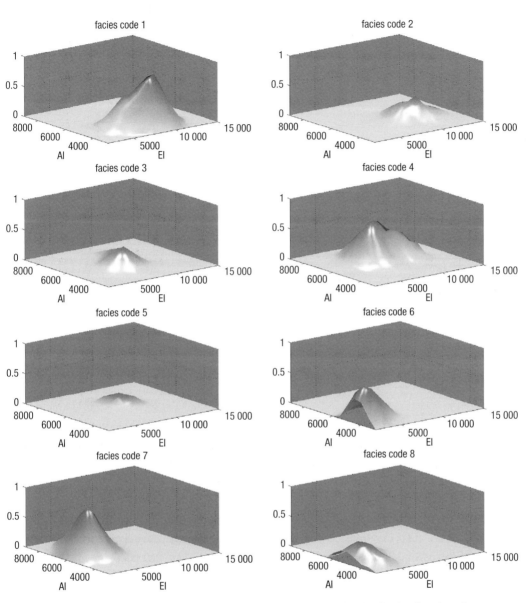

Figure 5.31 Surface plot of I_a–I_e bivariate conditional probability distribution functions for different facies. Facies coded 1 and 2 represent shales, 3 to 5 are brine sands, while 6 to 8 are the corresponding oil sands obtained by fluid substitution of facies 3 to 5.

account the small-scale variability seen in the well logs, they provide a statistical estimate of small-scale spatial heterogeneity beyond the resolution of the seismic data. In the indicator formalism, the known facies at the wells are coded as binary [0 1] indicator random variables. These are taken as "hard data." In the interwell region, the probabilities (between 0 and 1) obtained from the voxel-by-voxel classification

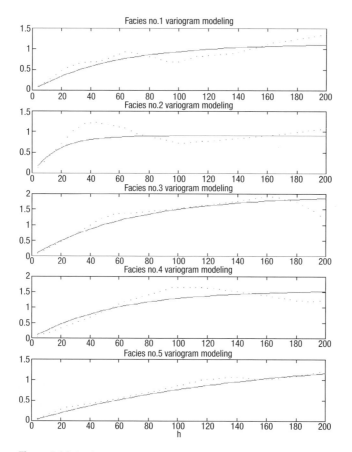

Figure 5.34 Facies experimental variograms from well data and model fits. The range varies from 54 m (facies 2) to 458 m (facies 5).

of the seismic attributes are taken as "soft indicators." The hard and soft indicator data are then combined using indicator simulation to provide multiple realizations of facies and fluid distributions in the reservoir. The first step in any geostatistical exercise is to model the spatial variability. We estimated the experimental variograms of the different facies from well data. Since there was sparse lateral control, we chose to use the horizontal spatial anisotropy ratio from the variograms of the seismic impedance which had much more exhaustive lateral coverage. The experimental variograms were then fit to functional forms by least-square minimization. The well variograms were modeled with a single spherical function (see e.g. Deutsch and Journel, 1996) while the near- and far-offset impedance variograms were modeled with a spherical function with two structures (Figures 5.34 and 5.35). The well variograms for the different facies had a range from 54 m (facies code 2) to 458 m (facies code 5). The experimental variogram for the acoustic impedance did not show a well-defined range or sill (i.e., approximate flattening out for large values of spatial lag distances).

5.2 Case 2: Lithofacies and fluid probabilities

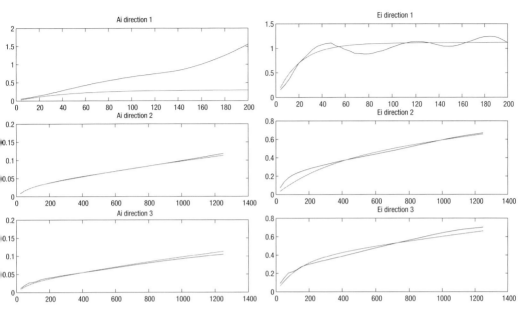

Figure 5.35 Acoustic and elastic impedance variograms. The acoustic impedance variograms do not show a well-defined sill. However, the variogram shapes look reasonable for the elastic impedance.

The variograms for the elastic impedance were well defined with a range of 60 m in the vertical direction, 861 m in the cross-line direction and 392 m in the inline direction.

After estimating and modeling the variograms, we carried out indicator kriging and simulation. Indicator random function models are ideally suited for estimating and simulating categorical variables. The Markov–Bayes indicator formalism (Deutsch and Journel, 1996) was used to obtain the posterior conditional pdfs, incorporating the spatial correlations of the facies as estimated from the facies indicator variograms. This updates the prior pdf, $P(\text{facies} \mid I_a, I_e)$ to give the posterior pdf, $P(\text{facies} \mid I_a, I_e,$ well indicator data). In the Markov–Bayes approach the covariance function of the soft seismic indicator is taken to be proportional to the covariance function of hard indicators obtained from well data. The proportionality constant is obtained from the collocated hard and soft indicator data as described by Deutsch and Journel (1996). The well coverage was very sparse, and the variogram model for facies indicators along the horizontal direction had to be completed by borrowing the spatial anisotropy ratio from variograms of the seismic impedance. Hence the updating by conditioning to hard well indicator data did not change the seismically obtained prior pdfs significantly. Plate 5.36 shows vertical sections of one stochastic realization of the simulated facies from the sequential indicator simulation. Figure 5.37 displays the probability maps (Markov–Bayes posterior probabilities) for shales, oil sands and brine sands as inlines through the cube.

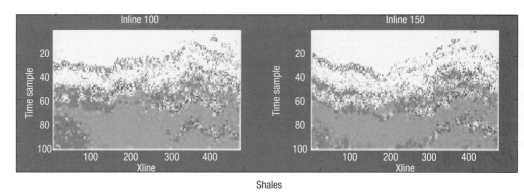

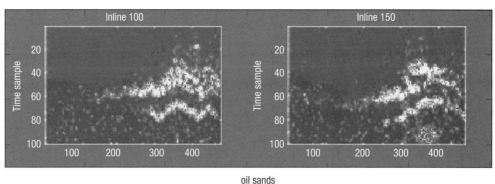

Figure 5.37 Vertical sections along two inlines showing probability of shales (top), oil sands (center), and brine sands (bottom). Light color indicates higher probability.

We now compare the predictions from this study with lithofacies actually observed in a well drilled after this study was done. Plate 5.38 shows a depth-averaged probability map of oil sands, with the location of the new well. The location of the well was independent of the probability maps from our analysis. The comparison between predictions from seismic inversion and classification with lithologies as interpreted from well-log analysis is shown in Figure 5.39. The three sand intervals and the shale just above the lowermost sand are all correctly predicted from the seismic data. In Figure 5.39 the

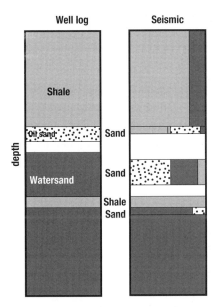

Figure 5.39 Comparison between predictions of lithologies and pore fluids from seismic inversion and classification (right) with lithologies and fluids as interpreted from well-log analysis (left). The shading indicates different lithofacies and fluids: light gray for shales, stippled for oil sands, and dark gray for brine sands. The horizontal widths of the bars on the right panel indicate the probabilities of the corresponding facies as estimated from seismic data.

shading indicates different lithofacies and fluids: light gray for shales, stippled for oil sands, and dark gray for brine sands. The horizontal widths of the bars on the right panel indicate the probabilities of the corresponding facies as estimated from seismic data. At the well location the middle sand was interpreted to be water-wet whereas the seismic interpretation assigned a somewhat higher probability to its being oil sand rather than brine sand (stippled bar slightly wider than dark gray bar at the middle sand). The upper sand had a higher probability for oil than brine, which matches with the observation at the well. The lower sand was also correctly predicted to be a brine sand (dark gray bar wider than stippled bar).

5.2.6 Discussion and conclusions

This case study shows how near- and far-offset seismic impedance attributes can be optimally combined with well-log petrophysical analysis, calibration, and statistical rock physics to classify and map the occurrence probabilities of reservoir lithofacies and fluids. Seismic impedance inversion in conjunction with lithofacies classification is applied to a North Sea reservoir data set to map out the iso-probability surfaces of shales, oil sands and brine sands within the reservoir. This strategy based on seismic inversions can complement the more traditional approach to seismic reservoir characterization, based on AVO gradient and reflectivity (see case study 1). Synthetic seismic modeling of AVO has shown that sometimes it can be difficult to use the seismic amplitudes quantitatively because of practicalities of picking, phase changes, resolution problems, and thin-layer effects. Lithofacies identification based on seismic impedance inversions can alleviate some of these problems as it uses information from the full waveform, not just the picked amplitudes. Impedance inversion is also less affected by the problems of horizon interpretation which can be subjective in heterogeneous reservoirs. Usually impedance inversion is applied to zero-offset or near-offset sections to estimate the acoustic impedance ρV, and therefore does not contain V_P/V_S information. Here we used a pseudo-impedance attribute, the elastic impedance, which is a far-offset equivalent of the more conventional zero-offset impedance. The elastic impedance contains information about the V_P/V_S ratio. This approach allows us to use the same algorithm for inversion of the far-offset stack as for the near-offset stack, and get an elastic impedance cube. The inversion for this pseudo-impedance parameter is therefore economical and simple, with no additional software required. A different inversion approach (not done in this case study) at considerably more computational cost is pre-stack full-waveform impedance inversions. This would give estimates of the P and S impedances (or P impedance and Poisson's ratio) directly, instead of the far-offset impedance-like attribute (elastic impedance) used in this study. If the objective is to identify and discriminate lithofacies and pore fluids, P and S impedances may not have higher classification success rate than the near- and far-offset impedances. However, the different impedance inversions (full pre-stack versus partial stacks) can have different errors associated with the estimated impedances.

Well-log training data should be used before doing any inversions to estimate the classification success rates for different attributes. This will help to decide whether it is worthwhile to do far-offset inversions, and how much improvement would be achieved by including far-offset impedance in the reservoir characterization strategy. For example, in this study we used well data to do a validation test which showed that the classification success rate with both near- and far-offset impedances was about 73%, whereas it was only about 49% with near-offset attribute alone. This clearly indicated that both attributes should be used.

We used discriminant analysis and Bayesian classification based on nonparametric pdfs. The advantage of the nonparametric approach is that, unlike a discriminant approach, it uses more than just the means and covariances of the data, and can capture nonlinear trends in the discriminant hyper-surface. The nonparametric Monte Carlo approach avoids restrictive, and sometimes erroneous, assumptions about the form of the underlying pdfs (e.g. Gaussianity). On the other hand, estimation of nonparametric pdfs may become unreliable and computationally intensive in very high-dimensional attribute spaces. For this study, where we have two attributes, the Bayesian nonparametric approach is a powerful method to obtain not only the most likely facies, but also the probability of each facies given the observed near- and far-offset seismic impedance attributes and assess the uncertainty of the interpretation.

As with any statistical analysis, there is always the issue of how representative of the whole population are the training data obtained from a few wells. With sparse data representativeness becomes problematic. In some situations it may be possible to borrow the population statistics from other similar reservoirs that have more data. Key wells have to be selected carefully so that they sample the important facies. One important aspect of the analysis is to use rock physics models (such as Gassmann's equations) to augment the training data by estimating rock properties for conditions not available in the initial training data (different pore-fluid saturations, different degrees of cementation etc.). The extended training data and the pdfs derived using physical models can capture more of the variability than the original data alone. This study lays the framework for an efficient strategy to optimally combine statistical techniques with physically based seismic-to-reservoir property transforms and apply them for reservoir characterization.

5.3 Case 3: Seismic lithology prediction and reservoir delineation using statistical AVO in the Grane field, North Sea

5.3.1 Introduction

In this case study we show how we can combine statistical rock physics, lithofacies interpretation, and AVO analysis to discriminate between lithologies and thereby improve detectability of hydrocarbons from seismic amplitudes in Grane field, North Sea. This Late Paleocene turbidite oil field has been problematic because of complex sand distribution and nonreservoir seismic anomalies. Plate 5.40 is a 3D visualization of the reservoir as delineated by conventional seismic interpretation. The reservoir is bounded by the Top Heimdal and Base Heimdal horizons. The figure includes seismic grids of the Top Chalk horizon and the overlying Base Balder Formation, which define the Late Paleocene target interval. Also shown are five wells, three of which (1, 2, and 3) penetrate reservoir sands. Wells 4 and 5 targeted possible satellite sands. However, neither encountered reservoir sands. This case study focuses on three 2D seismic lines that intersect Wells 1, 3, and 4 (Plate 5.41).

The sands of interest represent the Paleocene Heimdal Formation. The study area, on the eastern margin of the South Viking Graben, is complex in terms of lithology variation. In addition to sands and shales, carbonates and volcanic ash-fall deposits are relatively abundant. This is related to the particular setting and the local basin topography during deposition. Grane field is on the eastern flank of the South Viking Graben, near the Utsira High which had abundant limestone and marl deposition during Late Cretaceous and Early Paleocene, as siliciclastic sedimentation rates were low. The

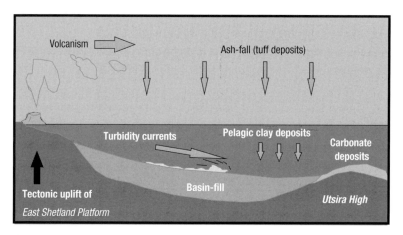

Figure 5.42 Structural setting and sedimentary processes in the South Viking Graben during the Paleocene, causing a great mix in lithologies in the Grane area, near the Utsira High.

complete Heimdal sequence is relatively thin (less than 100 m), and was deposited in the Late Paleocene. During deposition of the sands, limestones, marls, and shales were eroded and redeposited locally. The Paleocene also experienced repeated episodes of volcanic eruptions and ash-flow deposition, associated with the opening of the Norwegian Sea. Hence, the relatively thin Paleocene interval in the Grane area comprises a great mix of lithologies (Figure 5.42).

Our approach to discriminating between these lithologies from seismic data includes three major steps: facies analysis, rock physics analysis, and AVO analysis. The first step involves facies analysis of cores and well logs, the second explores site-specific rock physics trends in terms of lithology and diagenesis using well-log data, and the third uses statistical AVO analysis to predict lithofacies from seismic data. Probability density functions (pdfs) are derived for each facies in terms of zero-offset reflectivity and AVO gradient. The $R(0)$ and G estimated from real seismic data were used to predict the most likely facies distribution along selected 2D seismic lines. In this methodology, the rock physics analysis provides the critical link between facies interpretation and AVO analysis. For more detailed description of the methodology, which is the same as applied in case study 1, see Chapter 4.

5.3.2 Facies analysis

In Grane field, the following lithofacies can occur at seismic scale: clean sandstone, pure shale, tuff, marl, and limestone. All are identified in Well 1 (Figure 5.43), based on core observations available for the entire zone.

The reservoir sands in Well 1, representing Heimdal Formation, are very clean, high-porosity sandstones saturated with water. The sands are embedded in relatively pure shales that represent Lista Formation. Balder Formation, representing the top of

5.3 Case 3: Statistical AVO

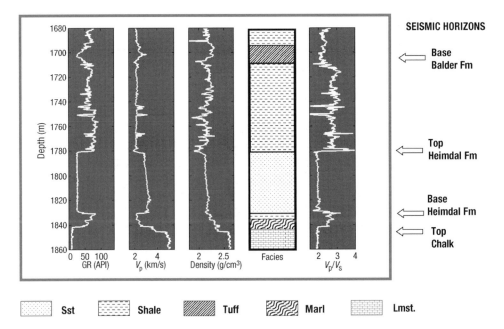

Figure 5.43 Various log data and facies in Well 1, the type-well. Facies observations are from cores. Key seismic horizons are noted.

the Paleocene interval, consists mainly of volcanic tuff or tuffaceous shales. Ekofisk Formation, of Cretaceous age, represents the base of the target interval. It consists of chalk deposits (limestones). The lower Paleocene Vaale Formation, directly over Ekofisk, consists of marl deposits. These are mixed deposits of limestone and shale.

Plate 5.44 shows the seismic signatures (zero-phase wavelet, peak frequency 30 Hz) along a 2D post-stack section intersecting Well 1. Here, we observe important seismic horizons that correspond with lithostratigraphic and facies boundaries. Balder Formation shows a prominent red reflector, indicating a positive stack response. Balder is about a wavelength thick; the black response below coincides with the base of the tuffaceous unit. The reservoir sands (Heimdal Formation) are also identified. The top reflector is prominent, but has an incoherent character. A black reflector that undulates in shape, just beneath the Top Heimdal reflector, represents the base of the reservoir. We also observe some subtle internal reflectors within the reservoir. The Base Heimdal horizon interferes with the sidelobe of the peak wavelet representing Top Chalk, which is the most prominent seismic reflector in the area. It shows a very strong positive reflectivity.

Well 1 was used as a type-well for a multivariate statistical classification of seismic lithofacies in other wells. The gamma-ray, density, and P-wave velocity logs were used as training data. Shear-wave velocity was not used, because it was not available in all the wells. Classification was done using quadratic discriminant analysis (see Chapter 3). Because significant mud-filtrate invasion has been documented in other

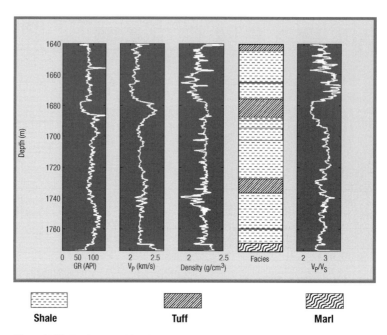

Figure 5.45 Facies classification results of Well 4. The well was drilled on a seismic anomaly at a depth of 1722 m (see Figure 5.50), but no sand was encountered. Note the tuff unit near the depth of this anomaly.

wells penetrating oil zones, we did not include oil-saturated facies in the classification of well logs. We expanded our training data to include oil-saturated facies in the prediction from seismic data, using the Gassmann equations (Chapter 1; see also the rock physics analysis section below).

Figure 5.45 shows classification results for Well 4. No sands are identified in the target zone. The Balder Formation tuff is identified around 1680 m and marl deposits about 1770 m. We identified a zone of tuff facies in the target zone between 1725–1735 m, embedded in shaly facies. Core data from the well confirm the presence of tuff at this level. The Intra-Paleocene tuff unit in Well 4 is of great interest, because it could explain the observed seismic anomaly on which Well 4 was targeted. The well that was drilled through the seismic anomaly, however, encountered no reservoir sands. In the next sections, we investigate the possibility that tuff is responsible for the seismic anomaly, and we will use statistical rock physics and AVO analysis to see if we are able to distinguish these tuffs from oil sands.

5.3.3 Rock physics analysis

We want to link the facies defined above, with rock physics properties. This is important for better understanding and interpretation of seismic amplitudes in terms of lithofacies, pore fluids, and the distributions of these (Chapter 2).

5.3 Case 3: Statistical AVO

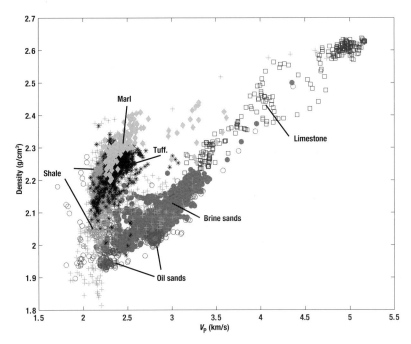

Figure 5.46 P-wave velocity versus density for different lithofacies.

Figure 5.46 plots the P-wave velocity versus density. In the V-shaped trend shales, marls, and tuffs have relatively low velocities and high densities, brine sands have intermediate velocities but relatively low densities, and limestones (chalks) have very high velocities and densities. Tuffs and marls overlap shales, but tend to have slightly higher velocities and densities. For instance, the tuff units in Well 4 (Figure 5.45) show a significant increase in P-wave velocity, and a subtle increase in density, compared with surrounding shales. The sandstones in Figure 5.46 have relatively long velocity and density ranges, and there are large overlaps with other facies. The variability within the sandstone cluster in terms of rock physics properties is related to sandstone texture, depositional sorting, and diagenetic cementation (Avseth *et al.*, 2000; see also Chapter 2).

The sandstones in the Grane area are either water-saturated or oil-saturated. Hence, we needed to expand on our training-facies data base to include oil-saturated sands. We applied Gassmann theory (Chapter 1) to calculate the rock physics properties of oil-saturated sands based on properties for the water-saturated sands in Well 1. Figure 5.46 shows that oil-saturated sands have slightly lower densities and velocities than water-saturated sands, but there are great overlaps between clusters. Oil is relatively heavy in Grane field (18 API), and the seismic properties do not change much as we go from brine-saturated to oil-saturated. Variability within the sandstone cluster is much larger than change related to pore fluids. This shows that the rock texture of the sands is seismically more important than pore fluids. Nevertheless, the oil saturation brings

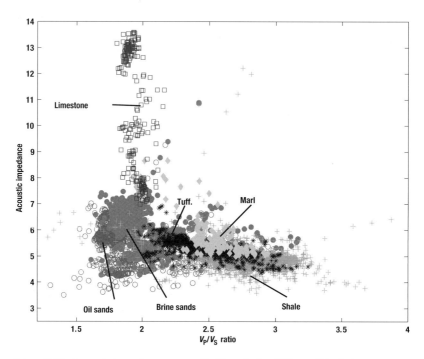

Figure 5.47 Acoustic impedance versus V_P/V_S ratio for different lithofacies.

the overall cluster of sands closer to the tuff and marl clusters in terms of P-wave velocity.

Figure 5.47 plots acoustic impedance versus V_P/V_S ratio for the various facies. Note the great overlap between different lithofacies in terms of acoustic impedance; V_P/V_S ratio is a much better facies discriminator. The exception is limestones. Limestones are easily distinguished in terms of acoustic impedance, whereas the V_P/V_S ratios are similar to those of sands ($V_P/V_S \sim 2$). The V_P/V_S ratio at the tuff units in Well 4 (Figure 5.45) is clearly dropping relative to the surrounding shales. In fact, the V_P/V_S curve mimics very well the facies classification results even though V_S information was not used in the classification procedure. The same is true for the sandstone and tuff unit in Well 1 (Figure 5.43). The observations in Figure 5.47 are important in order to assess seismic detectability in the area. The contrast in acoustic impedances between two layers controls the zero-offset reflectivity at the layer interface, whereas the contrast in V_P/V_S has a large effect on the offset-dependent reflectivity (see Chapter 4). Hence, the observations in Figure 5.47 indicate that AVO analysis must be conducted to predict lithofacies from seismic data in this case.

5.3.4 AVO analysis

We first conducted deterministic AVO analysis in Well 3 to study the offset-dependent reflectivity of oil sands in the area. We then did similar studies in Well 4 to see if tuffs

5.3 Case 3: Statistical AVO

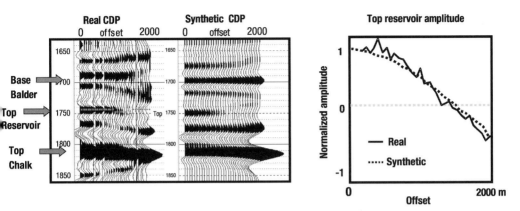

Figure 5.49 AVO analysis at Well 3. There is an excellent match between real and synthetic gathers. The top of the oil-saturated sands shows a prominent zero-offset reflectivity and a strong negative AVO gradient, resulting in a phase shift at an offset of approximately 1500 m.

in this well yield a characteristic AVO response. Finally, we did probabilistic AVO analysis along 2D seismic sections intersecting these wells to predict lithofacies and pore fluids away from the wells.

Plate 5.48 shows a seismic near-offset stack section intersecting Well 3. This well encountered thick reservoir sand saturated with oil. The oil–water contact is at 1765 m. There is a prominent positive near-offset stack response at the top of the reservoir. Figure 5.49 shows the real CDP gather and the modeled synthetic CDP gather at the well. The synthetic gather is based on log properties in Well 3. Shear-wave velocity information was not available so we used the V_P/V_S ratio in Well 1 to calculate V_S in this well. Because of the mud-filtrate invasion effect, fluid substitution using the Gassmann theory is done to calculate properties of the sands saturated with oil. The response of the resulting synthetic CDP gather is very similar to the real gather. Moreover, the picked amplitude of the top reservoir shows a phase shift at the same offset for the real and the synthetic case, when normalized at zero-offset reflectivity.

Figure 5.50 shows a seismic stack section intersecting Well 4. Note the prominent seismic reflector around the well. This anomaly was interpreted as potential reservoir sands before drilling Well 4. The main reservoir sands of Grane field are a significant seismic reflector around CDP 1350. The stack response of reservoir sands will vary as a function of the sand texture, and because of phase shifts (see Figure 5.49) the stack response of the sands can be very weak. Variation in amplitude can also be related to tuning effects or diffractions related to an uneven top reservoir. Figure 5.51 shows the real and the synthetic CDP gather in Well 4. The log data in Figure 5.45 were inputs for the AVO modeling. Shear-wave velocity is available in this well. There is a good match between the real AVO response and the synthetic one. The AVO modeling confirms that the tuff unit gives a significant seismic reflector that is also recognized in the real data. The tuff unit shows a prominent zero-offset reflectivity that decreases with offset.

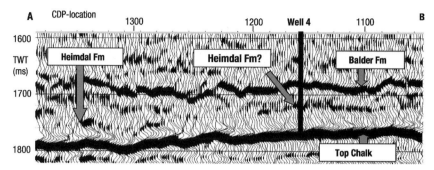

Figure 5.50 Seismic section intersecting Well 4 (for scale and location see Plate 5.41).

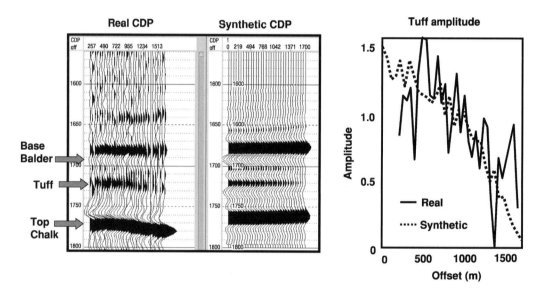

Figure 5.51 AVO analysis in Well 4. There is a good match between real and synthetic gathers. The top of the tuff unit shows a prominent zero-offset reflectivity and a negative AVO gradient, resulting in a weak far-offset reflectivity.

On the basis of the facies classification, we generated cumulative distribution functions (cdfs) of rock physics properties for each facies population. The cdfs are the basis for generation of the AVO pdfs (Chapter 4). We did Monte Carlo simulation (Chapter 3) of the seismic properties from cdfs, and calculated corresponding realizations of reflectivity versus offset, using Shuey's approximation of the Zoeppritz equations (see Chapter 4). Uncertainties in the properties of the cap-rock as well as of the reservoir zone are included in the simulations. On the basis of these simulations, we created bivariate pdfs of $R(0)$ and G for the different facies combinations (Figure 5.52). We assume shale as cap-rock. These pdfs create the probabilistic link between lithofacies and seismic properties, and they will be used below to predict lithofacies from seismic data. We observe that the various facies have different locations in terms of $R(0)$ and

5.3 Case 3: Statistical AVO

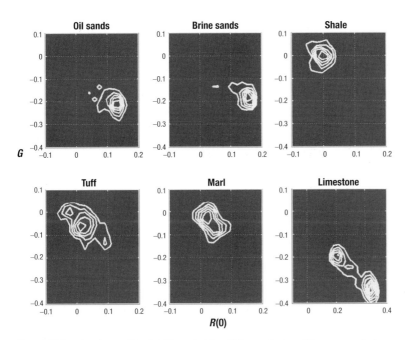

Figure 5.52 Bivariate pdfs of $R(0)$ and G for different facies. We assume shale as cap-rock.

G. Oil sands and brine sands have relatively large $R(0)$ and G values, and there is great overlap between the two. Hence, this plot indicates that seismic data can hardly discriminate between oil and brine sands. Shales have very low $R(0)$ and G values centering around 0, since the cap-rock is also shale. Tuffs and marls have intermediate $R(0)$ and G values. Finally, limestone has very large $R(0)$ and G values, and is easily separated from the other facies. Note that tuff capped by a shale produces a significant $R(0)$ and a negative AVO gradient. Shale–shale interfaces can also give some $R(0)$ response.

Let us focus again on the seismic anomaly around Well 4. Figure 5.53 illustrates the potential ambiguity between tuffs and oil sands. Intermediate positive $R(0)$ and negative G of tuff could give similar stack responses to the strong positive $R(0)$ and large negative G of oil sands. However, statistical AVO analysis should be able to distinguish between them, if both facies are included in the training data. Even statistical AVO would fail if tuff was not included as a facies in the training data.

The next step is to apply the bivariate AVO pdfs to predict seismic lithofacies from pre-stack seismic data. We selected two seismic lines from which we extracted $R(0)$ and G along the Top Heimdal horizon using commercial AVO inversion software. The inverted $R(0)$ and G values from the seismic data were calibrated to the log data and classified according to our bivariate pdfs of $R(0)$ and G.

One selected line intersects Well 3 (Plate 5.48). For this line the goal was to delineate the extent of the reservoir sands laterally, and see if results corresponded with the extent determined from the conventional seismic interpretation. The other selected line

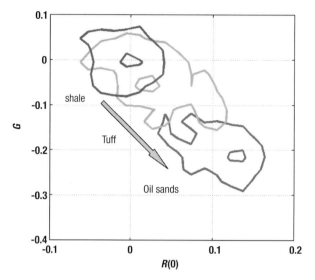

Figure 5.53 Iso-probability contours (50%, outer; 90%, inner) of shale, tuff, and oil sands. This figure illustrates the potential pitfall of tuff in the assessment of seismic amplitudes. The tuff data are between shales and oil sands. Hence, a tuff data point can easily be mistaken for an oil sand, if we ignore tuffs and only try to distinguish sands and shales.

intersects Well 4 (Figure 5.50). For this line, the goal was to do a blind test of the well. We wanted to determine if we could predict the presence of volcanic tuff based upon a calibration within the main reservoir sands. Figure 5.54 shows the calibrated $R(0)$ and G values along the Top Heimdal horizon in the line intersecting Well 3, where oil sands were encountered. The AVO pdfs in Figure 5.52 were used to predict the most likely facies underlying this horizon. We predicted both oil- and brine-saturated sands along the horizon. The total extent of the reservoir sands coincides nicely with the extent determined from conventional seismic interpretation (Plate 5.41).

Next, we conducted a blind test on the seismic anomaly around Well 4, and predicted seismic lithofacies along the Top Heimdal horizon. For each location along a horizon, we obtained $R(0)$ and G from inversion of pre-stack seismic data. These values were calibrated inside the main reservoir sands of Grane field. We calibrated the average of unscaled $R(0)$ and G values from a range of CDPs inside the reservoir (as defined by the map in Plate 5.41) with the mean values of $R(0)$ and G for oil-saturated sandstone facies as defined by the training data. Figure 5.55 includes the calibrated $R(0)$ and G as well as the predicted most likely lithofacies along this line. We confirmed the oil-filled sands of the main reservoir (CDP 1225–1375). At Well 4 we predicted the most likely facies to be tuff, present beneath the seismic anomaly in Figure 5.50. This prediction matches core observations and log classification results. A local water-saturated sand body is predicted just east of the well.

5.3 Case 3: Statistical AVO

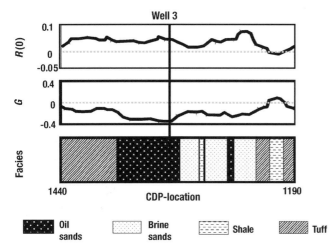

Figure 5.54 Seismic lithofacies prediction along Top Heimdal horizon inline intersecting Well 3. We predicted both oil and brine sands within the reservoir. The extent of the reservoir sands coincides well with the extent determined from the conventional seismic interpretation. Note that we predicted volcanic tuff as well as shales outside the reservoir.

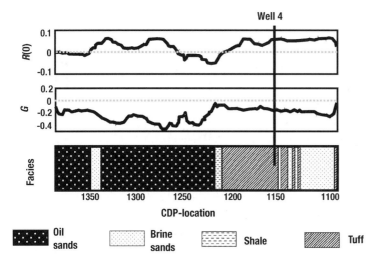

Figure 5.55 Seismic two-way traveltime, $R(0)$, and G along the Top Heimdal horizon extended to the anomaly around Well 4. Lowest bar indicates most likely facies and/or pore fluid predicted under the seismic horizon, assuming shale as cap-rock. We predicted tuff at Well 4.

5.3.5 Conclusions

Rock physics analysis shows that volcanic tuffs and marls with relatively low porosity can have similar acoustic impedances to high-porosity sandstones in the Grane area, especially if sands are saturated with oil. The ratio V_P/V_S is a better parameter to discriminate different types of lithofacies. Because of these ambiguities in acoustic impedance, tuffs and marls can cause seismic anomalies that are potential pitfalls in hydrocarbon exploration. We have shown how statistical AVO analysis can be applied to seismically discriminate sands from other lithofacies. Oil sands can be hard to discriminate from brine sands in this field, because of the relatively heavy oil. In general, this case study demonstrates how we can combine rock physics and facies analysis with statistical AVO to improve delineation of hydrocarbon reservoirs, identify possible nearby satellite reservoirs, and reveal potential pitfall anomalies caused by nonreservoir rock types.

5.4 Case 4: AVO depth trends for lithology and pore fluid classification in unconsolidated deep-water systems, offshore West Africa

5.4.1 Introduction

In this case study we apply statistical AVO classification constrained by rock physics depth trends to predict most likely lithofacies and pore fluids in an offshore West Africa deep-water system (Plate 5.56). First, we calculate expected depth trends in rock physics properties for different lithologies and pore fluids (see Section 2.6). These trends are calculated from empirical porosity–depth models representing the local burial and compaction history. Next, we calculate the corresponding AVO depth trends from the depth trends in rock properties (see Section 4.3.11). Different models are generated based on the knowledge of local geology and depositional environment. AVO uncertainties are included and take into account the expected or observed natural variability in the rock properties. In this way we can obtain AVO pdfs for any given depth of burial. Finally, the modelled AVO pdfs are used to predict the most likely lithology and pore fluids for different depth intervals from real seismic data.

The seismic signature of hydrocarbons can be very different from one depth to another owing to different compaction trends for different lithologies. Therefore, it is necessary to include depth as a parameter when we use AVO analysis to predict lithology and pore fluids from seismic data.

5.4.2 Rock physics depth trends

During early burial porosity is reduced mainly through packing change and ductile grain deformation. Ramm and Bjørlykke (1994) suggested a clay-dependent regression model for porosity versus depth (Z) of sands, owing to mechanical compaction:

$$\phi = \phi_c e^{-(\alpha + \beta Cl)Z} \tag{5.3}$$

where ϕ_c is the critical (i.e., depositional) porosity, whereas α and β are regression coefficients representing a framework grain stability factor for clean sandstones (Cl = 0)

5.4 Case 4: AVO depth trends

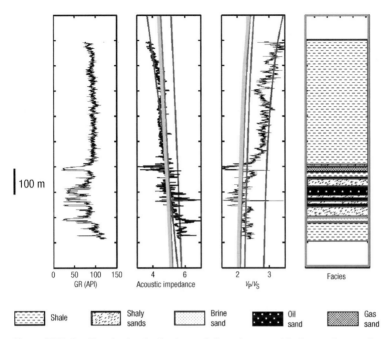

Figure 5.57 Predicted seismic depth trends based on empirical porosity trends, compared with well-log data from the Tertiary deep-water turbidite case in Plate 5.56.

and a factor describing the sensitivity towards increasing clay, respectively. The clay index, Cl, is defined as the content of total clays relative to the total content of stable framework grains. Porosity–depth trends are calibrated locally using reasonable critical porosity values (0.4–0.45 for sands and 0.6–0.8 for shales) at the surface, and inverted density logs for any burial depth.

Hertz–Mindlin (HM) theory (Mindlin, 1949) can be used to calculate elastic moduli of unconsolidated sediments as a function of porosity and pressure (Chapter 2). Densities are calculated from the empirical porosity–depth trends. From elastic moduli and densities we can calculate acoustic impedance (AI) and V_P/V_S ratios versus depth. We calculate depth trends for clean sands, shaly sands and shales. We assume 100% quartz and 0% clay for the clean sand trend and 80% quartz and 20% pore-filling clay (smectite) for the shaly sand trend. Effective mineral moduli are estimated using the Hill's average (see Mavko *et al.*, 1998). Bulk and shear moduli of quartz are 36.8 GPa and 44 GPa, respectively. The same parameters for smectite are 15 GPa and 5 GPa.

Figure 5.57 shows calculated trend lines of AI and V_P/V_S versus depth compared with observed well-log data from the well intersected by the seismic line in Plate 5.56. We observe a very good match between the shale trend and the log data in the shaly intervals (i.e., zones with high gamma-ray values), both in terms of acoustic impedance and V_P/V_S. Deviations from the modeled shale trend may reflect variation in silt content within the shales. The sandy reservoir zone nicely follows the shaly sand

308 Case studies

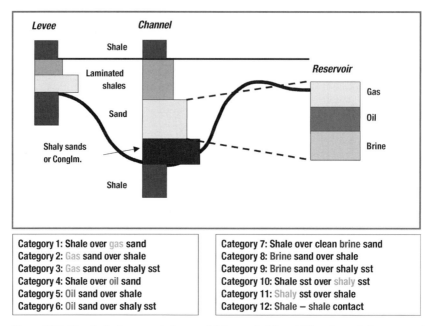

Figure 5.58 Simple facies association model for mud-rich turbidite channel–levee complex.

trend. Deviations from the clean sand and shaly sand trends reflect both presence of hydrocarbons (the trend lines are calculated for brine-saturated rocks) and variability in clay content. A few local peaks of high impedance anomalies may reflect local cementation.

5.4.3 Statistical AVO modeling constrained by rock physics depth trends

The estimated acoustic impedance and V_P/V_S trends in Figure 5.57 can be used to calculate the expected AVO response with depth, for sand–shale interfaces. The example in this study is from a deep-water turbidite setting, and we assume 12 different interface categories. These are based on realistic layer configurations in a turbiditic environment, and are depicted in Figure 5.58. This model of facies associations is rather simplified compared with the true sedimentologic observations in the area, but we attempt to reduce the amount of interface categories while still honoring geologic variations that may be seismically significant. If we include too many interface categories, we may introduce too much overlap between individual classes in a binary AVO cross-plot.

Next, we extract V_P, V_S and density for clean brine sand, pure shale, and shaly sand, from the calculated depth trends in the previous section. These are assumed to be the mean values for the different facies at the target level. We assume multi-Gaussian distributions where the variances are selected on the basis of information from analog

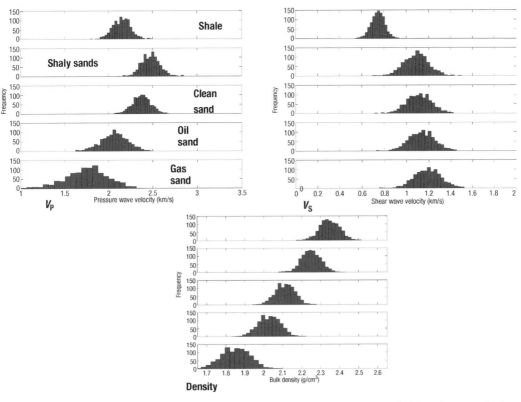

Figure 5.59 Histograms of V_P, V_S and density for different lithologies and fluids at the target depth level corresponding to the reservoir sands penetrated by the well in Figure 5.57. The mean values are determined by the depth trends, while the variances are assumed to be depth-independent and are taken from a nearby well.

areas, or from nearby wells. In this example we have used nearby wells to calculate the variances. Moreover, we use Gassmann theory (Chapter 1) to estimate the rock properties for gas- and oil-saturated sands. (Fluid properties used for the turbidite field: gas gravity = 0.7, oil reference density = 28 API, and brine salinity = 80 000 ppm). The resulting histograms of V_P, V_S and density for different facies and fluids are shown in Figure 5.59.

For each interface category, the expected AVO response at a target depth is calculated using Shuey's approximation to the Zoeppritz equation for P-wave reflectivity, valid for angles less than 30° (see Chapter 4). In Shuey's equation, $R(0)$ is controlled by the contrast in acoustic impedance across an interface. The gradient, G, is mainly controlled by the contrast in V_P/V_S ratio. We do a Monte Carlo (MC) simulation to estimate the distribution of $R(0)$ versus G, based on the mean and covariances in V_P, V_S, and ρ for the different interface categories. The structure of the covariance matrix determines the dependencies between the variables. Normally, there is a higher correlation

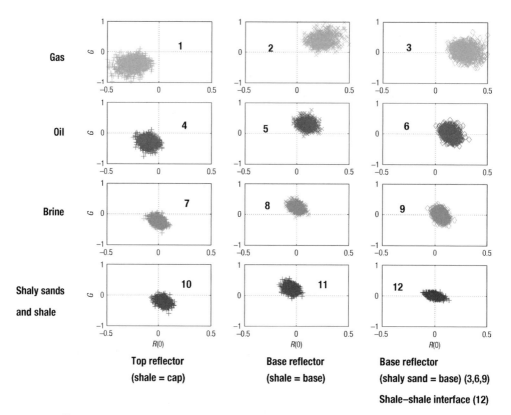

Figure 5.60 Modeled AVO scatter plots of $R(0)$ versus G for different interface categories for the target depth level. (See Figure 5.58 for explanations of the interface categories.)

between V_P and V_S than between V_P and ρ. The resulting AVO scatter plots representative of the target depth, from which the AVO pdfs can be estimated, are shown in Figure 5.60.

5.4.4 Seismic calibration and AVO classification

The final step in the AVO classification technique is to apply the modeled AVO pdfs to predict the most likely facies and pore fluid from seismic data. We did a blind test of the well intersecting the line in Plate 5.56, using the AVO pdfs derived from the modeled depth trends. We calibrate $R(0)$ and G estimated from pre-stack gathers along the line to the modeled AVO pdfs in Figure 5.60. We identify a background "window" in the seismic section near or around the target interval. For the studied turbiditic environment we assume the background trend to be characterized by interface categories 10–12, as most seismic horizons in a mud-rich turbiditic environment are made up of these categories. We calibrate the covariance matrix of $R(0)$ and G for the background trend in the seismic data with the covariance matrix of $R(0)$ and G for the background trend in the model, either by matching the covariances or by univariate variance matching. This calibration is then applied to seismic data in the target area (Plate 5.61).

After calibrating the seismic data with the modeled AVO pdfs, we perform the AVO classification. We use the Mahalanobis distance (see Chapter 3) to estimate the most likely layer category for each data sample in the data. The classification result is shown in Plate 5.62, where we lump the 12 layer interface categories into five facies and fluid groups (tops and bases together). We obtain a good match with the observations in the well (compare with Figure 5.57). The top reservoir is successfully identified as gas-bearing, while zones of oil sands are identified below the gas reservoir. However, significant parts of the reservoir are characterized as water-bearing. This could reflect the great overlap and ambiguities between oil and brine sands in terms of AVO properties. The cap-rock is predicted to be predominantly heterolithics and shales. Bear in mind, however, that the final result in Plate 5.62 represents classification of interfaces, not layers. However, the methodology used in this case study can also be applied to layer inversion results. For instance, elastic inversion could be classified using pdfs of AI versus V_P/V_S. In that case, the calibration of the seismic data is not necessary. Also note in Plate 5.62 that a few data points have been categorized as "no class," depicted in black. In the classification procedure, data points located a certain distance away from any of the modeled interface categories in the $R(0)$–G cross-plot are rejected. The unclassified units could either represent noise in the data, or lithologies/facies not included in the modeling. We suspect these to be thin units of cemented sands. This would be in accordance with the well-log data in Figure 5.57, where we observe a few anomalous high-velocity peaks in the sandy target interval.

5.4.5 Discussion and conclusions

The seismic signature of hydrocarbons can be very different from one depth to another because of different compaction trends for different lithologies. Therefore, it is necessary to include depth as a parameter when we use AVO analysis to predict lithology and pore fluids from seismic data. The depth-dependent probabilistic AVO technique presented in Chapter 4 enables us to predict the *most likely lithology and pore fluid* from seismic data, even in areas with sparse local well-log information. Nevertheless, the presence of local well-log data will improve the modeling of AVO pdfs, and give better control of the seismic calibration. The main limitations of the methodology include tuning and overburden effects, as well as the inherent ambiguities in rock physics properties and AVO response. Moreover, only unconsolidated siliciclastic sediments have been modeled. One future extension of the AVO technique used in this case study will be to include depth trends for cemented sandstone. It is also important to note that fluid properties will be depth-dependent. Pressure and temperature control the compressibility of fluids, but the chemical properties of fluids can also change with depth. In particular, oil reference density (API gravity) tends to be depth-dependent, where biodegradation of oil decreases with depth. Hence, shallow reservoirs will normally contain relatively thick oil, compared with deeper reservoirs. Trend lines of oil reference density versus depth would be valuable information to be included in this AVO classification technique. Another future extension will be to include facies transition probabilities and spatial statistics to improve the constraints on the classification of vertical and lateral geologic variations from seismic data.

5.5 Case 5: Seismic reservoir mapping using rock physics templates. Example from a North Sea turbidite system

(Courtesy of Aart-Jan van Wijngaarden, Susanne Lund Jensen, and Erik Ødegård, Norsk Hydro.)

Rock physics templates (RPTs) are locally calibrated rock physics models that can be used for interpretation of well-log and seismic data. In our case, the RPTs are plotted in terms of acoustic impedance (AI) versus V_P/V_S. The models used include the friable-sand model and the constant-clay model presented in Chapter 2. In Chapter 2 we showed how to construct the templates, and we presented an example of how to apply RPTs for well-log analysis. We also demonstrated how well-log data can be used to calibrate/validate the RPTs before the interpretation of seismic data. An example of seismic data analysis using RPTs was shown in Chapter 4. In this case study, we present an example of the full workflow of RPT analysis, starting with the selection of the most appropriate RPT using well-log cross-plot analysis followed by rock physics interpretation of elastic inversion results using the selected template. We also demonstrate how the RPTs can guide the classification of elastic inversion results. The example is from the Grane field in the North Sea, the same field as for case study 3. This is a producing oil field, and the goal is to characterize the reservoir in terms of lithofacies and depositional units.

5.5.1 Step 1: RPT selection and validation to well-log data

As mentioned in Chapter 2, we need to validate the RPT to well-log data before we apply it on seismic data. Plate 5.63 illustrates the template verification step. The log data are from Well 2 shown in Figure 2.13, Section 2.3.1. The lithologic interpretation is based on a suite of different log data (gamma-ray, V_P, V_S, density, and neutron porosity) as well as core information (see case study 3, Figure 5.43). Plate 5.63 also includes a cross-plot of AI versus V_P/V_S where we have superimposed the log data onto a selected RPT. We have identified certain populations in the data and marked these with colored polygons. The boundaries of the zones are selected by qualitative judgement. Still, we see that the defined zones match perfectly with different lithofacies identified in the well-log data. The AI and V_P/V_S logs have been color-coded according to the zones defined in the AI vs. V_P/V_S log cross-plot. The red zone includes two different shales, and the yellow zone contains mixed facies (shaly sand, silty shale and marls). The chalk constitutes a small and clearly separate cluster in the high-AI violet zone. The oil sand is highlighted by the blue zone, while the brine zone is represented by the green zone. Note that the oil–water contact coincides with a sudden change in porosity: the brine sand has higher AI values than the oil sand, as expected, but lower V_P/V_S values, which is the opposite of expected. Normally, we expect oil sands to have lower V_P/V_S values than brine sands at the same stratigraphic level. In Chapter 2 and in case study 3, we have demonstrated that the Grane sands exhibit great changes in porosity due to variation in sorting. Moreover, the oil in this example is very dense (18 API) and has a relatively low GOR, so a large deviation from the brine-sand trend is not expected. We can therefore conclude that the log data match the theoretical trends in the selected RPT fairly well. Bear in mind that the RPT in Plate 5.63 only includes the friable or unconsolidated sand model together with a shale model. However, as shown in Chapter 2, the sandy interval has a little bit of quartz cement. This may explain why the sands plot somewhat below the friable-sand line in the RPT plot. Cementation will cause a drop in V_P/V_S ratio and an increase in acoustic impedance (cf. Figure 2.44, Section 2.8.2). The exact porosity values of the friable-sand line are therefore not representative for the Grane sands. Nevertheless, the modeled trends are relevant, and the friable-sand model together with a shale model (constant-clay line) nicely explains the geologic trends in the target zone.

5.5.2 Step 2: RPT interpretation and classification of elastic inversion results

RPT analysis of elastic seismic inversion results

A sub-cube of 3D elastic inversion data exists around the well shown in Plate 5.63. Plates 5.64 and 5.65 show vertical sections of estimated AI and V_P/V_S across the Grane reservoir. AI is determined from near-stack inversion, while V_P/V_S ratio is determined from elastic impedance combining near- and far-offset stacks. For details about the methodology, see Chapter 4. Plate 5.66 shows a cross-plot of the AI versus V_P/V_S derived from the seismic data, superimposed onto the same RPT as selected in step 1. The cross-plot only contains the data within the limited time-window 1650–1850 ms, since the template is strictly valid only for a given depth. The templates should, however, be fairly robust within a window of a few hundred meters, as shown in the validation step in the previous section. The target time depth in this case is around 1750 ms. The selected time-window corresponds roughly to the depth range cross-plotted in Plate 6.63, for which the selected RPT is assumed to be valid.

The rock physics interpretation of Plate 5.66 appears to be straightforward. The population that sits along the theoretical shale trend is interpreted to represent shale. Note that the shale points appear to move closer to the sand trend for the highest AI values. This could reflect shales becoming increasingly more silty, and the points between the shale and brine-sand trends are interpreted to be silty shales and/or shaly sand. The points close to the theoretical brine-sand trend probably represent clean sand. Some of the data points plot below the sand line, with lower V_P/V_S ratios than we expect for brine sands. As discussed above, this could reflect the fact that most of the clean sands in the Grane field are slightly cemented (1–2%). We do not expect to see a clear oil-sand response, as the oil is fairly heavy in this case, but some of the data that plot significantly below the brine-sand trend may also be attributed to oil saturation.

Seismic facies classification and reservoir mapping guided by RPT analysis

Next, we want to use the RPT models in Plate 5.66 to guide the selection of certain populations or polygons of data that we assume represent various lithofacies or rock types. In Plate 5.66 we have defined the following facies populations:
- Shale (red polygon): high V_P/V_S and low AI values.
- Sands 1 (purple polygon): low V_P/V_S and low to intermediate AI values.
- Sands 2 (green polygon): intermediate V_P/V_S and intermediate to high AI values.
- Sands 3 (blue polygon): intermediate to high V_P/V_S and low to intermediate AI values.
- Chalk (yellow polygon): low V_P/V_S and high AI values.

The polygons made in Plate 5.66 are somewhat different from the polygons in Plate 5.63. This is because we want to capture the texture-related changes of the sands rather than the fluid changes. Also, the seismic data contain a larger variability in facies than we observe in one single well, and we want to include facies changes not encountered in the well. This is where the benefit of the rock physics template comes in. The underlying models make it easier to interpret facies not observed in the wells. Also note that the polygon colors in Plate 5.66 are different from those in Plate 5.63.

The sandstones in the Grane field have large variation in porosity associated with sorting (see Chapter 2 and case study 3, this chapter). This variation is associated with different depositional events and sub-environments within the Grane turbidite system. By grouping into different sandy facies we hope to delineate different depositional units from the seismic inversion results. As shown in Plate 5.63, the change from oil sands to brine sands followed a porosity trend, not a fluid trend. Indeed, for the well shown in Plate 5.63, the oil sands are better sorted than the brine zone. The oil is relatively dense and therefore difficult to differentiate seismically from brine. Nevertheless, we define Sands 1 to fall entirely within the hydrocarbon zone of the template. The low V_P/V_S ratios may or may not be related to hydrocarbons. As mentioned above, the sands in the Grane area are slightly cemented (case study 3), so the low V_P/V_S ratios may also reflect clean sands with contact cement. Consequently, in this case we assume the different sand populations to represent variation in sand quality, and not variation in pore fluids:

- Sands 1 = clean, well-sorted, high-porosity sandstone. Slightly cemented.
- Sands 2 = relatively clean sands, moderately well-sorted, with intermediate porosity values. Assumed to be less cemented than Sands 1.
- Sands 3 = relatively dirty "sands," more likely silty shales. Poorly sorted rocks. Possibly including some tuff deposits (see case study 3).

The classification result based on the RPT analysis in Plate 5.66 is shown in Plate 5.67. The cap-rock is classified as massive shales, just as we observe in the well-log data. The Grane reservoir sands are identified, and we observe the better-quality sands (Sands 1) in the center, with adjacent sands of lower quality (Sands 2 and Sands 3). As mentioned above, "Sands 3" are probably not true sands, but silty shales. This facies category may also include some volcanic tuff deposits (see case study 3). Silty shales and tuff deposits will indeed be located between the shale line and the clean sand line in the RPT plot, as we observe in Plate 5.66. This makes geologic sense, as we expect silty, hemipelagic shales to be present laterally besides the sandy turbidite channels and lobes. It is also interesting to note that we are able to identify the thin shale between the reservoir sands and the chalk beneath (compare with well-log observations in Plate 5.63). The sandy facies predicted within the chalk unit we suspect to be erroneously classified marl deposits (i.e., mixed facies of chalks and siliciclastics).

In Plate 5.68 we have plotted a time slice (at 1780 ms) of the classified 3D cube. The resulting map view of the classification results shows some nice geologic features. We identify a rather patchy but channelized outline of Sands 2 (green), surrounded by the "Sands 3" facies, more likely to represent silty shales and/or tuff deposits. The Sands 1 facies are almost absent in this time slice. However, a "movie" through the cube (not included here for practical reasons) reveals how the different facies are associated with different depositional units. This "movie" shows that Sands 1 is likely to be the youngest event of sands, sometimes located on top of Sands 2, and sometimes filling in depositional lows between geologically older units of Sands 2. The underlying shale and chalk facies are seen in the lower right of the map in Plate 5.68. This is due to the structural dip of the stratigraphy in the direction towards a structural high (the Utsira High; see case study 3).

5.5.3 Conclusions

We have demonstrated how RPT analysis can be used to guide interpretation and classification of pre-stack seismic inversion results into various lithofacies and depositional units in the Grane field. The procedure consists of two basic steps: (1) selecting the template that is consistent with the well-log data; and (2) applying the user-defined polygon boundaries in the template to classify the seismic data. This is a semi-quantitative method. The templates are based on quantitative rock physics models. But the validation with log data, and especially the choice of separation polygons, is based on qualitative judgement. RPTs help to combine qualitative expert knowledge with model curves.

The results show that we can potentially distinguish between different types of sands in the Grane field. We are also able to delineate reservoir sands from silty shales. In this way, this case study is a complementary technique to the statistical AVO technique applied in case study 3. The final classified cube of lithofacies can be used as input in well planning and field development during the production phase of the Grane field.

6 Workflows and guidelines

Damn the torpedoes, full speed ahead! *Admiral David Glasgow Farragut*

In this chapter we provide a summarizing workflow, or road map, explaining the major steps of the methodologies for seismic reservoir prediction and characterization presented in this book. *In the description of the workflows we consider the term AVO to represent all offset reflectivity-dependent seismic attributes.* These are not limited to the classical intercept–gradient attributes but also include other elastic parameters extracted from pre-stack data such as near- and far-offset impedances, elastic Lamé parameters, converted wave impedance, P-wave and S-wave impedances, and density. The workflows are general and are applicable to any quantitative seismic attribute that can be linked to rock properties.

A complete workflow of *quantitative* seismic interpretation should also include some necessary *qualitative* steps, including AVO reconnaissance, semi-quantitative feasibility studies based on well-log analysis, and qualitative interpretations of the inversion results. Below, we list our recommended techniques to be included in a combined qualitative and quantitative seismic interpretation workflow:

(1) AVO reconnaissance and seismic anomaly hunting (performed together with conventional seismic interpretation, not afterwards!).
(2) Well-log-based rock physics and AVO feasibility study (rock physics templates (RPTs) and cross-plot analysis, lithology and fluid substitution, forward seismic modeling).
(3) RPT interpretation of elastic inversion results.
(4) AVO lithology and fluid classification constrained by rock physics depth trends.
(5) AVO analysis and classification constrained by statistical rock physics and facies analysis of well logs.
(6) Elastic inversions and classification constrained by statistical rock physics and facies analysis of well logs.
(7) Quantifying the uncertainty associated in the interpretations in terms of probabilities.

Workflows and guidelines

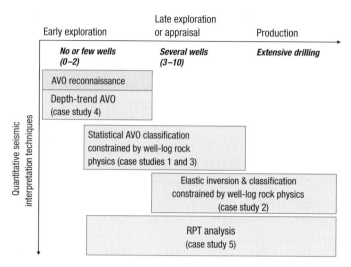

Figure 6.1 Quantitative seismic interpretation techniques performed at different stages of development of an oil field.

All the methodologies listed above are complementary to each other, yet they should be applied at different stages during exploration and production of an oil field. Figure 6.1 depicts a suggested flow-scheme for how the different techniques are tied together and performed at different stages during the development of an oil field. Below, we first go through the steps of each of the methodologies listed above. Finally, we will discuss how these technologies should be used together, and how some of the technologies are most appropriate for the early exploration stage, whereas others are more appropriate for later exploration, development, and production stages of an oil field.

6.1 AVO reconnaissance

AVO analysis should always be done (if pre-stack data are available) at an early stage of seismic exploration. The first thing to do is to create sub-stacks of seismic data, and/or AVO attributes that combine intercept and gradient, plus other attributes if possible (Poisson's reflectivity, fluid factor, etc.).

The primary reason for early use of AVO is to scan for anomalies that may represent hydrocarbon accumulations. In order to understand what is an anomaly, one must also understand what is the background "wet" response at the given target depth. At a very early exploration stage, the target depth may not be defined, and rock physics depth trends may be useful, especially if well-log data are lacking (see Sections 6.2 and 6.4).

A further motivation for early AVO analysis is understanding the seismic signatures of lithologies. It makes little sense to start interpreting a hard event as a potential reservoir sand, if the expected seismic response of a wet sand is relatively soft (or dim) compared with an overlying shale. Also, the appropriate AVO attribute must be selected on the basis of expected AVO response for a given case. Hence, early AVO screening should go along with rock physics feasibility studies and qualitative seismic interpretation.

To summarize the steps of AVO reconnaissance, we suggest the following workflow, assuming that the appropriate pre-processing scheme (see Chapter 4) has been carried out in advance:

- Create sub-stacks that are balanced correctly, so the near-stack response and far-stack response are consistent with each other, with relative amplitude levels intact.
- Create far versus near attributes (see Section 4.3.11). Make sure the far and near stacks are properly aligned, with no phase difference.
- If pre-stack CDP gathers are available, estimate $R(0)$ and G. Make various attributes that combine these parameters. ($R(0)$ and G may also be estimated from partial stacks.)
- If well logs are available in the area, make rock physics cross-plots to improve your understanding of the expected seismic signatures in the area, both in terms of background "wet" response and anomalous hydrocarbon response. Go to Section 6.2. The results will tell you which attributes are most appropriate (see Chapter 4). These results can also be valuable input information to the conventional seismic interpreter.
- If well logs are not available, no attempt should be made to interpret the AVO results quantitatively. However, if well logs from nearby or regional wells are available, rock physics depth trends may be very helpful in improving the understanding of expected seismic signatures. Rock physics depth trends can also be used to calibrate the AVO parameter estimates. Go to Section 6.2 and/or 6.4.
- Having found the optimal AVO screening attribute, start "hunting" for anomalies. If an anomaly is detected, investigate the AVO data from a time-limited window around the observed anomaly, in an attribute cross-plot, for example $R(0)$ versus G.
- Identify the background trend in the cross-plot and in the seismic section. Do you observe geologic elements/layers that have been confirmed or interpreted to be water-saturated sands at the same depth? Investigate where the known/interpreted water sands locate in an AVO cross-plot compared with the observed anomaly.
- If the anomaly is first detected in a cross-plot, then identify where on the stacked section the anomaly plots. Compare the anomaly with qualitative seismic interpretation results. Does the anomaly occur in a location that is geologically plausible for hydrocarbon accumulation (structural and/or stratigraphic trap)?
- Investigate the CDP gathers where the anomaly occurs. Do you observe an AVO response at the same location where the AVO attributes indicate an AVO anomaly?
- Do noise analysis of the cross-plot. If the fluid anomaly is still seen in a cross-plot of $R(0)$ versus full-stack amplitude, the fluid anomaly observed in $R(0)$ versus G is more likely to be real (see Section 4.3.10).
- Analyze potential geologic pitfalls that could have caused the observed anomaly. Does the anomaly occur at a pinch-out zone, so that it could reflect tuning effects?

Any signs of overburden effects (gas clouds, complex geology with lateral velocity shifts, faulting, extensive thin-bed effects, etc.)?
- Integrate the AVO results with conventional seismic interpretation results.

Complementary to the above recommended workflow for AVO screening, we also include the AVO checklist of Castagna (1993), which contains some additional important points to be aware of during qualitative AVO analysis.

> **Castagna's AVO checklist**
>
> - Is the expected variation in rock properties sufficient to produce a detectable AVO anomaly? What are the petrophysical signal and background?
> - Can the same seismic response result from other (nonpay) Earth models? What are the chances that one of these other models is the real one?
> - If AVO correctly predicts the occurrence of hydrocarbons, what are the chances that the saturations will be commercial? Do all parties involved understand that AVO cannot generally be used to distinguish between commercial and noncommercial saturations?
> - Is there sufficient angular coverage for the event of interest?
> - How do I know that processing has preserved and isolated the "true" relative AVO response? What quality control (QC) displays have been used to verify this?
> - What is the seismic data quality? Has coherent noise been adequately suppressed? If so, has the procedure corrupted the relative AVO signal?
> - Have lateral variations in overburden effects been properly compensated?
> - Has an appropriate amplitude-preserving migration algorithm adequately corrected for structural effects?
> - Is the AVO algorithm being used sensitive to velocity errors? Do the NMO velocities need to be repicked?
> - Can the AVO anomaly be verified by eye on the CDP gathers? Is the anomaly robust or must it be coaxed from the data?
> - Does the AVO anomaly confirm to structure? Do I really believe the result or am I ignoring my instincts because I want to drill the prospect? Am I using AVO to better characterize the subsurface or just as a flashy sales device?
> - Do I understand what "red" on the AVO display really means in physical terms?

6.2 Rock physics "*What ifs*" and AVO feasibility studies

In conjunction with the qualitative AVO screening summarized above, one should perform rock physics and AVO feasibility studies. This feasibility study should include rock physics analysis of well-log data, rock physics modeling, and AVO reflectivity

6.2 Rock physics "*What ifs*" and AVO feasibility studies

modeling as well as seismic modeling and analysis of CDP gathers at well locations. Doing this, we can find out if AVO will work in our case. Also, different rock physics "*What if*" questions can be modeled in terms of AVO response. Sensitivity studies of both fluid and lithology changes should be performed. Statistical rock physics analysis (see Chapter 3) can also help to determine what seismic attributes are optimal for fluid and facies discrimination in our case.

The feasibility studies can be done to help the reconnaissance, but may also help us to decide which quantitative seismic techniques would be most appropriate. Is inversion of near-stacks and estimation of acoustic impedance enough to discriminate hydrocarbons from lithologies, or do we need to carry out full-offset inversion and estimate elastic parameters?

We can summarize the steps of the rock physics and AVO feasibility studies as follows, assuming that thorough quality control and corrections of well logs have been performed:

- Make cross-plots of different well-log data, including V_P–porosity, V_S–porosity, V_P–V_S, AI–V_P/V_S, AI–EI, etc. When shear logs are missing, perform V_S prediction using Greenberg–Castagna's formulas or other locally calibrated methods that give you estimated V_S (see Chapter 1).
- Create color-coded scatter plots with color indicating a third dimension of gamma-ray, saturation, clay content, or other petrophysical logs or estimates. The third dimension may help to identify the reasons for trends observed in the cross-plots. What do the cross-plots tell you in terms of seismic detectability? Are fluids discriminated mainly in terms of V_P/V_S ratios (soft rocks) or in terms of AI (stiff rocks)? (See Chapter 2.)
- If necessary, apply rock physics models to understand the trends seen in the cross-plots. RPT analysis of well-log data is a means of comparing well-log data with rock physics models for interpretation of observed trends, and quick determination of various "*What if*" scenarios, including porosity, lithology and fluid saturation changes not encountered in the data (see Chapters 1 and 2).
- In the case that well logs only contain data from brine zones, perform fluid substitution to find out where oil and gas sands will plot. Also check for different saturation scenarios, including patchy versus uniform saturation, and commercial versus residual saturations. Plot the fluid-substituted data in the same cross-plots as above (see Chapter 1).
- Perform lithology substitution to capture lithofacies or textural scenarios not encountered in the wells, but likely to occur away from the wells. If we observe a given type of sand (in terms of porosity, clay content, sorting, and cementation) at a drilled location, how do we expect this sand to change away from the well? Add the new modeled data in the same cross-plots as above, in order to determine the seismic detectability of the "*What if*" scenarios (see Chapters 1 and 2).
- Make synthetic seismograms (CDP gathers) at the well locations, and investigate the seismic signatures (including AVO responses) of the target zone. Also, make

synthetic seismograms of the *"What if"* scenarios, including both fluid and lithology substitutions (see Chapters 1 and 4).
- Compare the original synthetic seismograms with real CDP gathers at the well location. Then, compare CDP gathers away from well location with the various *"What if"* synthetic seismograms. Are the observed lateral changes in AVO response more in agreement with the lithology-substituted seismograms, or with the fluid-substituted seismograms? (See Chapters 1 and 4.)
- Perform 2D (and/or 3D) forward seismic modeling, and conduct the same fluid and lithology *"What if"* changes as mentioned above. In addition, geometric *"What if"* changes may have to be modeled as well (including layer thickness, faulting patterns, net–gross, etc.). (See Section 4.5.)

6.3 RPT analysis

The rock physics template (RPT) analysis methodology is described in Chapter 2 (theory and well-log applications) and in Chapter 4 (application to seismic data analysis). Also see case study 5 in Chapter 5. The main goal of RPT analysis is to use rock physics models constrained by local geology to interpret and classify well-log and seismic data. The RPT plots are also very intuitive and allow for better communication between geologists and geophysicists. The workflow of a complete RPT analysis is summarized in Plate 6.2.

Step 1: Well-log analysis and RPT validation

- The RPTs are based on the rock physics models described in Chapter 2. The RPTs can also be made from other, alternative rock physics models. A compilation of RPTs should be generated for a given basin, as a catalog of rock physics charts describing various lithology, mineralogy, diagenesis, and depth scenarios.
- One needs well-log data to validate and calibrate the models used in the RPT plots. However, if a few well logs exist in an area, one can make RPT plots to cover a range of *"What if"* scenarios away from the wells. In this way, the RPT plots can be used in feasibility studies (go back to Section 6.2), to predict expected lithology and fluid trends in terms of seismic properties.
- If no existing RPT plot matches the well-log data in a new well, one should update the RPT models. Alternatively, one should check the quality of the well-log data. Wash-out effects and mud-filtrate invasions are common causes of mismatch between observed well-log data and the models in the RPT plots.
- Select the appropriate RPT that matches with the well-log data in the target zone. The same RPT is applied to interpret seismic inversion results (go to step 2).

Step 2: Seismic data analysis using selected RPT plot

- Perform elastic seismic inversion and obtain acoustic impedance (AI) and elastic impedance. From elastic impedance, calculate V_P/V_S ratios (see Section 4.4).
- Cross-plot the AI versus V_P/V_S data onto the RPT selected in step 1 (see Section 4.4.6). Other attributes such as AI versus EI, $\lambda-\mu$ etc. could also be investigated in RPT plots.
- Interpret the trends in the elastic inversion results using the underlying models in the RPT plots. If there is a great mismatch between inversion results and RPT models, evaluate the quality of the inversion procedure. Also, evaluate possible scale effects that can cause differences between the seismic data and the well-log data, especially if there are a lot of thin-bedded layers in the target zone (see Chapter 4).
- Use the RPT plot to guide manual classification of characteristic data populations, representative of lithology and/or fluid changes (see case study 5, Chapter 5).
- Evaluate the RPT classification results along seismic sections and/or 3D cubes.

The wide range of applications of RPT analysis

- Interpretation of well-log data and assessing seismic detectability of observed facies and fluid variations.
- Petrophysical quality control of well-log data.
- "Quick and dirty" qualitative feasibility studies of various rock physics "*What if*" scenarios, both in terms of lithology and fluid substitutions.
- Calibration and validation of rock physics models to local conditions.
- Interpretation and classification of elastic inversion results.
- The RPT plots are intuitive and link rock physics properties to geologic trends. Hence, the RPT plots, and the interpretations of these, are perfect for round-table, cross-disciplinary discussions of inversion results.

6.4 AVO classification constrained by rock physics depth trends

During early exploration, when few or no well-log data are available, AVO classification can be constrained by rock physics depth trends. Plate 6.3 shows a flowchart of this technique. The methodology is thoroughly explained in Chapter 4.

Step 1: Calibration/estimation of rock physics depth trends

- Using nearby or regional wells, establish the local or regional compaction trends for different lithologies expected to occur in the area of investigation. Porosity–depth trends are calibrated to density logs and/or direct measurements of porosity (for

example helium porosities measured on core samples). For siliciclastic systems, we suggest that you make depth trends for sands, shales, and shaly sands (see Chapter 2).
- Estimate corresponding seismic properties (V_P, V_S, and density) for brine-saturated rocks as a function of depth, using rock physics models. If unconsolidated sediments, use Hertz–Mindlin models. Use Gassmann's theory to obtain brine-saturated values from models that give dry rock properties. If consolidated rocks, use rock physics models for cemented rocks. Empirical regression trends may be used instead of theoretical rock physics models, if sufficient well-log data are available. It is important to discuss with a geologist to find out if and at what depth cementation is expected to occur. Geochemical modeling of cement volume as a function of depth is very valuable input information in the modeling of seismic depth trends (see Chapter 2).
- In the lack of any well-log data, velocity–depth trends extracted from seismic travel-time inversion may be used as input.

Step 2: Modeling AVO pdfs at a target depth

- Select a target depth at which to pick V_P, V_S, and density from the depth trends derived in step 1. These represent the mean values for the different lithologies.
- Define variances and covariances of the same properties. The variances are assumed to be depth-independent and may be selected from an analog area or from well logs nearby. Now we have the estimated distributions of V_P, V_S, and density, for the various lithologies, representative for the target depth.
- For the depositional system in the area of interest, define likely facies models and corresponding interface categories (see Chapter 2). This procedure should be done in collaboration with an experienced sedimentologist who has worked in the area, or in analog systems. Also, include various pore-fluid scenarios. However, avoid making too many interface categories. This will make the following classification unstable. Fluid–fluid contacts should be avoided, because these often exhibit a relatively weak AVO response, and may be misclassified as shale over brine-sand. Flat spots and fluid contacts should be investigated using qualitative AVO combined with rock physics sensitivity studies (Sections 6.1 and 6.2).
- For each interface category, perform Monte Carlo simulations to create AVO pdfs of $R(0)$ versus G representative of the target depth (see Chapter 3). Use Gassmann's theory to extend the training data to include pdfs of categories not encountered in the well, for example, oil sand or gas sand below shales.

Step 3: AVO parameter estimation

- Estimate AVO parameters from CDP gathers along 2D lines or for a 3D cube of pre-stack seismic data (see Chapter 4).

- Extract the uncalibrated $R(0)$ and G values only for a target window, representative for the selected target depth in step 2. (A time interval of 300–400 ms should be fine, since very little depth-related variation in the seismic properties will generally occur within this window.) Include all samples, not only trough and peak values, from this target window. We want to include potential class II AVO anomalies, which are easily left out if only maximum amplitude values are selected. Be aware, though, that sidelobe energy will be classified as an interface category.

Step 4: Calibration of AVO attributes to modeled AVO pdfs

- Identify a "background-trend" window in the seismic data, located at the same depth as the target anomaly.
- Extract the variances or covariances of $R(0)$ and G for the uncalibrated background-trend data window.
- Define which modeled AVO pdfs shall constitute the background trend. For a siliciclastic sand–shale system, we suggest that the background trend includes shale–shale and brine-saturated shaly sand–shale interfaces. Clean brine sands should be avoided as part of the background trend, because these are actually rare, and they can have significant AVO effects off the shaly background trend. Extract variances or covariances of the modeled background trend.
- Calibrate the real background trend with the modeled background trend.
- Do the same calibration of the AVO data around the target anomaly as for the background window. Now you have obtained calibrated $R(0)$ and G values for the target zone.

Step 5: Lithology and fluid classification

- Use discriminant analysis or other classification techniques to classify the calibrated AVO data with the modeled AVO pdfs as the training data (see Chapter 3).
- Plot the resulting most likely lithofacies and fluids as sections or maps.
- Plot the occurrence probabilities of various facies and fluids (see Chapter 3).
- Use the seismic reservoir prediction results as input in prospect evaluation and risk assessment.

6.5 Seismic reservoir characterization constrained by lithofacies analysis and statistical rock physics

In this section we present in detail a general workflow that can be applied to map 3D seismic data attributes into 3D lithological cubes and evaluate the uncertainty associated with the prediction (Plate 6.4). The workflow describes the steps of the techniques listed as 5, 6 and 7 at the beginning of this chapter. These techniques combine rock physics theory (Chapter 1) with local geologic parameters or trends (Chapter 2) and statistical techniques (Chapter 3) to predict and map the most likely lithofacies and pore fluids, as well as their occurrence probabilities, from seismic AVO or impedance inversion data (Chapter 4). This workflow, which has been used in case studies 1, 2, and 3 in Chapter 5, is best suited for situations where there are at least a few key wells available.

Step 1: Seismic and rock physics lithofacies analysis

Mapping seismic volumes into lithology cubes is not a unique process, and a mechanism for incorporating a-priori geological information is needed. The first step of the workflow is dedicated to identifying textural effects, depth trends, and composition for the target interval, defining the range of seismic lithofacies, to which the seismic data volume will be mapped, and identifying key trends and rock physics models that will be used in transforming the seismic data into lithofacies. These goals imply the following specific tasks:

- Gather prior geological/sedimentary knowledge for the selected zone of interest from key wells. Identify depositional and compactional trends. What is the depositional environment and what are the associated facies? What are the dominant mineralogies? What types of rock are expected? Thin-section and core analysis will provide important information about sorting, clay content and cement volume. What types of pore fluids are expected? Gather information about fluid properties, including API, GOR, salinity, etc.
- Map well logs, core data, and geological knowledge into the rock physics domains, and make cross-plots of V_P–porosity, V_P–V_S, AI–V_P/V_S, etc. Be sure to correct the well logs for mud-filtrate invasion if necessary. Make V_S prediction if V_S log is not available (see Chapter 1).
- Select appropriate rock physics models and constraints to describe the well-log data. We suggest using the rock physics diagnostic procedure introduced by Dvorkin and Nur (1996) (see Chapter 2) to generate velocity/porosity/depth relations from all available data, as well as identifying key textural trends in the data (sorting, cementation, sand–shale–clay, etc.). Rock physics models can help us to discover subtle but important textural or mineralogical variations within the rocks.
- Define seismic lithofacies. On the basis of the well-log facies analysis and the rock physics diagnostic, we define seismic lithofacies by one of the following criteria. (1) It has a distinct lithologic/geologic definition, or (2) it has distinct acoustic properties. For example, shales and gas sands are two distinct lithofacies, although these may have the same seismic velocity. Gas-saturated sands, oil-saturated sands and brine-saturated sands are also distinct seismic lithofacies since each of them may have a different seismic signature.

Step 2: Derive attribute probability density functions (pdfs) from well-log pdfs, rock physics models and surface seismic data at the well location

In this step we define the uncertainty associated with the deterministic trends we have established in step 1 and analyze different attributes at the well location. The rock physics models and trends will be characterized in terms of probability density functions (pdfs). This step comprises the following tasks:
- Estimate the pdfs of rock properties for different lithofacies (Chapter 3).
- Extend the original data set where necessary by Monte Carlo simulations and rock physics models (Chapter 3).
- Derive the distributions (pdfs) of sets of possible seismic attributes. The seismic attributes to be tested are physical attributes that have specific models relating them to the rock properties. Among these are V_P, V_S, R_0, G, I_P, I_S, EI, Poisson ratio, $\lambda-\mu$, etc. Assessment of the surface seismic quality at the well location will be made. The quality of the data will determine to some extent which attributes can be extracted from it.
- Use the class-conditioned probabilities for statistical classification in step 3.

Step 3: Classification and validation tests at the well locations

- Estimate the separability of the lithofacies using the different attributes computed from the extended training data in step 2.
- Validate facies classification of different sets of computed attributes at the well locations. Different statistical algorithms that can be used include Bayesian classification based on pdfs, discriminant analysis, and neural networks. The output of a Bayesian classification procedure is the full posterior facies pdf, $P(\text{facies}|\text{seismic attributes})$ so an assessment of the uncertainty in the prediction is inherent in this procedure. The uncertainty is estimated in terms of the classification confusion matrix, which gives the conditional probability of the true lithofacies given the predicted facies.
- Choose an optimal set of attributes. This selection depends not only on the rock physics benefit as estimated by the classification success rate, but also on the economical constraints such as the cost of estimating that attribute from seismic data, and the quality of the seismic data. The classification validation also helps in selecting additional information that can help to improve the classification and reduce the uncertainty.

Step 4: Extract the optimal set of attributes from the seismic volume in 3D

- Extract the optimal set of attributes identified in step 3 from seismic data. The actual methodology will depend on the specific attributes, e.g. AVO analysis or impedance inversion.

- Calibrate the pdfs generated from the well data to the seismic and field scale. In general, the calibration procedure accounts for the upscaling of the well-data velocities and derivation of appropriate pdfs from analyses and/or inversion of seismic data available.

Step 5: Statistical classification of the seismic lithofacies using all available information

This step integrates the previous steps, and should include "all available information." We generate the final lithocube in the following way:
- Generate appropriate 3D attributes from the seismic data.
- Map/classify them into lithology cubes using the pdfs we have generated in the previous steps. Any number of 3D attributes extracted from the seismic data volume can be used and classified into most likely lithological units using the extended training data.
- We strongly recommend incorporating spatial correlation and seismically derived structure to constrain the lithocube further. This can be done using a conventional geostatistical framework (e.g. variogram-based co-kriging, simulation, indicator statistics) or using more recent multiple-point geostatistical methods. Geostatistics should certainly be used when there are a few wells (see Chapter 3).

Step 6: Visualization of lithocube and uncertainty in lithological prediction

Visualization of the output is an important part of the workflow. There are various ways of visualizing the final results. We recommend generating a maximum likelihood cube that will represent the most likely lithological unit at each point of the subsurface, as well as showing iso-probability surfaces which are essentially confidence intervals presented in 3D space. Such a display is presented in Plate 6.5 (Mukerji *et al.*, 2001). Once we obtain the posterior conditional pdfs [P(lithofacies|seismic attribute and training data)] other displays of critical values can also be made. For example, it might be of interest to display the probability that the depth-averaged sand/shale ratio is greater than some critical threshold.

6.6 Why and when should we do quantitative seismic interpretation?

In this book, we have shown the potential of various quantitative techniques to predict hydrocarbons, to determine lithologies, and to resolve ambiguities between different types of facies and fluid scenarios. All these methodologies take advantage of offset-dependent reflectivity information, via pre-stack or partial-stack seismic inversions, or AVO least-square estimates. Hence, in a complete industrial workflow of quantitative

6.6 Why and when?

seismic interpretation, conventional AVO modeling and analysis should be one of the first points on the task list. AVO analysis is also a cheap and rapid technology which can be very effective during early exploration. However, we have also emphasized the long list of pitfalls associated with AVO techniques.

Three very important questions to be raised before a new area is explored for potential hydrocarbon prospects are these.
(1) Does AVO analysis work in this area?
(2) Do we really need AVO in this case? (Do we expect a detectable AVO anomaly, given the known rock physics properties?)
(3) Do we observe any AVO anomalies in the real seismic data that can be related to hydrocarbons?

These questions are solved by AVO reconnaissance and AVO feasibility studies. These two tasks should go hand in hand, because the feasibility and modeling studies may tell us to go back and re-evaluate the observed AVO anomalies. Also, feasibility studies combined with AVO reconnaissance will tell us if there is any hope for more quantitative AVO analysis.

If we find that AVO analysis will work and will be needed in order to detect hydrocarbons in the area of investigation, a new, important question arises:
(4) When should we do AVO analysis? During seismic exploration, should it be done before, at the same time as, or after the conventional seismic interpretation and prospect evaluation?

We claim, as Castagna did more than 10 years ago (Castagna, 1993), that AVO analysis should always be included in the workflow of a seismic interpreter, from an early stage! It should be used for anomaly hunting rather than to validate already existing prospects. Unfortunately, in the oil industry, AVO analysis is often put at the end of the workflow of seismic exploration, after the qualitative seismic interpretation has been completed and prospects have already been defined. Also, the AVO analysis is often done by geophysicists who are not familiar with the local geology. While the seismic interpreter sits on valuable information about the area of investigation, the AVO analyst often must make vague assumptions about geologic input parameters. Often shortage of time results in lack of good communication and interaction between the seismic interpreters and the AVO analysts.

We would like to stress, as one of the final statements of this book, that AVO technologies, qualitative as well as quantitative, do not belong to the geophysicists, but should represent integrated technologies used by geophysicists and geologists together. Too often, hydrocarbon prospects are defined on the basis of conventional geologic interpretation of seismic geometries. Before these geologically (geometry) driven prospects are risk-evaluated, it is common to do late-stage AVO analysis to strengthen the prospect, making it an *AVO-supported prospect*. First of all, if AVO is implemented as a technical support, done after seismic interpretation, there is no real decision impact to be made by the AVO technology. The risk number may be adjusted up or down a little, depending on the AVO results. If the well turns out to be dry, it is not rare to see that the AVO analysis becomes the scapegoat. One easily forgets that the AVO analyst joined the game at a late stage, and probably did not receive too much information about local geologic parameters (if the AVO work is done after the seismic interpreter has finished his or her job, this makes interaction and collaboration between AVO analyst and seismic interpreter very challenging). Furthermore, defining the prospects before doing AVO analysis means that potential prospects that would be detected only using AVO techniques can be missed (like AVO class II anomalies causing dim spots on stacked sections). Fortunately, it is becoming more common for seismic interpreters to do interpretation on partial stacks.

There are also pitfalls in doing AVO analysis *before* conventional seismic interpretation. Defining a prospect based predominantly on an AVO anomaly would create an *AVO-driven prospect*. An AVO-driven prospect needs a geologic model that can explain the observed AVO anomaly. If the AVO work is done before there exists a thorough geologic interpretation in the area, it probably means that the geophysicist has made vague assumptions about the geologic input parameters in the first place. And there will always be some positive scenario that can explain an observed AVO anomaly. An AVO-driven prospect can easily make the interpreter blind to pitfalls.

If AVO techniques are *integrated* with geologic interpretations of seismic data during prospect evaluation (call that *geology-controlled AVO* analysis), it allows for more collaboration between the conventional seismic interpreter and the AVO analyst. The seismic interpreter can gain important input from the AVO analysis during the geometric interpretation, while the AVO analyst can gain important input information to better constrain the rock physics models behind the AVO analysis. If we want to discover increasingly subtle oil fields in the future, there is a need to establish more interaction between conventional seismic interpreters and quantitative seismic interpreters. This also means that the conventional seismic interpreters must become more knowledgeable about AVO analysis and other quantitative seismic techniques, whereas the rock physics and AVO analyst must become more knowledgeable about geologic aspects of seismic interpretation. This integration aspect is one of the main motivations of this book: to make quantitative seismic interpretation techniques more available to seismic interpreters and practitioners, and to improve interaction between the geologically inclined interpreters and the geophysically inclined interpreters. We hope this will result in improved drilling success for oil companies.

That being said, it is important to note that while some of the techniques presented in this book should only be used during early exploration, others can also be used during appraisal and production stages of an oil or gas field. A final question to be asked is therefore the following:

(5) What quantitative methodologies should we select to use, and at what stage?

AVO classification constrained by depth trends (see case study 4, Chapter 5) would be used only during early exploration, when few wells are available, or during appraisal

6.6 Why and when?

where satellite fields are explored at different depths from an existing field where wells are available. AVO classification constrained by well-log observations (case studies 1 and 3 in Chapter 5) can be performed during appraisal stage, where a few wells have been drilled and the goal is to predict the reservoir properties in the interwell areas. At the same stage or later, given that enough wells are available, impedance inversions ought to be carried out. Statistical rock physics models can then be used to classify the layer-based inversion results (case study 2). Classification of impedance inversions should also be the selected methodology during production stage, and the same classification techniques as presented here can be applied for pressure and saturation mapping from time-lapse (4D) seismic data. RPT analysis (case study 5) can in principle be carried out at any stage from early exploration to late production, depending on the goal of the analysis. However, some wells must be available in order to calibrate the models used in the rock physics templates. Also, if the templates are in terms of AI and V_P/V_S as shown in this book, impedance inversions are required and therefore some wells should be drilled before reliable seismic data analysis can be done using RPT plots. That being said, the RPTs are semi-quantitative, and one can use the RPT plots to analyze well-log data, and then extrapolate and predict expected seismic properties away from a given well. This can be done as part of feasibility studies during early or late exploration stages.

Again we emphasize that the methodologies and workflows are applicable to all quantitative offset-dependent attributes extracted from seismic data such as intercept and gradient, impedances, and elastic parameters.

7 Hands-on

For the things we have to learn before we can do them, we learn by doing them.

Aristotle

7.1 Introduction

This chapter provides problem sets and an extended reservoir characterization project based on an example seismic data set and well logs provided at the book website. The website also has additional resources in the form of downloadable MATLAB function files that may be helpful in solving the problems. We used the MATLAB statistics toolbox and neural net toolbox for solving the problems involving pdf estimation and statistical classification.

Well-log data from five wells are provided in flat ascii text files. Well 2 is taken as the type-well. Six different lithofacies have been identified from Well 2. For the purposes of the problem sets and the project the facies are: clean sand, cemented sand, silty-sand1, silty-sand2, silty-shale, and shale. The depth zones containing data for these facies have been extracted from Well 2 and provided in separate text files. The seismic data consist of one 2D section of NMO-corrected pre-stack CDP gathers, and two 3D cubes of near- and far-offset partial stacks. Details about the seismic data and well locations within the seismic cubes are provided below in Section 7.3.

7.2 Problems

Rock physics modeling

Load the well-log data for Well 2 into your software of choice (MATLAB, Excel or any other suitable software). The ascii file contains five columns: depth, V_P, density, gamma ray, and V_S.

(1) Make cross-plots of V_P versus porosity. Derive porosity from density assuming mineral density of 2.65 and fluid density of 1.05 g/cm^3. What can you say about

the trends of shales versus sands? How do you explain porosity variation for each lithology class? Compare with upper and lower Hashin–Shtrikman bounds. Compare with Han's empirical sandstone line, assuming 5% clay content (Chapter 1). Is there a good match? If not, how do you explain the mismatch? Model a constant-cement line (using modified Hashin–Shtrikman lower bound combined with Dvorkin–Nur cement model) to match the trend of the reservoir sands, assuming 100% quartz. What is the volume of cement in the reservoir sands? Use modified Hashin–Shtrikman lower bound to model the cap-rock shales, assuming a critical porosity of 0.6. What effective bulk and shear mineral moduli are needed to match the shale data? What can these values say about the mineralogical composition of the shales?

(2) Make cross-plots of V_P versus V_S. Compare with famous mudrock line (Chapter 1) and "dry rock" line ($V_P/V_S = 1.5$). What do V_P–V_S relationships tell you about lithology and fluids?

(3) Make cross-plots of AI versus V_P/V_S. Superimpose on appropriate rock physics template (RPT). Interpret the trends you see in terms of fluid, porosity, lithology, sorting, cement or other effects. What can you say about seismic contrasts between reservoir sands and cap-rock shales from the RPT analysis?

(4) Use Greenberg–Castagna or any other shear-wave prediction tool to predict V_S from V_P for Well 2. Assume 80% quartz, 15% feldspar and 5% clays in the sands, and 80% clays and 20% silt (quartz) in the shales. Compare predicted V_S with measured V_S.

(5) Use empirical porosity trends from Well 2 together with Hertz–Mindlin model to create V_P–depth trends for brine-saturated sands. Assume the same mineralogical composition as in Problem (4). Water depth is 100 m. To create the depth trend, first fit an exponential porosity versus depth trend to the sand points in Well 2. For each depth compute the effective pressure assuming no overpressure. Finally use the porosity and effective pressure as inputs in the Hertz–Mindlin model to compute the velocity–depth trend. What can you say about the reservoir sands?

(6) Create normal-incidence synthetic seismograms for the case of brine-saturated sands. Use a 35-Hz zero-phase wavelet. What is the seismic character and polarity of the top and base sand?

(7) Use Gassmann's relations to do fluid substitution in Well 2, first from brine to oil, then from brine to gas in the upper 50 m of the reservoir sands. The fluid properties are as follows: brine salinity 80 000 ppm, oil API 19, GOR 100 l/l, temperature 70 °C, pore pressure 16 MPa, gas gravity 0.6.

(8) Create AI–V_P/V_S cross-plots after fluid substitution and compare with the AI–V_P/V_S for brine sands using rock physics templates (RPTs). What can you say about the seismic contrast between gas sands and shales versus oil sands and shales?

(9) Create synthetic seismograms for the oil and gas cases. What is the seismic character of the top and base sand? Do you observe any fluid contact (gas–water or oil–water contact)?

Statistical rock physics

(10) Plot histograms, boxplots, cumulative distribution functions and quantile–quantile plots of gamma ray, V_P, V_S, density. Explore the variability in the different rock properties for the different facies.

(11) Estimate and compare univariate pdfs of P-wave impedance for the different facies. A simple estimate of the pdf is obtained by a kernel-based smoothing of the data points. Explore the impact of the width of the smoothing kernel on the estimated pdfs.

(12) Make a color-coded cross-plot of gamma ray versus P-wave impedance for the different facies. Assign a different color to each facies. Explore qualitatively whether the clean sands (low gamma ray) can be separated from the shales (high gamma ray) using P-wave impedance.

(13) Estimate 2D pdfs for P- and S-wave impedance for the clean sand, cemented sand and shale facies. A simple way to estimate the 2D pdf is to smooth the 2D histogram. Visualize the 2D pdfs using contour plots or surface plots.

(14) Monte Carlo simulation: compute the nonparametric univariate cdf of V_P for the clean sand facies. A simple estimate of the cdf is obtained by a cumulative sum of the histogram frequencies. Draw 1000 uniform random numbers between 0 and 1, and interpolate the inverse of the cdf to obtain 1000 Monte Carlo simulations of V_P. Check that the histogram of the simulated V_P is similar to the histogram of the original data. Establish a linear regression between the well-log V_P and V_S of the form: $V_S = a + bV_P$. Estimate the coefficients a and b by standard least-squares curve fitting. Using the linear regression between V_P and V_S draw correlated Monte Carlo realizations of (V_P, V_S) pairs. Be careful not to make the Monte Carlo (V_P, V_S) pairs perfectly correlated. Check that the Monte Carlo simulations of (V_P, V_S) have about the same correlation as the original log data.

(15) Derive the distributions of P-wave velocity and impedance for oil-saturated clean sand. For this problem, the Well 2 data for clean sands represent brine-saturated sands. Draw 1000 correlated Monte Carlo realizations of V_P, V_S, and density from the log data. Make sure that the V_P–V_S and V_P–density correlations are honored. Propagate the simulated points through Gassmann's relation to get 1000 realizations of V_P, V_S, and density for oil-saturated clean sands. Use the same fluid properties as in Problem (7). Compare the histograms and pdfs of P-wave velocity and impedance for the original brine-saturated clean sand, and the derived

oil-saturated sands. Using simulations we can derive the pdfs and extend our training data beyond the conditions encountered in the well.

(16) Classify Well 3 into different facies based on the training data from Well 2. The facies in Well 2 are: clean sand, cemented sand, silty-sand1, silty-sand2, silty-shale, and shale. Use a minimum Mahalanobis distance discriminant classification. Gamma ray and V_P will be the input attributes for the classification. Test the classification error rate in the training data and then classify Well 3. Compare the classification with one attribute (gamma ray alone) and with two attributes (gamma ray and V_P). Remember to normalize the gamma-ray values in Well 3 to be within the range observed in the type-well 2. A color-coded cross-plot of gamma ray versus V_P will help in understanding qualitatively some of the causes for misclassification.

(17) Estimate 2D pdfs of gamma ray and V_P from the training data in Well 2 for the different facies. Use the pdfs in Bayes classification of Well 3. Compare with the classification obtained by discriminant analysis in Problem (16).

(18) Using the training data from Well 2 train a feed-forward back-propagation neural network to classify the different facies. The neural net will have two inputs: gamma ray and V_P. The output will be the facies class. There are six facies classes: clean sand, cemented sand, silty-sand1, silty-sand2, silty-shale, and shale. One way to represent the desired net output is by a six-element indicator vector with '1' in the jth position representing the jth class, and zeros in all other positions. Thus [0 1 0 0 0 0] would represent facies class 2. Use the trained neural net to classify the data in Well 3. Compare with discriminant and Bayesian classification obtained in Problems (16) and (17).

(19) Compute the 2D pdfs of AVO intercept, $R(0)$, and gradient, G, for the following pairs of cap-rock over reservoir rock: shale/clean sand (brine); shale/clean sand (oil); shale/cemented sand (brine); shale/cemented sand (oil); shale/silty-sand1 (brine); shale/silty-sand1 (oil); shale/silty-shale; shale/shale. The data for the six different facies in Well 2 represent brine-saturated conditions. Derived distributions for oil-saturated conditions will have to be obtained using Monte Carlo simulation as in Problem (15). For each cap-rock/reservoir-rock pair, draw 1000 correlated Monte Carlo points of (V_P, V_S, ρ) for the cap-rock as well as the reservoir rock. Propagate the simulated point through the equations for $R(0)$ and G (Chapter 4) to obtain 1000 simulated $R(0)$–G pairs. Estimate the pdfs from the simulated point for each class.

(20) From the simulations of $R(0)$ and G in Problem (19), compute the classification error rate and confusion matrix using a leave-one-out jackknife (Chapter 3). Use any classification method (discriminant, Bayesian, neural network). Combine all the oil sands into one group, the brine sands into a second group and the silty-shales and shales into a third group. Now compute the classification error rate

and confusion matrix for the three groups instead of the original six groups. Do a bootstrap analysis of the uncertainty of the error rate by drawing a large number of bootstrap samples of the training data and computing classification error rates for all of the bootstrap samples.

(21) Compute classification error rate and confusion matrix for different pairs of attributes for the different facies classes: oil sands, brine sands, and shales (including silty shales). Oil-sand distributions will have to be derived using Monte Carlo simulations and fluid substitution. Compute the following different attribute sets: P-impedance and S-impedance; AI–EI($30°$) (Chapter 4); $\lambda - \mu$, P-impedance, S-impedance, and density. For each attribute set estimate and compare the classification confusion matrix for different classification methods.

7.3 Project

Objective

Using quantitative interpretation methods seismically characterize the reservoir. Create maps of most likely facies and fluids. Estimate the uncertainties and create maps of probability of occurrence for different lithofacies and pore fluids.

Reservoir information for North Sea oil field

The field to be investigated is located in the South Viking Graben in the North Sea. It is of Paleocene age, and represents turbidite sand deposits. The sands were eroded off the Scottish Mainland and East Shetland Platform, and transported to the "deep sea" between Scotland and Norway, into the graben basins of the North Sea.

The sediments are today buried at a depth of about 2200 m in the area of study. But they are still loosely consolidated sediments. The episodes of sand deposition were separated by longer periods of high-stand shale deposition. Hence, the lithology variation can be complex and variable both vertically and laterally in these systems.

Hydrocarbons were encountered in Well 2 and Well 5 in these sands, often referred to as the Heimdal Formation sands. In Well 2, the Top Heimdal is located at a depth of 2153 m. The OWC (oil–water contact) in this well is at 2183 m. Hence, the oil column is 30 m. The traveltime topography to the Top Heimdal horizon is given in the text file Top_Heimdal_subset.txt in inline and cross-line coordinates. You will have to focus on the depth zone of about 2100–2300 m in all wells.

Well logs, rock and fluid properties

When you start recognizing clusters of data, we recommend you to use Well 2 as a type-well or reference well to create a database of training data. This is because of

7.3 Project

all the direct information available here, including cores and thin sections, and helium porosity and permeability measurements. Wells 2 and 5 are the only wells with shear-wave information (V_S). If you choose to use V_S as a parameter in your classification, you should carry out V_S prediction (based on V_P) in the other wells, before you do classification. The gamma-ray log is a very good clay indicator in the North Sea. So V_P and gamma ray together will work well in order to classify lithologies in the area. Remember to normalize your gamma-ray values to be within the range observed in the type-well. Alternatively, you can calculate clay content empirically from gamma-ray values using the formula:

$$\text{clay content} = (GR_{log} - GR_{min})/(GR_{max} - GR_{min})$$

This assumes a linear relationship between clay content and gamma ray, which is not necessarily true. If you want you can use other empirical relations that exist in the literature. Clay content can also be calculated from relations between density and neutron porosity; this can be done in Well 2, and the gamma-ray method can be controlled or calibrated. The P-wave velocity log should be very reliable in all the wells available. However, keep in mind that the P-wave velocity log will probably read the velocity of mud-filtrate-invaded zones even in wells that encountered hydrocarbons. It would be most reasonable to define your training data in a zone where we know we have brine, so this invasion effect would be negligible. Check the shallow and deep saturations in Well 2. By knowing what lithology we have from cores, we can check the invasion effect by using rock diagnostics. The density log is available in all the wells, and can be used in the classification procedure. The density logging tool is sensitive to rough borehole surfaces. This is why we have to provide you with a corrected density log in Well 2. It turns out that the cap-rock is a silty shale with thinly laminated beds that cause reading errors. This effect has not been corrected for in other wells, so the density log values there may not always be reliable. In practice, one would check the caliper log to identify zones where the density tool is reliable or not. The density is needed because we want to calculate acoustic impedance (AI), which is the product of velocity and density. Porosity can be directly related to density by the following formula:

$$\text{Porosity} = (\rho_{matrix} - \rho_{log})/(\rho_{matrix} - \rho_{fluid})$$

Parameters and rock properties that you need in your North Sea project

Effective pressure at reservoir depth $\approx$ 20 MPa or assuming hydrostatic pressure we can calculate:

$$P_{eff} = (\rho - \rho_{fluid})g \times \text{depth}$$

where $g = 9.8$ m/s^2 is the acceleration due to gravity.
Temperature at reservoir depth = 77.2 °C.

Fluid properties:
   brine density 1.09 g/cm^3
   brine bulk modulus 2.8 GPa
   oil density 0.78 g/cm^3
   oil gravity 32 API
   GOR 64 Sm^3/Sm^3

Rock properties:
   quartz mineral bulk modulus 36.8 GPa
   quartz mineral shear modulus 44 GPa
   clay bulk modulus 15 GPa
   clay shear modulus 5 GPa

You will need to calculate the fluid bulk modulus of oil before you do pore-fluid substitution. This can be done using Batzle–Wang relations. Mineral moduli will also be needed in Gassmann fluid substitution modeling. Furthermore, they will be needed in rock diagnostics.

Three-dimensional seismic information

Near- and far-offset partial stacks: sub-cubes from full 3D cubes in SEGY format. The survey was processed for true amplitude recovery. The maximum fold is 30, which corresponds to a maximum source receiver offset of approximately 2500 m. The near-offset stack of 10 traces has an average incidence angle of 8° at the target level, while the far-offset stack of the 10 last traces in each CDP gather has an average incidence angle of 26°.

   inline: 1300–1500, every 2nd line (50 m spacing)
   xline: 1500–2000, every 2nd line (50 m spacing)
   time: 1500–2500 ms
   total 25 351 traces for each sub-cube

xline number stored in bytes 21–26; inline number (multiplied by 1000) stored in bytes 41–44.

The file cdps_line2.sgy contains NMO corrected pre-stack CDP gathers. Well 2 corresponds to CDP 2232.

Well locations (inline and xline coordinates):
   Well 1: inline 1448, xline 945 (outside the sub-cube)
   Well 2: inline 1376, xline 1776 (inside)
   Well 3: inline 1468, xline 1847 (inside)
   Well 4: inline 1154, xline 1128 (outside)
   Well 5: inline 1242, xline 1636 (outside)

The sub-cube is defined within inlines 1300–1500 and xlines 1500–2000.

7.3 Project

Tasks

From the well logs identify the different facies. You can use the six different facies already identified in Well 2, or form your own facies clusters. Make V_S predictions in well with missing shear-wave sonics.

Derive distributions of V_P, V_S, and density for the different facies. Consider both brine and hydrocarbon saturations for the sands.

Using simulations derive the distributions for different seismic attributes: $R(0)$–G, P- and S-impedances, near- and far-offset elastic impedance, etc. Using the extended training data perform a statistical analysis of classification success rates. Compute the classification confusion matrix. How many classes should you keep? Should some of the initial classes be grouped together?

Perform AVO analysis (using any commercially available or in-house software) of the 2D line of CDP gathers and extract $R(0)$ and G from the pre-stack seismic data. Calibrate the $R(0)$ and G to the log-derived $R(0)$ and G distributions. Statistically classify the seismically extracted $R(0)$–G attributes and estimate the most likely facies and fluids along the 2D section. (As an alternative, already extracted $R(0)$–G attributes along the Top Heimdal horizon are provided in a data file at the website.) Classify $R(0)$–G along the horizon to estimate the most likely facies, and the probabilities of occurrence of different facies. Make plots of the Top Heimdal showing the most likely oil sands.

Perform impedance inversion (using any commercially available or in-house software) of the near- and far-offset partial stacks. Calibrate the inverted impedances to log-derived near- and far-offset impedance. Statistically classify the cubes of near- and far-offset impedances to get the probabilities of different facies, and the most likely facies at each point within the cube. Make 3D iso-probability plots showing occurrence of oil sands. Compute and plot vertically averaged probability of oil sands. As an alternative, already inverted near- and far-offset impedance cubes have been provided at the website. These inverted impedance cubes can be used for statistical classification.

Estimate and model the spatial correlation of different facies from the experimental variogram of facies indicators at the well-log locations. Perform geostatistical simulations of different facies conditioned to the well facies indicators, and constrained by the probabilities obtained from statistical classification of the seismic attributes. From the multiple geostatistical simulations, estimate the joint spatial uncertainty of occurrence of oil sands.

References

Adler, F., 1999, Robust estimation of dense 3D stacking velocities from automated picking. 61st Mtg Eur. Assoc. Geosci. Eng. (EAGE) Abstract.

Aki, K., and Richards, P. G., 1980, *Quantitative Seismology: Theory and Methods*. San Francisco: W. H. Freeman and Co.

Al-Chalabi, M., 1997, Instantaneous slowness versus depth functions. *Geophysics*, **62**, 270–273.

Al-Chalabi, M., and Rosenkranz, P. L., 2002, Velocity–depth and time–depth relationships for a decompacted uplifted unit. *Geophys. Prospecting*, **50**, 661–664.

Allen, J. L., and Peddy, C. P., 1993, *Amplitude Variation with Offset: Gulf Coast Case Studies*. Geophysical Development Series, Vol. 4, Soc. Expl. Geophys.

Anselmetti, F. S., and Eberli, G. P., 1997, Sonic velocity in carbonate sediments and rocks. In *Carbonate Seismology* (ed. Palaz, I. and Marfurt, K. J.), 53–75, Soc. Expl. Geophys.

Avseth, P., 2000, Combining rock physics and sedimentology for seismic reservoir characterization of North Sea turbidite systems. Unpublished Ph.D. thesis, Stanford University.

Avseth, P., and Mukerji, T., 2002, Seismic lithofacies classification from well logs using statistical rock physics. *Petrophysics*, **43**, 70–81.

Avseth, P., Mavko, G., and Mukerji, T., 1998a, Seismic lithofacies prediction using AVO-analysis: Application to a North Sea deep-water clastic system. Am. Assoc. Petr. Geol. 82nd Ann. Mtg, Expanded Abstracts, 1888.

Avseth, P., Mukerji, T., Mavko, M., and Veggeland, T., 1998b, Statistical discrimination of lithofacies from pre-stack seismic data constrained by well log rock physics. Application to a North Sea turbidite system. Soc. Expl. Geophys. 68th Ann. Mtg, Expanded Abstracts, 890–893.

Avseth, P., Dvorkin, J., Mavko, G., and Rykkje, J., 2000, Rock physics diagnostic of North Sea sands: Link between microstructure and seismic properties. *Geophys. Res. Lett.*, **27**, 2761–2764.

Avseth, P., Mukerji, T., Mavko, G., and Tyssekvam, J. A., 2001a, Rock physics and AVO analysis for lithofacies and pore fluid prediction in a North Sea oil field. *The Leading Edge*, **20**, 429.

Avseth, P., Mukerji, T., Jørstad, T., Mavko, G., and Veggeland, T., 2001b, Seismic reservoir mapping from 3-D AVO in a North Sea turbidite system. *Geophysics*, **66**, 1157–1176.

Avseth, P., Mavko, G., Dvorkin, J., and Mukerji, T., 2001c, Rock physics and seismic properties of sands and shales as a function of burial depth. Soc. Expl. Geophys. 71st Ann. Mtg, Expanded Abstracts.

Avseth, P., Flesche, H., and van Wijngaarden, A.-J., 2003, AVO classification of lithology and pore fluids constrained by rock physics depth trends. *The Leading Edge*, **22**, 1004–1011.

Bachrach, R., and Dutta, N., 2004, Joint estimation of porosity and saturation using stochastic rock physics modeling. Soc. Expl. Geophys. 74th Ann. Mtg, Expanded Abstracts.

Bachrach, R., Mallick, S., Dutta, N., and Perdomo, J., 2003, Propagating seismic data quality into rock physics analysis and reservoir property estimation: Case study of lithology prediction using full waveform inversion in clastic basins. Soc. Expl. Geophys. 73rd Ann. Int. Mtg, Expanded Abstracts.

Bachrach, R., Beller, M., Liu, C. C. et al., 2004, Combining rock physics analysis, full waveform prestack inversion and high-resolution seismic interpretation to map lithology units in deep water: A Gulf of Mexico case study. *The Leading Edge*, **23**, 378–383.

Backus, G. E., 1962, Long-wave elastic anisotropy produced by horizontal layering. *J. Geophys. Res.*, **67**, 4427–4440.

Bacon, M., Simm, R., and Redshaw, T., 2003, *3-D Seismic Interpretation*. Cambridge: Cambridge University Press.

Bakke, N., and Ursin, B., 1998, Thin-bed AVO effects. *Geophys. Prospecting*, **46**, 571–587.

Baldwin, B., and Butler, C. O., 1985, Compaction curves. *AAPG Bull.*, **69**, 622–626.

Baldwin, J. L., Otte, D. N., and Wheatley, C. L., 1989, Computer emulation of human mental process: Application of neural network simulations to problems in well-log interpretation. *Soc. Petroleum Engineers (SPE) Paper* 19619, 481–493.

Baldwin, J. L., Bateman, A. R. M., and Wheatley, C. L., 1990, Applications of neural networks to the problem of mineral identification from well-logs. *Log Analyst*, **3**, 279–293.

Banik, N. C., 1987, An effective anisotropy parameter in transversely isotropic media. *Geophysics*, **52**, 1654–1664.

Batzle, M., and Wang, Z., 1992, Seismic properties of pore fluids. *Geophysics*, **57**, 1396–1408.

Benabentos, M., Mallick, S., Sigismondi, M., and Soldo, J., 2002, Seismic reservoir description using hybrid seismic inversion: A 3D case study from the María Inés Oeste Field, Argentina. *The Leading Edge*, **21**, 1002.

Berger, J. O., 1985, *Statistical Decision Theory and Bayesian Analysis*, 2nd edn. New York: Springer-Verlag.

Berkhout, A. J., 1977, Least squares inverse filtering and wavelet deconvolution. *Geophysics*, **42**, 1369–1383.

  1985, *Seismic Migration: Imaging of Acoustic Energy by Wave Field Extrapolation. A. Theoretical Aspects*, 3rd edn. New York: Elsevier Science.

Berryman, J. G., 1980, Long-wavelength propagation in composite elastic media. *J. Acoust. Soc. Am.*, **68**, 1809–1831.

  1995, Mixture theories for rock properties. In *A Handbook of Physical Constants* (ed. Ahrens, T. J.) Washington: American Geophysical Union.

Berryman, J. G., and Milton, G. W., 1991, Exact results for generalized Gassmann's equation in composite porous media with two constituents. *Geophysics*, **56**, 1950–1960.

Biot, M. A., 1956, Theory of propagation of elastic waves in a fluid saturated porous solid. I. Low-frequency range. *J. Acoust. Soc. Am.*, **28**, 168–178.

  1962, Mechanics of deformation and acoustic propagation in porous media, *J. Appl. Phys.*, **33**, 1482–1498.

Bishop, C., 1995, *Neural Networks for Pattern Recognition*. Oxford: Clarendon Press.

Bjørlykke, K., and Egeberg, 1993, Quartz cementation in sedimentary basins. *AAPG Bull.*, **77**, 1538–1548.

Blangy, J. P., 1992, Integrated seismic lithologic interpretation: the petrophysical basis. Unpublished Ph.D. thesis, Stanford University.

  1994, AVO in transversely isotropic media: An overview. *Geophysics*, **59**, 775–781.

Bleistein, N., 1987, On the imaging of reflectors in the earth. *Geophysics*, **52**, 931–942.

Boggs, S. Jr., 1987, *Principles of Sedimentology and Stratigraphy*. London: Merrill Publishing Company.

Bortoli, L. J., Alabert, F., Haas, A., and Journel, A. G., 1993, Constraining stochastic images to seismic data. In *Geostatistics Troia '92* (ed. Soares, A.) Dordrecht: Kluwer, 325–337.

Bourbie, T., Coussy, O., and Zinszner, B., 1987, *Acoustics of Porous Media*. Houston: Gulf Publishing Co.

Brevik, I., 1996, Inversion and analysis of Gassmann skeleton properties of shaly sandstones using wireline log data from the Norwegian North Sea. Soc. Expl. Geophys. 66th Ann. Int. Mtg, Expanded Abstracts, 130–133.

Brie, A., Pampuri, F., Marsala, A. F., and Meazza, O., 1995, Shear sonic in gas-bearing sands. SPE 30595, 701–710.

Brown, A. R., 1992, *Interpretation of Three-Dimensional Seismic Data*. AAPG Memoir **42**, 3rd edn.
   1996, Seismic attributes and their classification. *The Leading Edge*, **15**, 1090.
   1998, Interpreter's corner – Picking philosophy for 3-D stratigraphic interpretation. *The Leading Edge*, **17**, 1198–1200.
   1999, *Interpretation of Three-dimensional Seismic Data*. 5th edn, Soc. Expl. Geophys.
   2001a, Data polarity for the interpreter. *The Leading Edge*, **20**, 549.
   2001b, Color in seismic display. *The Leading Edge*, **20**, 549.

Brown, R., and Korringa, J., 1975, On the dependence of the elastic properties of a porous rock on the compressibility of the pore fluid. *Geophysics*, **40**, 608–616.

Brown, A. R., Dahm, C. G., and Graebner, R. T., 1981, A stratigraphic case history using three-dimensional seismic data in the Gulf of Thailand. *Geophys Prospecting*, **29**, 327–349.

Buland, A., and Omre, H., 2003, Bayesian linearized AVO inversion. *Geophysics*, **68**, 185–198.

Buland, A., Landroe, M., Anderssen, M., and Dahl, T., 1996, AVO inversion of Troll Field data. *Geophysics*, **61**, 1589–1602.

Bulat, J., and Stoker, S. J., 1987, Uplift determination from interval velocity studies, UK, southern North Sea. In *Petroleum Geology of North West Europe* (ed. Brooks, J., and Glennie, K. W.) London: Graham & Trotman, 293–305.

Cadoret, T., 1993, *Effet de la Saturation Eau/Gaz sur les Propriétés Acoustiques des Roches*. Unpublished Ph.D. dissertation, University of Paris VII.

Caers, J., Avseth, P., and Mukerji, T., 2001, Geostatistical integration of rock physics, seismic amplitudes, and geologic models in North Sea turbidite systems. *The Leading Edge*, **20**, 308.

Cambois, G., 2000, Can *P*-wave AVO be quantitative? *The Leading Edge*, **19**, 1246.
   2001, AVO processing: Myths and reality. *CSEG Recorder*, March 2001, 30–33.

Cambois, G., and Magesan, M., 1997, Seismic wavelet phase stability: Problems and solutions. EAGE 59th Mtg Abstracts, Session A026.

Carcione, J., 1999, Effects of vector attenuation on AVO of offshore reflections. *Geophysics*, **64**, 815–819.
   2001, Amplitude variations with offset of pressure-seal reflections. *Geophysics*, **66**, 283–293.

Carcione, J., Helle, H. B., and Zhao, T., 1998, Effects of attenuation and anisotropy on reflection amplitude versus offset. *Geophysics*, **63**, 1652–1658.

Carcione, J., Helle, H. B., Pham, N. H., and Toverud, T., 2003, Pore pressure estimation in reservoir rocks from seismic reflection data. *Geophysics*, **68**, 1569–1579.

Carter, D. C., 1989, Downhole geophysics as an aid to the interpretation of an evaporite sequence: Examples from Nova Scotia. *CIM Bull*. (1974), **82**, no. 925, pp. 58–64.

Castagna, J. P., 1993, Petrophysical imaging using AVO. *The Leading Edge*, **12**, 172.
   2001, Recent advances in seismic lithologic analysis. *Geophysics*, **66**, 42–46.

Castagna, J. P., and Backus, M. (eds.), 1993, *Offset-Dependent Reflectivity – Theory and Practice of AVO Analysis*. Investigations in Geophysics, no. 8. Tulsa: Soc. Expl. Geophys.

References

Castagna, J. P., and Smith, S. W., 1994, Comparison of AVO indicators: A modeling study. *Geophysics*, **59**, 1849–1855.

Castagna, J. P., and Swan, H. W., 1997, Principles of AVO crossplotting. *The Leading Edge*, **16**, 337–342.

Castagna, J. P., Batzle, M. L., and Eastwood, R. O., 1985, Relationships between compressional-wave and shear-wave velocities in clastic silicate rocks. *Geophysics*, **50**, 571–581.

Castagna, J. P., Batzle, M. L., and Kan, T. K., 1993, Rock physics: The link between rock properties and AVO response. In *Offset-Dependent Reflectivity: Theory and Practice of AVO Analysis* (ed. Castagna, J. P. and Backus, M.) *Investigations in Geophysics*, No. 8. Tulsa: Soc. Expl. Geophys., 135–171.

Castagna, J. G., Swan, H. W., and Foster, D. J., 1998, Framework for AVO gradient and intercept interpretation. *Geophysics*, **63**, 948–956.

Cerveny, V., 2001, *Seismic Ray Theory*. Cambridge: Cambridge University Press.

Chemingui, N., and Biondi, B., 2002, Seismic data reconstruction by inversion to common offset. *Geophysics*, **67**, 1575–1585.

Chen, H., Castagna, J. P., Brown, R. L., and Ramos, A. C. B., 2001, Three-parameter AVO crossplotting in anisotropic media. *Geophysics*, **66**, 1359–1363.

Chen, W., 1995, AVO in azimuthally anisotropic media. Unpublished Ph.D. thesis, Stanford University.

Cheng, C. H., 1978, Seismic velocities in porous rocks: Direct and inverse problems. Unpublished Sc.D. thesis, MIT.

1993, Crack models for a transversely anisotropic medium. *J. Geophys. Res.* **98**, 675–684.

Chiburis, E. F., 1984, Analysis of amplitude versus offset to detect gas-oil contacts in the Arabia Gulf. Soc. Expl. Geophys. 54th Ann. Mtg, Expanded Abstracts, 669–670.

1993, AVO applications in Saudi Arabia. In *Offset-Dependent Reflectivity: Theory and Practice of AVO Analysis* (ed. Castagna, J. P. and Backus, M.) Tulsa: Soc. Expl. Geophys.

Chiles, J. P., and Delfiner, P., 1999, *Geostatistics: Modeling Spatial Uncertainty*. New York: Wiley.

Connolly, P., 1998, Calibration and inversion of non-zero offset seismic. *Soc. Expl. Geophys. 68th Ann. Mtg, Expanded Abstracts*.

1999, Elastic impedance. *The Leading Edge*, **18**, 438–452.

Cover, T., and Thomas, J., 1991, *Elements of Information Theory*. New York: Wiley.

Crampin, S., 1984, Effective anisotropic elastic constants for wave propagation through cracked solids. *J. Roy. Astr. Soc.*, **76**, 135–145.

Cressie, N. A. C., 1993, *Statistics for Spatial Data*. New York: Wiley.

Dahl, T., and Ursin, B., 1992, Non-linear AVO inversion for a stack of anelastic layers. *Geophys. Prospecting*, **40**, 243–265.

Daley, P. T., and Hron, F., 1977, Reflection and transmission coefficients for transversely isotropic media. *Bull. Seism. Soc. Am.*, **67**, 661–675.

Davis, J. C., 2002, *Statistics and Data Analysis in Geology*, 3rd edn, Chichester: Wiley.

Deutsch, C. V., and Journel, A. G., 1996, *GSLIB: Geostatistical Software Library and User's Guide*, 2nd edn. Oxford: Oxford University Press.

Dey-Sarkar, S. K., and Svatek, S. V., 1993, Prestack analysis: An integrated approach for seismic interpretation in clastic basins. In *Offset-Dependent Reflectivity: Theory and Practice of AVO Analysis* (ed. Castagna, J. P. and Backus, M.) Tulsa: Soc. Expl. Geophys., 57–77.

Dickinson, G., 1953, Geological aspects of abnormal reservoir pressures in Gulf Coast Louisiana. *AAPG Bull.* **37**, 410–432.

Digby, P. J., 1981, The effective elastic moduli of porous granular rocks. *J. Appl. Mech.*, **48**, 803–808.

Domenico, S. N., 1976, Effect of brine–gas mixture on velocity in an unconsolidated sand reservoir. *Geophysics*, **41**, 882–894.

Dong, W., 1998, AVO detectability against tuning and stretching artifacts. *Soc. Expl. Geophys., 68th Ann. Mtg, Expanded Abstracts*, 236–239.

    1999, AVO detectability against tuning and stretching artifacts. *Geophysics*, **64**, 494–503.

Dorn, G. A., 1998, Modern 3-D seismic interpretation. *The Leading Edge*, **17**, 1262–1272.

Doveton, J. H., 1994, Geologic log analysis using computer methods. *AAPG Computer Applications in Geology*, No. 2, Tulsa: AAPG.

Dowla, F. U., and Rogers, L. L., 1995, *Solving Problems in Environmental Engineering and Geosciences with Artificial Neural Networks*. Cambridge, MA: MIT Press.

Doyen, P. M., 1988, Porosity from seismic data: A geostatistical approach. *Geophysics*, **53**, 1263–1275.

Doyen, P. M., and Guidish, T. M., 1992, Seismic discrimination of lithology: A Monte Carlo approach. In *Reservoir Geophysics, Investigations in Geophysics*, no. 7 (ed. Sheriff, R. E.), Tulsa: Soc. Expl. Geophys., 243–250.

Duda, R. O., and Hart, P. E., 1973, *Pattern Classification and Scene Analysis*. New York: Wiley.

Duda, R. O., Hart, P. E., and Stork, D. G., 2000, *Pattern Classification*. New York: John Wiley & Sons.

Duffaut, K., Alsos, T., Landrø, M., Rognø, H., and Al-Najjar, N. F., 2002, Shear-wave elastic impedance. *The Leading Edge*, **19**, 1222–1229.

Dutta, N., Mukerji, T., Prasad, M., and Dvorkin, J., 2002, Seismic detection and estimation of overpressures. Part II: Field applications. *CSEG Recorder*, **27**, 58–73.

Dutton, S. P., and Diggs, T. N., 1990, History of quartz cementation in the Lower Cretaceous Travis Peak Formation, East Texas. *J. Sedim. Petrol.*, **60**, 191–202.

Dvorkin, J., and Gutierrez, M., 2001, Textural sorting effect on elastic velocities, Part II: Elasticity of a bimodal grain mixture. Soc. Expl. Geophys. 71st Ann. Int. Mtg, 1764–1767.

    2002, Grain sorting, porosity, and elasticity. *Petrophysics*, **43**, 185–196.

Dvorkin, J., and Nur, A., 1993, Dynamic poroelasticity: A unified model with the squirt and the Biot mechanisms. *Geophysics*, **58**, 524–533.

    1996, Elasticity of high-porosity sandstones: Theory for two North Sea datasets. *Geophysics*, **61**, 1363–1370.

Dvorkin, J., Nur, A., and Yin, H., 1994, Effective properties of cemented granular material. *Mech. Mater.*, **18**, 351–366.

Efron, B., 1979, Computers and the theory of statistics: Thinking the unthinkable. *SIAM Review*, **21**, 460–480.

Efron, B., and Tibshirani, R. J., 1993, *An Introduction to the Bootstrap*. New York: Chapman & Hall.

Eide, A. L., Ursin, B., and Omre, H., 1997, Stochastic simulation of porosity and acoustic impedance conditioned to seismic data and well data. Soc. Expl. Geophys. 67th Ann. Int. Mtg, Extended Abstracts, 1614–1617.

Eidsvik, J., Omre, H., Mukerji, T., Mavko, G., and Avseth, P., 2002, Seismic reservoir prediction using Bayesian integration of rock physics and Markov random fields: A North Sea example. *The Leading Edge*, **21**, 290–294.

Eidsvik, J., Mukerji, T., and Switzer, P., 2004, Estimation of geological attributes from a well log: an application of hidden Markov chains. *J. Math. Geol.*, **36**, 379–397.

Enachescu, M. E., 1993, Amplitude interpretation of three-dimensional reflection data. *The Leading Edge*, **12**, 678–685.

References

Eshelby, J. D., 1957, The determination of the elastic field of an ellipsoidal inclusion, and related problems. *Proc. Roy. Soc. London*, A241, 376–396.

Fatti, J. L., Smith, G. C., Vail, P. J., Strauss, P. J., and Levitt, P. R., 1994, Detection of gas in sandstone reservoirs using AVO analysis: A 3-D seismic case history using the Geostack technique. *Geophysics*, **59**, 1362–1376.

Fertl, W. H., Chapman, R. E., and Hotz, R. F., 1994, *Studies in Abnormal Pressures*. Amsterdam: Elsevier.

Foster, D. J., Keys, R. G., and Reilly, J. M., 1997, Another perspective on AVO crossplotting. *The Leading Edge*, **16**, 1233–1236.

Fouquet, D. F., 1990, Principles of AVO processing. Soc. Expl. Geophys. 60th Ann. Mtg, Expanded Abstracts, 1486.

Fournier, F., 1989, Extraction of quantitative geologic information from seismic data with multidimensional statistical analysis: Part I, methodology, and Part II, a case study. Soc. Expl. Geophys. 59th Ann. Int. Mtg, Expanded Abstracts, 726–733.

Fukunaga, K., 1990, *Introduction to Statistical Pattern Recognition*. Boston: Academic Press.

Gardner, G. H. F., Gardner, L. W., and Gregory, A. R., 1974, Formation velocity and density: The diagnostic basics for stratigraphic traps. *Geophysics*, **39**, 770–780.

Gassaway, G. S., Brown, R. A., and Bennett, L. E., 1986, Pitfalls in seismic amplitude versus offset analysis: Case histories. Soc. Expl. Geophys. 56th Int. Mtg, 332–334.

Gassmann, F., 1951, Uber die elastizitat poroser medien. *Vier. Natur Gesellschaft*, **96**, 1–23.

Gastaldi, C., Roy, D., Doyen, P., and Boer, L. D., 1998, Using Bayesian simulations to predict reservoir thickness under tuning conditions. *The Leading Edge*, **17**, 539–543.

Gluck, S., Juve, E., and Lafet, Y., 1997, High-resolution impedance layering through 3-D stratigraphic inversion of poststack seismic data. *The Leading Edge*, **16**, 1309.

Godfrey, R., Muir, F., and Rocca, F., 1980, Modeling seismic impedance with Markov chains. *Geophysics*, **45**, 1351–1372.

Gonzalez, E. F., Mukerji, T., Mavko, G., and Michelena, R. J., 2003a, Far offset P-to-S "elastic impedance" for lithology and partial gas saturation (fizz water) identification: Applications with well logs. Soc. Expl. Geophys. 73rd Ann. Int. Mtg, Expanded Abstracts.

2003b, Near and far offset P-to-S elastic impedance for discriminating fizz water from commercial gas. *The Leading Edge*, **22**, 1012.

Goodway, B., Chen, T., and Downton, J., 1997, Improved AVO fluid detection and lithology discrimination using Lamé petrophysical parameters; "$\lambda\rho$", "$\mu\rho$", and "λ/μ" fluid stack from P and S inversions. Soc. Expl. Geophys. 67th Ann. Int. Mtg, Expanded Abstracts, 183–186.

Goovaerts, P., 1997, *Geostatistics for Natural Resources Evaluation*. New York: Oxford University Press.

Gouveia, W., 1996, Bayesian seismic waveform inversion: parameter estimation and uncertainty analysis. Unpublished Ph.D. thesis, Colorado School of Mines.

Gouveia, W., and Scales, J. A., 1998, Bayesian seismic waveform inversion: Parameter estimation and uncertainty analysis. *J. Geophys. Res.*, **103**, 2759–2779.

Gray, D., 2002, Elastic inversion for Lamé parameters. Soc. Expl. Geophys. 72nd Ann. Int. Mtg, Expanded Abstracts, 213–216.

Gray, D., Goodway, B., and Chen, T., 1999, Bridging the gap: Using AVO to detect changes in fundamental elastic constants. Soc. Expl. Geophys. 69th Ann. Int. Mtg, Expanded Abstracts, 852–855.

Greenberg, M. L., and Castagna, J. P., 1992, Shear-wave velocity estimation in porous rocks: Theoretical formulation, preliminary verification and applications. *Geophys. Prospecting*, **40**, 195–209.

Grubb, H., Tura, A., and Hanitzsch, C., 2001, Estimating and interpreting velocity uncertainty in migrated images and AVO attributes. *Geophysics*, **66**, 1208–1216.

Gueguen, Y., and Palciauskas, V., 1994, *Introduction to the Physics of Rocks*. Princeton, NJ: Princeton University Press.

Gutierrez, M., 2001, Rock physics and 3-D seismic characterization of reservoir heterogeneities to improve recovery efficiency. Unpublished Ph.D. thesis, Stanford University.

Haas, A., and Dubrule, O., 1994, Geostatistical inversion: a sequential method of stochastic reservoir modeling constrained by seismic data. *First Break*, **12**, 561–569.

Hamilton, E. L., 1956, Low sound velocities in high porosity sediments. *J. Acoust. Soc. Am.*, **28**, 16–19.

Hammond, A. L., 1974, Bright Spot: Better seismological indicators of gas and oil. *Science*, **185**, 515–517.

Hampson, D., 1986, Inverse velocity stacking for multiple elimination. *J. Can. Soc. Expl. Geophys.*, **22**, 44–55.

Hampson, D., and Russell, B., 1995, *AVO UNIX Software* (manual/tutorial). Calgary: Hampson-Russell Software Services Ltd.

Han, D., 1986, Effects of porosity and clay content on acoustic properties of sandstones and unconsolidated sediments. Unpublished Ph.D. dissertation, Stanford University.

Han, D., and Batzle, M., 2002, Fizz water and low gas-saturated reservoirs. *The Leading Edge*, **21**, 395.

Hanitzsch, C., 1997, Comparison of weights in prestack amplitude-preserving Kirchhoff depth migration. *Geophysics*, **62**, 1812–1816.

Harbaugh, J. W., Davis, J. C., and Wendebourg, J., 1995, *Computing Risk for Oil Prospects: Principles and Programs*. New York: Pergamon.

Hardage, B. A., 1985, Vertical seismic profiling. *The Leading Edge*, **4**, 59.

Harris, D. A., Lewis, J. J. M., and Wallace, D. J., 1993, The identification of lithofacies types in geological imagery using neural networks. *Conf. Proc. EUROCAIPEP*, **93**, Aberdeen.

Hashin, Z., and Shtrikman, S., 1963, A variational approach to the elastic behavior of multiphase materials. *J. Mech. Phys. Solids*, **11**, 127–140.

Hastie, T., Tibshirani, R., and Freidman, J., 2001, *The Elements of Statistical Learning: Data Mining, Inference, and Prediction*. New York: Springer-Verlag.

Hawkins, K., Roberts, G., Leggott, R., and Williams, G., 2001, Rock properties from longer seismic offsets. EAGE Ann. Mtg, Expanded Abstract.

Helseth, H. M., Mathews, J. C., Avseth, P., and van Wijngaarden, A.-J., 2004, Combined diagenetic and rock physics modeling for improved control on seismic depth trends. 66th Ann. Mtg EAGE.

Herring, E. A., 1973, North Sea abnormal pressures determined from logs. In *Oil and Gas Production Handbook* (ed. Kastrop, J. E.) Dallas: Energy Communic., Inc., Petrol. Eng. Publ. Co., 9–13.

Herrmann, P., Mojesky, T., Magesan, M., and Hugonnet, P., 2000, De-aliased, high-resolution Radon transforms. Soc. Expl. Geophys. 70th Ann. Mtg, Expanded Abstract, SP2.3.

Hill, K. B., and Halvatzis, G. J., 2001, Sand thickness prediction from 3-D seismic data: A case study of the Upper Jurassic Frisco City Sand of southwest Alabama. *The Leading Edge*, **20**, 950.

Hilterman, F., 1990, Is AVO the seismic signature of lithology? A case history of Ship Shoal – South Addition. *The Leading Edge*, **9**, 15–22.

  2003, Predicting fizz saturation with AVO. Presentation at Dallas Geophys. Soc., Abstract.

Hilterman, F. J., 2001, *Seismic Amplitude Interpretation*. Tulsa: Soc. Expl. Geophys.

Hilterman, F., Sherwood, J. W. C., Schellhorn, R., Bankhead, B., and DeVault, B., 1998, Identification of lithology in the Gulf of Mexico. *The Leading Edge*, **17**, 215.

References

Hilterman, F., van Schuyver, C., and Sbar, M., 2000, AVO examples of long-offset 2-D data in the Gulf of Mexico. *The Leading Edge*, **19**, 1200.

Hornby, B. E., Schwartz, L. M., and Hudson, J. A., 1994, Anisotropic effective-medium modeling of the elastic properties of shales. *Geophysics*, **59**, 1570–1583.

Houck, R. T., 1999, Estimating uncertainty in interpreting seismic indicators. *The Leading Edge*, **18**, 320–325.

2002, Quantifying the uncertainty in an AVO interpretation. *Geophysics*, **67**, 117–125.

Hudson, J. A., 1980, Overall properties of a cracked solid. *Math. Proc. Cambridge Phil. Soc.*, **88**, 371–384.

1981, Wave speeds and attenuation of elastic waves in material containing cracks. *Geophys. J. Roy. Astron. Soc.*, **64**, 133–150.

1990, Overall elastic properties of isotropic materials with arbitrary distribution of circular cracks. *Geophys. J. Int.* **102**, 465–469.

Isaaks, E. H., and Srivastava, R. M., 1989, *An Introduction to Applied Geostatistics*. New York: Oxford University Press.

Jakobsen, M., Hudson, J., and Johansen, T. A., 2003, T-matrix approach to shale acoustics. *Geophys. J. Int.*, **154**, 533–558.

Japsen, P., 1993, Influence of lithology and Neogene uplift on seismic velocities in Denmark. *AAPG Bull.*, **77**, 192–211.

1998, Regional velocity-depth anomalies, North Sea Chalk: A record of overpressure and Neogene uplift and erosion. *AAPG Bull.*, **82**, 2031–2074.

Japsen, P., Mukerji, T., and Mavko, G., 2001, Constraints on velocity–depth trends from rock physics models. *Stanford Rock Physics and Borehole Geophysics, Ann. Report*, **84**, paper A3.

Jizba, D., 1991, Mechanical and acoustical properties of sandstones and shales. Unpublished Ph.D. thesis, Stanford University.

Johansen, T. A., Jakobsen, M., and Ruud, B. O., 2002, Estimation of the internal structure and anisotropy of shales from borehole data. *J. Seism. Explor.*, **11**, 363–381.

Jolliffe, I. T., 2002, *Principal Component Analysis*. New York: Springer-Verlag.

Jones, T. D., 1983, Wave propagation in porous rocks and models for crystal structure. Unpublished Ph.D. thesis, Stanford University.

Jørstad, A., Mukerji, T., and Mavko, G., 1999, Model-based shear-wave velocity estimation versus empirical regressions. *Geophys. Prospecting*, **47**, 785–797.

Journel, A. G., and Huijbregts, C. J., 1978, *Mining Geostatistics*. New York: Academic Press.

Juhlin, C., and Young, R., 1993, Implications of thin layers for amplitude variation with offset (AVO) studies. *Geophysics*, **58**, 1200–1204.

Kabir, N., Lavaud, B., and Chavent, G., 2000, Estimation of the density contrast by AVO inversion beyond the linearized approximation: An indicator of gas saturation. Soc. Expl. Geophys. 70th Ann. Mtg, Expanded Abstracts.

Kan, T. K., and Sicking, C. J., 1994, Pre-drill geophysical methods for geopressure detection and evaluation. In *Studies in Abnormal Pressures* (ed. Fertl, W. H., Chapman, R. E., and Hotz, R. F.) New York: Elsevier.

Katahara, K. W., 1996, Clay mineral elastic properties. Soc. Expl. Geophys. Ann. Mtg, Expanded Abstracts, **66**, 1691–1694.

Keho, T., 2000, The AVO hodogram: Using polarization to identify anomalies. Soc. Expl. Geophys. 70th Ann. Mtg. 118–121.

Keho, T. H., Lemanski, S., Ripple, R., and Tambunan, B. R., 2001, The AVO hodogram. *The Leading Edge*, **20**, 1214–1224.

Kim, K. Y., Wrolstad, K. H., and Aminzadeh, F., 1993, Effects of transverse isotropy on P-wave AVO for gas sands. *Geophysics*, **58**, 883–888.

Kitanidis, P. K., 1997, *Introduction to Geostatistics: Applications in Hydrology*. Cambridge: Cambridge University Press.

Klimentos, T., 1991, The effects of porosity–permeability–clay content on the velocity of compressional waves. *Geophysics*, **56**, 1930–1939.

Koefoed, O., 1955, On the effect of Poisson's ratios of rock strata on the reflection coefficients of plane waves. *Geophys. Prospecting*, **3**, 381–387.

Krail, P. M., and Shin, Y., 1990, Deconvolution of a directional marine source. *Geophysics*, **55**, 1542–1548.

Krynine, P. D., 1948, The megascopic study and field classification of sedimentary rocks. *J. Geol.*, **56**, 130–165.

Kuster, G. T., and Toksöz, M. N., 1974, Velocity and attenuation of seismic waves in two-phase media. Part I. Theoretical formulations. *Geophysics*, **39**, 587–606.

Kyburg, H. E. Jr., and Smokler, H. E. (eds), 1980, *Studies in Subjective Probability*. New York: Robert E. Krieger Publ. Co.

Lander, R. H., and Walderhaug, O., 1999, Predicting porosity through simulating sandstone compaction and quartz cementation. *AAPG Bull.*, **83**, 433–449.

Landrø, M., 2001, Discrimination between pressure and fluid saturation changes from time-lapse seismic data. *Geophysics*, **66**, 836–844.

Le Meur, D., and Magneron, C., 2000, Quality check of automatic velocity picking. EAGE Abstract, B-36.

Leinbach, J., 1995, Wiener spiking deconvolution and minimum-phase wavelets: A tutorial. *The Leading Edge*, **14**, 189.

Lin, C. T., and Lee, C. S. G., 1996, *Neural Fuzzy Systems: A Neuro-Fuzzy Synergism to Intelligent Systems*. New Jersey: Prentice-Hall.

Lin, L., and Phair, R., 1993, AVO tuning. Soc. Expl. Geophys. 63rd Ann. Mtg, Expanded Abstracts, 727–730.

Liu, J. S., 2001, *Monte Carlo Strategies in Scientific Computing*. New York: Springer-Verlag.

Loertzer, G. J. M., and Berkhout, A. J., 1992, An integrated approach to lithologic inversion: Part 1, Theory. *Geophysics*, **57**, 233–244.

1993, Linearized AVO inversion of multicomponent seismic data. In *Offset-Dependent Reflectivity: Theory and Practice of AVO Analysis* (ed. Castagna, J. and Backus, M.) Tulsa: Soc. Expl. Geophys., 317–332.

Lowe, D., Hickson, T., and Guy, M., 1995, Slurry flows; an important class of sediment gravity flows with examples of slurry-flow deposits from the Lower Cretaceous of the North Sea. Geol. Soc. Am. 1995 Ann. Mtg, Abstracts with Programs, **27**, no. 6, 128–129.

Lucet, N., and Mavko, G., 1991, Images of rock properties estimated from a crosswell tomogram. Soc. Expl. Geophys. 61st Ann. Int. Mtg., Expanded Abstracts, **61**, 363–366.

Luh, P., 1993, Wavelet attenuation and AVO. In *Offset-Dependent Reflectivity: Theory and Practice of AVO Analysis* (ed. Castagna, J. and Backus, M.) Tulsa: Soc. Expl. Geophys., 190–198.

MacLeod, M. K., and Martin, H. L., 1988, Amplitude changes due to bed curvature. Soc. Expl. Geophys. 58th Ann. Mtg, Expanded Abstracts, 1209–1212.

Magara, K., 1980, Comparison of porosity–depth relationships of shale and sandstone. *J. Petrol. Geol.*, **3**, 175–185.

Mahob, P. N., and Castagna, J. P., 2003, AVO polarization and hodograms: AVO strength and polarization product. *Geophysics*, **68**, 849–862.

Makse, H. A., Gland, N., Johnson, D. L., and Schwartz, L. M., 1999, Why effective medium theory fails in granular materials. *Phys. Rev. Lett.*, **83**, 5070–5073.

Mallick, S., 1999, Some practical aspects on implementation of prestack waveform inversion using a genetic algorithm: An example from east Texas Woodbine gas sand. *Geophysics* **64**, 326–336.

2001, AVO and elastic impedance. *The Leading Edge*, **20**, 1094–1104.

Mallick, S., and Dutta, N., 2002, Shallow water flow prediction using prestack waveform inversion of conventional 3D seismic data and rock modeling. *The Leading Edge*, **21**, 675.

Marion, D., 1990, Acoustical, mechanical and transport properties of sediments and granular materials. Unpublished Ph.D. thesis, Stanford University.

Marion, D., and Nur, A., 1991, Pore-filling material and its effect on velocity in rocks. *Geophysics*, **56**, 225–230.

Marion, D., Nur, A., Yin, H., and Han, D., 1992, Compressional velocity and porosity in sand–clay mixtures. *Geophysics*, **57**, 554–563.

Martinez, R. D., 1993, Wave propagation effects on amplitude variation with offset measurements: A modeling study. *Geophysics*, **58**, 534–543.

Martinsen, O., Holmefjord, I. B., Midtbø, R. E., Torkildsen, G., and Rykkje, J. M., 1996, Standard core description (Glitne field, well name confidential). Norsk Hydro Internal Report.

Mavko, G., 1980, Velocity and attenuation in partially molten rocks. *J. Geophys. Res.*, **85**, 5173–5189.

Mavko, G., and Jizba, D., 1991, Estimating grain-scale fluid effects on velocity dispersion in rocks. *Geophysics*, **56**, 1940–1949.

Mavko, G., and Mukerji, T., 1995, Pore space compressibility and Gassmann's relation. *Geophysics*, **60**, 1743–1749.

1998a, Bounds on low-frequency seismic velocities in partially saturated rocks. *Geophysics*, **63**, 918–924.

1998b, A rock physics strategy for quantifying uncertainty in common hydrocarbon indicators. *Geophysics*, **63**, 1997–2008.

Mavko, G., and Nur, A., 1975, Melt squirt in the asthenosphere. *J. Geophys. Res.*, **80**, 1444–1448.

1978, The effect of nonelliptical cracks on the compressibility of rocks. *J. Geophys. Res.*, **83**, 4459–4468.

Mavko, G., Mukerji, T., and Godfrey, N., 1995, Predicting stress-induced velocity anisotropy in rocks. *Geophysics*, **60**, 1081–1087.

Mavko, G., Mukerji, T., and Dvorkin, J., 1998, *The Rock Physics Handbook*. Cambridge: Cambridge University Press.

Meckel, L. D. Jr., and Nath, A. K., 1977, Geologic considerations for stratigraphic modeling and interpretation. In *Seismic Stratigraphy; Applications to Hydrocarbon Exploration* (ed. Payton, C. E.) American Association of Petroleum Geologists, Memoir no. 26, 417–438.

Menke, W., 1989, *Geophysical Data Analysis: Discrete Inverse Theory*. San Diego: Academic Press, 289.

Middleton, G. V., 1973, Johannes Walther's Law of the correlation of facies. *Geol. Soc. Am. Bull.*, **84**, 979–988.

Milholland, P., Manghnani, M. H., Schlanger, S. O., and Sutton, G. H., 1980, Geoacoustic modeling of deep-sea carbonate sediments. *J. Acoust. Soc. Am.*, **68**, 1351–1360.

Mindlin, R. D., 1949, Compliance of elastic bodies in contact. *J. Appl. Mech.*, **16**, 259–268.

Mitchum, R. M. Jr., Vail, P. R., and Thompson, S. III, 1977, Seismic stratigraphy and global changes of sea level, Part 2: The depositional sequence as a basic unit for stratigraphic analysis. *AAPG Memoir*, **26**, 53–62.

Mosher, C. C., Keho, T. H., Weglein, A. B., and Foster, D. J., 1996, The impact of migration on AVO. *Geophysics*, **61**, 1603–1615.

Mukerji, T., Jørstad, A., Mavko, G., and Granli, J., 1998a, Applying statistical rock physics and seismic inversions to map lithofacies and pore fluid probabilities in a North Sea reservoir. Soc. Expl. Geophys. 68th Ann. Mtg Expanded Abstracts.

　1998b, Near and far offset impedances: Seismic attributes for identifying lithofacies and pore fluids. *Geophys. Res. Lett.*, **25**, 4557–4560.

Mukerji, T., Jørstad, A., Avseth, P., Mavko, G., and Granli, J. R., 2001, Mapping lithofacies and pore-fluid probabilities in a North Sea reservoir: Seismic inversions and statistical rock physics. *Geophysics*, **66**, 988–1001.

Mukerji, T., Dutta, N., Prasad, M., and Dvorkin, J., 2002, Seismic detection and estimation of overpressures. Part I: The rock physics basis. *CSEG Recorder*, **27**, 34–57.

Murphy, W. F. III, 1982, Effects of microstructure and pore fluids on the acoustic properties of granular sedimentary materials. Unpublished Ph.D. thesis, Stanford University.

Neri, P., 1997, Revolutionary software for seismic analysis. *World Oil*, **218**, 90–93.

Newendorp, P. D., and Schuyler, J., 2000, *Decision Analysis for Petroleum Exploration*, 2nd edn., Aurora, CO: Planning Press, 606.

Newman, P., 1973, Divergence effects in a layered earth. *Geophysics*, **38**, 481–488.

Norris, A. N., and Johnson, D. L., 1997, Nonlinear elasticity of granular media. *ASME J. Appl. Mech.*, **64**, 39–49.

Nur, A., 1971, Effects of stress on velocity anisotropy in rocks with cracks. *J. Geophys. Res.*, **76**, 2022–2034.

　1992, Critical porosity and the seismic velocities in rocks. *EOS, Trans. Am. Geophys. Un.*, **73**, 43–66.

Nur, A., and Simmons, G., 1969, Stress-induced velocity anisotropy in rocks: An experimental study. *J. Geophys. Res.*, **74**, 6667.

O'Connell, R. J., and Budiansky, B., 1974, Seismic velocities in dry and saturated cracked solids. *J. Geophys. Res.*, **79**, 4626–4627.

　1977, Viscoelastic properties of fluid-saturated cracked solids. *J. Geophys. Res.*, **82**, 5719–5735.

Ødegaard, E., and Avseth, P., 2003, Interpretation of elastic inversion results using rock physics templates. EAGE Ann. Mtg Extended Abstract.

　2004, Well log and seismic data analysis using rock physics templates. *First Break*, **22**, 37–43.

Ostrander, W. J., 1984, Plane-wave reflection coefficients for gas sands at non-normal angles of incidence. *Geophysics*, **49**, 1637–1648.

Parker, R. L., 1994, *Geophysical Inverse Theory*. Princeton, NJ: Princeton Univ. Press.

Petersen, S. A., 1999, Compound modeling, a geological approach to the construction of shared earth models. 61st Mtg EAGE.

Pettijohn, F. J., 1975, *Sedimentary Rocks*. New York: Harper & Row.

Pickett, G. R., 1963, Acoustic character logs and their applications in formation evaluation. *J. Petrol. Technol.*, **15**, 650–667.

Priestley, M. B., 1983, *Spectral Analysis and Time Series*. New York: Academic Press.

Pyrak-Nolte, L. J., Myer, L. R., and Cook, N. G. W., 1990a, Transmission of seismic waves across single natural fractures. *J. Geophys. Res.*, **95**, 8617–8638.

　1990b, Anisotropy in seismic velocities and amplitudes from multiple parallel fractures. *J. Geophys. Res.*, **95**, 11345–11358.

Raiga-Clemenceau, J., Martin, J. P., and Nicoletis, S., 1988, The concept of acoustic formation factor for more accurate porosity determination from sonic transit time data. *Log Analyst*, **219**, 54–60.

References

Ramm, M., and Bjørlykke, K., 1994, Porosity/depth trends in reservoir sandstones: assessing the quantitative effects of varying pore-pressure, temperature history and mineralogy, Norwegian Shelf data. *Clay Miner.*, **29**, 475–490.

Ramm, M., Martinsen, O. J., Holmefjord, I., Johnsen, A., and Rykkje, J., 1992, Standard core description (Grane field, well name confidential). Norsk Hydro Internal Report.

Ratcliffe, A., and Adler, F., 2000, Combined velocity and AVO analysis. EAGE Abstract C-36.

Raymer, L. L., Hunt, E. R., and Gardner, J. S., 1980, An improved sonic transit time-to-porosity transform. Soc. Professional Well Log Analysis (SPWLA), 21st Ann. Logg. Symp., Paper P July 1980.

Reading, H. G., and Richards, M., 1994, Turbidite systems in deep-water basin margins classified by grain-size and feeder system. *AAPG Bull.*, **78**, 792–822.

Resnick, J. R., Ng, P., and Larner, K., 1987, Amplitude versus offset analysis in the presence of dip. Soc. Expl. Geophys. 57th Ann. Mtg, Expanded Abstracts, 617–620.

Reuss, A., 1929, Berechnung der Fliessgrenzen von Mischkristallen. *Z. Angew. Math. Mech.*, **9**, 49–58.

Rider, M. H., 1996, *The Geological Interpretation of Well Logs*, 2nd edn., Houston: Gulf Publishing Co.

Rieke, H. H. III, and Chilingarian, G. V., 1974, *Compaction of Argillaceous Sediments, Developments in Sedimentology*, vol. 16. Amsterdam-Oxford-New York: Elsevier.

Rijks, E. J. K., and Jauffred, J. C. E. M., 1991, Attribute extraction: An important application in any detailed 3-D interpretation study. *The Leading Edge*, **10**, 11–19.

Roberts, G., 2000, Wide angle AVO. EAGE Ann. Mtg, Expanded Abstract.

Roberts, G., Went, D., Hawkins, K., *et al.*, 2002, Case study of long-offset acquisition, processing and interpretation over the Harding Field, North Sea. EAGE 64th Ann. Mtg, Expanded Abstract, G 104.

Rogers, S. J., Fang, J. H., Karr, C. L., and Stanley, D. A., 1992, Determination of lithology from well-logs using a neural network. *Am. Assoc. Petrol. Geol. (AAPG) Bull.*, **76**, 792–822.

Ross, C., 1997, AVO and nonhyperbolic moveout: A practical example. *First Break*, **15**, 43–48.

2000, Effective AVO crossplot modeling: A tutorial. *Geophysics*, **65**, 700–711.

2002, Interpreter's Corner: Comparison of popular AVO attributes, AVO inversion, and calibrated AVO predictions. *The Leading Edge*, **21**, 244.

Ross, C. P., and Kinman, D. L., 1995, Non-bright spot AVO: Two examples. *Geophysics*, **60**, 1398–1408.

Rowbotham, P. S., Lamy, P., Swaby, P. A., Dubrule, O., and Cadoret, T., 1998, Geostatistical inversion for reservoir characterization. Soc. Expl. Geophys. 68th Ann. Int. Mtg, Expanded Abstracts, 886–889.

Rubey, W. W., and Hubbert, M. K., 1959, Mechanics of fluid-filled porous solids and its application to overthrust faulting. Part 1 of Role of fluid pressure in mechanics of overthrust faulting. *Geo. Soc. Am. Bull.*, **70**, 115–166.

Rubin, Y., 2003, *Applied Stochastic Hydrogeology*. Oxford: Oxford University Press.

Rüger, A., 1995, P-wave reflection coefficients for transversely isotropic media with vertical and horizontal axis of symmetry. Soc. Expl. Geophys. 65th Int. Mtg, Expanded Abstracts.

1996, Variation of P-wave reflectivity with offset and azimuth in anisotropic media. Soc. Expl. Geophys. 66th Int. Mtg, Expanded Abstracts.

1997, P-wave reflection coefficients for transversely isotropic models with vertical and horizontal axis of symmetry. *Geophysics*, **62**, 713–722.

Rutherford, S. R., and Williams, R. H., 1989, Amplitude-versus-offset variations in gas sands. *Geophysics*, **54**, 680–688.

Ryseth, A., Fjellbirkeland, H., Osmundsen, I. K., Skålnes, Å., and Zachariassen, E., 1998, High-resolution stratigraphy and seismic attribute mapping of a fluvial reservoir; Middle Jurassic Ness Formation, Oseberg Field. *AAPG Bull.*, **82**, 1627–1651.

Sams, M., 1998, Yet another perspective on AVO crossplotting. *The Leading Edge*, **17**, 911–917.

Sams, M., and Andrea, M., 2001, The effect of clay distribution on the elastic properties of sandstones. *Geophys. Prospecting*, **49**, 128–150.

Sava, D., Mukerji, T., Diaz, M., and Mavko, G., 2000, Seismic detection of pore fluids: Pitfalls of ignoring anisotropy. Soc. Expl. Geophys., 70th Ann. Int. Mtg, Expanded Abstracts, 1842–1845.

Savage, S. L., 2002, The flaw of averages. *Harvard Business Review*, November, 20–21.

2003, *Decision Making with Insight*. Belmont: Brooks/Cole Publ. Co.

Sayers, C. M., 1988, Stress-induced ultrasonic wave velocity anisotropy in fractured rock. *Ultrasonics*, **26**, 311–317.

Schleicher, J., Tygel, M., and Hubral, P., 1993, 3-D true-amplitude finite-offset migration. *Geophysics*, **58**, 1112–1126.

Schoenberg, M., 1983, Reflection of elastic waves from periodically stratified media with interfacial slip. *Geophys. Prospecting*, **31**, 265–292.

Sclater, J. G., and Christie, P. A. F., 1980, Continental stretching: An explanation of the post-mid-Cretaceous subsidence of the central North Sea basin. *J. Geophys. Res.*, **85**, 3711–3739.

Sen, M., and Stoffa, P. L., 1995, *Global Optimization Methods in Geophysical Inversion*. New York: Elsevier.

Sengupta, M., 2000, Integrating rock physics and flow simulation to reduce uncertainties in seismic reservoir monitoring. Unpublished Ph.D. dissertation, Stanford University.

Shang, Z., McDonald, J. A., and Gardner, G. H. F., 1993, Automated extraction of AVA information in the presence of structure. In *Offset-Dependent Reflectivity: Theory and Practice of AVO Analysis* (ed. Castagna, J., and Backus, M.) Tulsa: Soc. Expl. Geophys., 199–210.

Shanmugam, G., Bloch, R., Mitchell, S. *et al.*, 1995, Basin-floor fans in the North Sea; sequence stratigraphic models vs. sedimentary facies. *AAPG Bull.*, **79**, 477–512.

Shapiro, S., Hubral, P., and Zien, H., 1994a, Frequency-dependent anisotropy of scalar waves in a multilayered medium. *J. Seism. Expl.*, **3**, 37–52.

1994b, A generalized O'Doherty–Anstey formula for waves in finely layered media. *Geophysics*, **59**, 1750–1762.

Sheriff, R. E., and Geldhart, L. P., 1995, *Exploration Seismology*, 2nd edn. Cambridge: Cambridge University Press.

Shinn, E. A., and Robbin, D. M., 1983, Mechanical and chemical compaction in fine-grained shallow-water limestones. *J. Sedim. Petrol.*, **53**, 595–618.

Shuey, R. T., 1985, A simplification of the Zoeppritz equations. *Geophysics*, **50**, 609–614.

Sick, C. M. A., Müller, T. M., Shapiro, A. A., and Buske, S., 2003, Amplitude corrections for randomly distributed heterogeneities above a target reflector. *Geophysics*, **68**, 1497–1502.

Silverman, B. W., 1986, *Density Estimation for Statistics and Data Analysis*. New York: Chapman and Hall.

Simm, R., White, R., and Uden, R., 2000, The anatomy of AVO crossplots. *The Leading Edge*, **19**, 150–155.

Simmons, J., and Backus, M. M., 1994, AVO modeling and the locally converted shear wave. *Geophysics*, **46**, 1237–1248.

Sintubin, M., 1994, Clay fabrics in relation to the burial history of shales. *Sedimentology*, **41**, 1161–1169.

Sinvhal, A., Khattri, K. N., Sinvhal, H., and Awasthi, A. K., 1984, Seismic indicators of stratigraphy. *Geophysics*, **49**, 1196–1212.

Smith, G. C., and Gidlow, P. M., 1987, Weighted stacking for rock property estimation and detection of gas. *Geophys. Prospecting*, **35**, 993–1014.

Smith, G. C., and Sutherland, R. A., 1996, The fluid factor as an AVO indicator. *Geophysics*, **61**, 1425–1428.

Spratt, S., 1987, Effect of normal moveout errors on amplitude. Soc. Expl. Geophys. 57th Int. Mtg, Expanded Abstracts, 634–637.

Stoll, R. D., and Bryan, G. M., 1970, Wave attenuation in saturated sediments. *J. Acoust. Soc. Am.*, **47**, 1440–1447.

Surdam, R. S., Dunn, T. L., MacGowan, D. B., and Heasler, H. P., 1989, Conceptual models for the prediction of porosity. Evolution with an example from the Frontier Sandstone, Bighorn Basin, Wyoming. In *Petrogenesis and Petrophysics of Selected Sandstone Reservoirs of the Rocky Mountain Region* (ed.-in-chief Coalson, E. B.) Denver: Rocky Mountain Association of Geologists.

Swan, H., 1991, Amplitude-versus-offset measurement errors in a finely layered medium. *Geophysics*, **56**, 41–49.

Swan, H. W., 1993, Properties of direct AVO hydrocarbon indicators. In *Offset-Dependent Reflectivity– Theory and Practice of AVO Analysis* (ed. Castagna, J. P. and Backus, M.) Investigations in Geophysics, No. 8. Tulsa: Soc. Expl. Geophys., 78–79.

Takahashi, I., 2000, Quantifying information and uncertainty of rock property estimation from seismic data. Unpublished Ph.D. thesis, Stanford University.

Takahashi, I., Mukerji, T., and Mavko, G., 1999, A strategy to select optimal seismic attributes for reservoir property estimation: Application of information theory. Soc. Expl. Geophys. 69th Ann. Mtg, Expanded Abstracts, 1584–1587.

Tarantola, A., 1987, *Inverse Problem Theory: Methods of Data Fitting and Model Parameter Estimation*. New York: Elsevier.

Thomsen, L., 1986. Weak elastic anisotropy. *Geophysics*, **51**, 1954–1966.

   1993, Weak anisotropic reflections. In *Offset-Dependent Reflectivity: Theory and Practice of AVO Analysis* (ed. Castagna, J. P., and Backus, M.) Tulsa: Soc. Expl. Geophys., 103–111.

Towner, B., and Lindsey, R., 2000, Subsalt success in the Republic of Yemen, using 3-D AVO and integrated exploration. *The Leading Edge*, **19**, 1064.

Tyler, N., and Finley, R. J., 1991, Architectural controls on the recovery of hydrocarbons from sandstone reservoirs. In *Three-Dimensional Facies Architecture of Terrigenous Clastic Sediments and its Implications for Hydrocarbon Discovery and Recovery* (ed. Miall, A. D. and Tyler, N.) Tulsa: Soc. Sed. Geol., 1–5.

Ursin, B., 1990, Offset-dependent geometrical spreading in a layered medium. *Geophysics*, **55**, 492–496.

Ursin, B., and Ekren, B., 1994, Robust AVO analysis. *Geophysics*, **60**, 317–326.

Ursin, B., and Stovas, A., 2002, Reflection and transmission responses of a layered isotropic viscoelastic medium. *Geophysics*, **67**, 307–323.

Vail, P. R., Mitchum, R. M., and Thompson, S., 1977, Seismic stratigraphy and global changes of sea level. Part 3: Relative changes of sea level from coastal onlap. In *Seismic Stratigraphy – Applications to Hydrocarbon Exploration* (ed. Clayton, C. E.) Tulsa: Am. Assoc. Petrol. Geol. Memoir 26, 63–81.

Vanorio, T., Prasad, M., and Nur, A., 2003, Elastic properties of dry clay mineral aggregates, suspensions and sandstones. *Geophys. J. Int.*, **155**, 319–326.

Varsek, J. L., 1985, Lithology prediction and discrimination by amplitude offset modeling. *Geophysics*, **50**, 1377.

Velzeboer, C. J., 1981, The theoretical seismic reflection response of sedimentary sequences. *Geophysics*, **46**, 843–853.

Verm, R., and Hilterman, F., 1995, Lithology color-coded seismic sections: The calibration of AVO crossplotting to rock properties. *The Leading Edge*, **14**, 847.

Vernik, L., 1992, Petrophysical classification of siliciclastics for lithology and porosity prediction from seismic velocities. *AAPG Bull.*, **76**, 1295–1309.

1997, Predicting porosity from acoustic velocities in siliciclastics: A new look. *Geophysics*, **62**, 118–128.

Vernik, L., and Liu, X., 1997, Velocity anisotropy in shales: A petrophysical study. *Geophysics*, **62**, 521–532.

Vernik, L., and Nur, A., 1992, Petrophysical classification of siliciclastics for lithology and porosity prediction from seismic velocities. *AAPG Bull.*, **76**, 1295–1309.

Vernik, L., Fisher, D., and Bahret, S., 2002, Estimation of net-to-gross from P and S impedance in deepwater turbidites. *The Leading Edge*, **21**, 380.

Voigt, W., 1910, *Lehrbuch der Kristallphysik*. Leipzig: Teubner.

Walden, A. T., 1991, Making AVO sections more robust. *Geophys. Prospecting*, **40**, 483–512.

Walker, R., 1978, Deep-water sandstone facies and ancient submarine fans: Models for exploration for stratigraphic traps. *AAPG Bull.*, **62**, 932–966.

Walsh, J. B., 1965, The effect of cracks on the compressibility of rock. *J. Geophys. Res.*, **70**, 381–389.

Walther, J., 1893–4, *Einleitung in die Geologie als historische Wissenschaft*. 3 vols. Jena: Gustav Fisher.

Walton, K., 1987, The effective elastic moduli of a random packing of spheres. *J. Mech. Phys. Solids*, **35**, 213–226.

Weglein, A. B., 1992, What can inverse scattering really do for you today? In *Geophysical Inversion* (ed. Bednar, J. B. *et al.*) Philadelphia: Soc. Industrial Appl. Math (SIAM), 20–45.

Weimer, P., and Link, M. H., 1991, Seismic facies and sedimentary processes of ancient submarine fans and turbidite systems. New York: Springer.

Whitcombe, D. N., 2002, Elastic impedance normalization. *Geophysics*, **67**, 59–61.

Whitcombe, D. N., Connolly, P. A., Reagan, R. L., and Redshaw, T. C., 2002, Extended elastic impedance for fluid and lithology prediction. *Geophysics*, **67**, 62–66.

Widess, M. B., 1973, How thin is a thin bed? *Geophysics*, **38**, 1176–1180.

Widmaier, M., Müller, T., Shapiro, S., and Hubral, P., 1995, Amplitude-preserving migration and elastic P-wave AVO corrected for thin layering. *J. Seism. Expl.*, **4**, 169–177.

Widmaier, M. T., Shapiro, S. A., and Hubral, P., 1996, AVO correction for scalar waves in the case of a thinly layered reflector overburden. *Geophysics*, **61**, 520–528.

Wiggins, R., Kenny, G. S., and McClure, C. D., 1983, *A method for determining and displaying the shear-velocity reflectivities of a geologic formation*. European Patent Application 0113944.

Woeber, A. F., Katz, S., and Ahrens, T. J., 1963, Elasticity of selected rocks and minerals. *Geophysics*, **28**, 658–663.

Wright, J., 1987, The effects of transverse isotropy on reflection amplitude versus offset. *Geophysics*, **52**, 564–567.

Wrolstad, K., and Lefebure, F., 1996, Clastic reservoir description from pre- and post-stack seismic data. Am. Assoc. Petrol. Geol. 1996 Ann. Convention, Abstracts, 154.

Wu, Y., 2000, Estimation of gas saturation using P-to-S converted waves. Soc. Expl. Geophys. 70th Ann. Int. Mtg, 158–161.

References

Wyllie, M. R. J., Gregory, A. R., and Gardner, L. W., 1956, Elastic wave velocities in heterogeneous and porous media. *Geophysics*, **21**, 41–70.

1958, An experimental investigation of factors affecting elastic wave velocities in porous media. *Geophysics*, **23**, 459–493.

Xu, S., and White, R. E., 1995, A new velocity model for clay-sand mixtures. *Geophys. Prospecting*, **43**, 91–118.

Yilmaz, O., 2001, *Seismic Data Processing*. Tulsa: Soc. Expl. Geophys.

Yin, H., 1992, Acoustic velocity and attenuation of rocks: Isotropy, intrinsic anisotropy, and stress induced anisotropy. Unpublished Ph.D. thesis, Stanford University.

Yin, H., Nur, A., and Mavko, G., 1993, Critical porosity: A physical boundary in poroelasticity. *Int. J. Rock Mech. Mining Sciences & Geomech. Abstracts*, **30**, 805–808.

Zeng, H., Backus, M. M., Barrow, K. T., and Tyler, N., 1996, Facies mapping from three-dimensional seismic data: Potential and guidelines from a Tertiary sandstone-shale sequence model, Powderhorn Field, Calhoun County, Texas. *AAPG Bull.*, **80**, 16–46.

Zhu, H., and Journel, A. G., 1993, Formatting and integrating soft data: Stochastic imaging via the Markov–Bayes algorithm. *Geostatistics*, 4th Int. Geostat. Congress, Troia 1992 (ed. Soares, A.) Dordrecht: Kluwer, 1–12.

Zhu, F., Gibson, R. L., Atkins, J., and Yuh, S. H., 2000, Distinguishing fizz gas from commercial gas reservoirs using multicomponent seismic data. *The Leading Edge*, **19**, 1238–1245.

Zimmerman, R. W., 1991, *Compressibility of Sandstones*. New York: Elsevier.

Zoeppritz, K., 1919, Erdbebenwellen VIIIB, Ueber Reflexion and Durchgang seismischer Wellen durch Unstetigkeitsflaechen. *Goettinger Nachrichten*, **I**, 66–84.

Index

acoustic impedance 104, 236, 278–295
   see also rock physics templates
adiabatic modulus *see* fluid properties
Aki–Richards AVO approximation *see* AVO, Aki–Richards approximation
American polarity standard *see* polarity
anisotropy, elastic 62, 183–185, 222–223, 248–251, 275
   AVAZ 248
   Brown and Korringa relations for fluid substitution 21
   elastic impedance, anisotropic 248–251
   Thomsen's parameters 184, 248
aspect ratio *see* inclusion models
attenuation, seismic 190–191
   layered media 188–189
   scattering 188–190
   see also Biot model; Bisq model; squirt model
autopicking 170
AVO 30, 180–230, 258–278, 306–312, 317–331
   Aki–Richards approximation 182, 283
   attributes 211–218, 327
   classes 202–204
   depth trends 227–229
   fluid factor 215–216
   inversion 201–202
   parameters 200–201
   Poisson reflectivity 214
   pre-processing 191–195, 320
   probabilistic 225–230
   Shuey approximation 182, 265, 275, 276, 302, 309
   three-term AVO 221–223
   trends 204–210
   ultra-far AVO 223–225
   see also shear attributes, value of

Backus average 117, 129–130, 275
Batzle–Wang relations 19, 29
Bayes classification 147–159, 163, 277, 285, 294, 327
   Bayes decision rule 149
   confusion matrix 141
   minimum Bayes error 151–152, 156
Biot model 19, 43
Bisq model 44
bootstrap 141
bounds, elastic 3–8, 21, 45
   Hashin–Shtrikman 6–8, 11, 13, 43, 56, 58, 61, 62
   modified lower bound 13, 56, 61, 62, 63, 66, 103
   modified upper bound 11, 58, 90
   Reuss bound 5–6, 7, 8, 11, 22, 60, 62, 63, 70
   Voigt bound 5–6, 7, 63, 90
Brie equation 24
bright spot 169, 174–175
Brown and Korringa 21, 22
bubble point 26

case studies 70–81, 83–90, 96–101, 106–107, 258–316
Castagna 320, 329
   see also empirical relations
cementation 11, 84, 92, 324
   cementation trend 11–15, 43
   constant-cement model 13, 58–59, 72, 80
   contact-cement model 12, 44, 52, 57–58, 72, 88, 96, 98, 104
   see also bounds, modified upper bound; diagenesis; Jizba's cemented shaly sand model
chemical compaction *see* diagenesis
classification 323–328
   see also Bayes classification; facies interpretation; statistical classification
clay 17, 30, 60–62
   clay minerals 60
   see also empirical relations; Han's velocity–porosity–clay relations; models, rock physics
compaction 11, 13, 25, 28, 90–95, 306
confusion matrix 141, 145
constant-cement model 13, 58–59, 72, 80
constant-clay model 13, 52, 53, 62, 312
contact-cement model 12, 44, 52, 57–58, 72, 88, 96, 98, 104
contact models 44, 55
converted waves 245–248, 317
coordination number 55, 57, 92, 98
critical porosity 2, 9, 54, 60, 62, 63, 98, 306

Index

cross-plot analysis 101–107, 202–211, 319, 321, 326
cumulative distribution functions 113

deposition 8
    depositional trends 11
    see also sorting
depth trends 50, 90–95, 227–229, 306–312, 319, 323–325, 330
    see also diagenesis
derived distributions 117, 124
diagenesis 1, 11, 28, 51–70, 90–95
    diagenetic trends 9, 10, 11–15
    pressure solution 11, 13, 92
    see also cementation; compaction
dim spot 169, 174–175, 318
discriminant analysis 140–145, 163
dispersion 20–21
    Biot model 19, 43
    Bisq model 44
    squirt model 20, 44
    see also attenuation
drainage 23
drained condition 19
dry rock 19
Dvorkin–Gutierrez models
    shaly sand model 53, 66
    silty shale model 53, 61

effective-medium models *see* models, rock physics
effective stress *see* pressure effects
elastic impedance *see* far-offset impedance
empirical relations
    Batzle–Wang relations 19, 29
    Brie equation 23
    coordination number 55
    Gardner relations 215
    Greenberg–Castagna relations 38, 321
    Han relations 2, 9–10, 11, 13, 30, 63
    Raiga-Clemenceau relation 2, 11
    Raymer–Hunt–Gardner relation 2, 11
    Wyllie time average 2, 11
    see also models, rock physics; V_P–V_S relations
entropy 132–135
European polarity standard *see* polarity

facies interpretation 50, 81–90, 125, 177, 258–316, 317, 323–328
far-offset impedance 30, 41, 104, 236–243, 278–295, 317
field examples *see* case studies
flat spot 175–176, 220, 324
flaw of averages 117–119
fluid factor 215–216
fluid properties 19, 28, 311
    adiabatic modulus 19
    Batzle–Wang relations 19, 29

isothermal modulus 19
    see also saturation; suspension
fluid substitution 15–24, 42–43, 258–316, 321
    Brie equation 24
    Brown and Korringa relations 21, 22
    Gassmann relations 3, 15–24, 34, 43, 64, 324
focusing *see* overburden effects
friable-sand model 52, 58, 60, 63, 72, 84, 88, 263, 312

Gardner relations 215
Gassmann 3, 15–24, 34, 43, 64, 324
    see also fluid substitution
geostatistics 124, 125, 131–132, 233, 288–293, 328
granular-medium models 4, 44, 51–70
    see also models, rock physics
Greenberg–Castagna 38, 321

Han 2, 9–10, 11, 13, 30, 63
hard event 169, 171–172, 318
Hashin–Shtrikman bounds *see* bounds, elastic
Hertz–Mindlin theory 55, 95, 100, 103, 307, 324
    see also contact models
heuristic models 46
    see also bounds, modified lower bound; bounds, modified upper bound

imbibition 23
impedance inversion 195, 234–243
    see also inversion
inclusion models 2, 4, 15, 20, 44, 54
increasing-cement model 52
information theory 41, 132
    value of additional information 113, 119–123
interference *see* tuning
intrinsic variability 126, 153
inversion 230–252
    hybrid inversion 240
    impedance inversion 234–243, 278–295
isostrain 5
isostress 5, 22
isothermal modulus (*see* fluid properties)

jackknife 141
Jizba's cemented shaly sand model 53, 58, 67, 81

kernel density estimators 114, 128
Kuster and Toksöz model 44

lambda–mu analysis 30, 41, 104, 243–245
Lamé parameters 41, 243–245, 317
laminated sand-shale model 53, 69, 84
lithofacies *see* facies interpretation
lithology substitution 42–43, 321

Mahalanobis distance 140, 264, 272, 285, 297, 311
Markov chains 129, 130, 136

Index

models, rock physics 43–45, 51–70
   Biot model 19, 43
   Bisq model 44
   Brown and Korringa relations 21, 22
   computational models 45
   constant-cement model 13, 58–59, 72, 80
   constant-clay model 13, 52, 53, 62, 312
   contact-cement model 12, 44, 52, 57–58, 72, 88, 96, 98, 104
   contact models 44, 55
   critical porosity model 2, 9
   Dvorkin–Gutierrez shaly sand model 53, 66
   Dvorkin–Gutierrez silty shale model 53
   friable-sand model 52, 58, 60, 63, 72, 84, 263, 312
   Gassmann relations 3, 15–24, 34, 43, 64, 324
   granular-medium models 4, 51–70
   Hertz–Mindlin model 55, 95, 100, 103, 307, 324
   inclusion models 2, 4, 15, 20, 44, 54
   increasing-cement model 52
   Jizba's cemented shaly sand model 53, 58, 67, 81
   Kuster–Toksöz model 44
   laminated sandy shale model 69, 84
   shales 60–62, 78, 79
   see also soft sediment model
   soft sediment model 52, 58, 60, 63, 72, 84
   squirt model 20, 44
   Yin–Marion shaly sand model 53, 64, 260
   Yin–Marion silty shale model 53, 260
   see also empirical relations
modified lower bound see heuristic models
modified upper bound see heuristic models
Monte Carlo simulation 119, 124, 127, 136–138, 283, 309, 324
   see also statistical rock physics

net-to-gross 179–180
neural networks 159–162, 163, 165
nonparametric density estimation 114, 115

O'Doherty–Anstey formula 188–189
overburden effects 187, 268, 275
overpressure 90, 221
   see also pressure effects

partial saturation see saturation
patchy saturation 23
penny-shaped cracks see inclusion models
Poisson reflectivity 214
polarity 169
   American polarity standard 170
   European polarity standard 170
polarization attributes 217–218
pore-fluid properties (see fluid properties)
pore stiffness 3, 15
pressure effects 24–29, 34
pressure solution see diagenesis
principal component analysis 145–146

probability 113
   frequentist 113
   subjective 113
   see also statistical rock physics
probability density functions 113–117, 147
probability mass function 113
P-to-S impedance see converted waves

Raiga-Clemenceau 2, 11
random variables 113
Raymer–Hunt–Gardner 2, 11
reflection coefficient 182, 185, 245, 248
residual gas 218–219, 246
resolution 172–173
Reuss see bounds, elastic
rock physics bottle-neck 39
rock physics diagnostics 50, 326
rock physics templates 50, 101–107, 251, 312–316, 317, 321, 322–323, 331
rock physics What ifs 42–43, 320–322
   see also fluid substitution; lithology substitution
RPT see rock physics templates

saturation 34, 321
   Brie equation 24
   partial saturation 22–24, 246
   patchy saturation 23
   uniform saturation 22
scattering 188–190
   O'Doherty–Anstey formula 188–189
sedimentation 1, 11, 51–70
   see also sorting
seismic modeling 252–256
shale–sand cross-over 91, 100, 170
shales 60–62, 78, 91
   see also facies interpretation; models, rock physics
shaly sands 62–70, 78
   see also facies interpretation; models, rock physics
shear attributes, value of 30–42, 119–123, 261
Shuey approximation see AVO, Shuey approximation
simulation, stochastic 131, 136–138
soft event 169, 171–172, 318
soft sediment model 52, 58, 60, 63, 72, 84
sorting
   sorting trend 11–15, 54–57, 58, 75, 104
squirt model 20, 44
statistical classification 124, 130, 138–165
   discriminant analysis 140–145
   Bayesian classification 147–159
   neural networks 159–162
statistical rock physics 111–167, 263–266, 269–274, 277–312, 321, 324, 325–328
   see also Bayes classification; statistical classification
stratigraphic interpretation 177–179
   see also facies interpretation
suspension 5, 8, 13, 60

Thomsen's parameters 184, 248
transmissivity 188
tuning 173–174, 179, 185–187, 268, 269, 275, 301, 319
turbidite systems 1, 70–81, 83–90, 258–306, 312–316

uncertainty, uses of 112
upscaling 129–130, 147

velocity–porosity 2–15
   see also empirical relations; models, rock physics
Voigt see bounds, elastic

V_P–V_S relations 34–38
   see also empirical relations

Walther's law 82, 127, 276
wavelet 169, 192, 234
   see also polarity
What ifs see rock physics *What ifs*
workflows 123–124, 317–331
Wyllie time average 2, 11

Yin–Marion models 260
   shaly sand model 53, 64
   silty shale model 53